Programming and
Customizing the
OOPic Microcontroller

**Other Books of Interest in the McGraw-Hill
Programming and Customizing Series**

Predko/Programming and Customizing PIC Microcontrollers, 2nd Edition, 0-07-136172-3

Edwards/Programming and Customizing the Basic Stamp Computer, 2nd Edition, 0-07-137192-3

Gadre/Programming and Customizing the AVR Microcontroller, 0-07-134666-X

Programming and Customizing the OOPic Microcontroller

The Official OOPic Handbook

Dennis Clark

McGraw-Hill

New York   Chicago   San Francisco   Lisbon   London
Madrid   Mexico City   Milan   New Delhi   San Juan
Seoul   Singapore   Sydney   Toronto

The McGraw·Hill Companies

Cataloging-in-Publication Data is on file with the Library of Congress.

Copyright © 2003 by The McGraw-Hill Companies, Inc. All rights reserved. Printed in the United States of America. Except as permitted under the United States Copyright Act of 1976, no part of this publication may be reproduced or distributed in any form or by any means, or stored in a data base or retrieval system, without the prior written permission of the publisher.

1 2 3 4 5 6 7 8 9 0   DOC/DOC   0 9 8 7 6 5 4 3

P/N 142085-1
Part of
ISBN 0-07-142084-3

The sponsoring editor for this book was Judy Bass and the production supervisor was Sherri Souffrance. It was set in Helvetica by MacAllister Publishing Services, LLC.

Printed and bound by RR Donnelley.

McGraw-Hill books are available at special quantity discounts to use as premiums and sales promotions, or for use in corporate training programs. For more information, please write to the Director of Special Sales, McGraw-Hill Professional, Two Penn Plaza, New York, NY 10121-2298. Or contact your local bookstore.

For Shannon, who is now happy I'm not burning the midnight oil anymore, and Brendan, who I'm sure is a unique two year old, since he not only knows what a robot is, he can pronounce it too.

CONTENTS

FOREWORD

This is a very exiting time for me. As a long-time fan of hobbyist robotics, I watched the movement grow from a hopeful beginning in the early 1980s to what it is now. It had its ups and downs, but overall, computers got faster and smaller, batteries got stronger and lighter, and sensors were developed. Before I knew it, the robotics hobbyist movement took off and was going full force by the mid-1990s.

But as much as I liked building robots, it was still a tedious job to program them. Programming microchips was just plain time-consuming, and there wasn't any way out of it. Even the basic compilers that made it easier to write programs left huge complexities when programming robots. They still required you to focus your programming efforts on understanding and controlling the hardware interface, instead of focusing on programming the robot itself. What I wanted was an easy way to program the robot to spin its wheels without spinning them myself.

Another big problem was the issue of multitasking. How could one control two motors, monitor two wheel encoders, and read a SONAR unit all at the same time when the available languages could only do one thing at a time?

The answer was *objects*.

As you can probably guess, I am a big fan of object-oriented programming. I have used a lot of programming languages, and I find object-oriented programming to be the easiest and fastest way of programming I have ever used. However, I thought that what would be really useful to the world of robotics would be to have an object-oriented microchip in which all the complexities of the hardware would be handled by objects, and the programmer would only need to tell the objects what to do with the hardware.

The result was *Object-Oriented Programmable integrated circuit* (OOPic).

When I started the OOPic project, the only goal I had in mind was to create a robotic controller that used objects to make programming my robots easier. But recently I attended the 2003 Trinity Fire-Fighting Home Robot Contest and was pleased to see that the first and second place awards were awarded to robots that used OOPic.

What's in store for future OOPic models? I am not at liberty to say, but new design ideas are constantly being tossed around and implemented. Who knows what the future OOPic models will be able to do?

Scott M. Savage
President and CEO
Savage Innnovations, Inc.

ACKNOWLEDGMENTS

This book would not have been possible without the intesive assistance given by Scott Savage of Savage Innovations, Inc. As the creator of the OOPic, only he could answer some of the questions that I kept pestering him about.

Thanks go out to Tony Brenke for translating the Basic code into C for me—there was no way that I could write two sets of code for everything that is in this book.

I would also like to thank my publisher, Judy Bass, for supporting me when I was pulling my hair out.

Finally, I would like to thank for the following who lent aid, parts, and inspiration—true heros everyone: Ron Nickolson, John Taylor, Tran Duong Dung, Charlie Payne, Nick Donaldson, Jeff Richeson, Tim Rohaly, and finally, Dan Michaels, who will never let anyone get a "big head."

INTRODUCTION

Welcome to the *OOPic*™, the *Hardware Object*™!

The OOPic microcontroller development system is arguably one of the most unique areas of the robotics hobbyist's embedded processor world. It stands alone as a proponent of the object-oriented design and an implementation paradigm for embedded programming. The OOPic also supports event-based programming as an integral part of its system definition.

At the core of the OOPic programming paradigm is the *virtual circuit* (VC). The VC is a means for connecting various hardware and processing capabilities of the PICmicro® microcontroller together (the hardware core of the OOPic and an obvious part of its name) to create both simple and complex embedded control systems. By the time you finish a few chapters of this book, you will be intimately familiar with the VC concept. The OOPic has so many interesting objects built into it that an OOPic project can be seen not so much as programming, but as solving a puzzle. Each of the pieces fits together in a certain way, and although the rules are explicit, the combinations are nearly endless. It's a game, but an elaborate one and, in the end, a very *useful* game.

The OOPic bases its networking on the Philips Inter-IC (*integrated circuit*) specification, usually called I²C or I2C. For more information on the I2C specification, see the Phillips web site at www.phillipslogic.com/i2c.

This book lists all the objects from the currently available versions of the OOPic microcontroller. I also detail many experiments that will help you learn the basics of programming the OOPic and understanding its capabilities. Also, several entire projects are included that take you through the process of designing and implementing an OOPic-based embedded control system. Although not all OOPic objects are represented by examples, all the *types* of OOPic objects are presented that will allow you to select and use other objects correctly. I will also show you how to connect and use many of the common hardware *input/output* (I/O) devices on the market, such as keypads, *liquid crystal displays* (LCDs), and PC terminal interfaces. Because I reside on many of the embedded discussion lists, including the Yahoo™ OOPic group, I've seen (and responded to) many of the questions out there. These questions formed the basis for the experiments and projects detailed in this book.

What Makes the OOPic So Special?

Several easy-to-use embedded controller platforms are available on the market today. Many of them are based on the Microchip PICmicro™ microcontroller (MCU). The list is so large that I don't want to detail it here because I'd have to keep typing all those confounded ® and ™ symbols. Get out a copy of your favorite embedded controller or hobbyist electronics magazine and you'll swiftly get an idea of what I'm talking about. The OOPic is unique among them because it focuses on *object-oriented* (OO) programming techniques instead of the more common procedural programming ones.

Procedural programming stresses a linear code process that starts at point A and proceeds to point B in a well-defined manner. OO programming, on the other hand, stresses the well-planned integration of data and processes, called attributes and methods, into self-contained units called objects. Objects deal with other objects by sending them *messages*. When an event occurs, a message is sent to the OOPic object that is configured to handle it.

In Chapter 4, "Your First OOPic Program, OOPic I/O (Project #1: Das Blinken Light)," I compare an OOPic program to the 800-pound gorilla of the hobbyist embedded world, the Parallax Stamp™. This should help you to understand the differences between procedural and OO-style programming.

Event programming is simply the realization that all activity is in response to some internal or external stimulus or event. A popular example of event programming is the user interface of your favorite operating system or graphic program. These programs respond to mouse clicks and keyboard events that signal when an operation is supposed to occur. The OOPic responds to internal and external events according to how the programmer *links* objects together.

The OOPic takes these concepts one step further in that it allows the programmer to combine these objects simply by linking them together, and the operating system takes care of the message passing from that point on. It is object-orientation at its most elegant and is known as the VC. I know of no other embedded hobbyist or industry-oriented platform that defines MCU internal activities in this manner. Because all the VC objects behave in similar ways, it's simple to use them once you learn how the entire system works. How to use these objects properly is explained in detail in subsequent chapters.

The Language of the OOPic: Basic, C, and Java™—What's the Difference?

When Savage Innovations created the OOPic microcontroller environment, they wanted anyone to be able to use it. At the time, Microsoft's Visual Basic™ was the most commonly known programming language used by nonprofessional and professional programmers alike. Thus, it was the natural language on which to base the OOPic microcontroller environment. However, other programmers out there were more familiar with C and Java, so those languages were supported as well, but with quirks. Those quirks will be discussed in the sections on the different languages.

It doesn't matter which language syntax you use; the OOPic program will act the same and take up the same space in memory. The syntax you choose will simply make it easier for you to use the environment. It has no effect on the outcome of the program itself. The OOPic runs on *bytecode* (also called *pcode*) that is interpreted by the operating system that is programmed into the OOPic chip itself. That bytecode is the same no matter which syntax you use to write the program.

Take, for instance, the following code sequence written in Basic and C. It's a data declaration and a loop construct in both languages, so it will do the same thing in the two languages. The Basic language syntax is as follows:

```
Dim LED as New oDio1

Sub Main()
 LED.IOLine = 31
 LED.Direction = cvOutput
 While True
 LED.Invert
 OOPic.Wait = 100
 Wend
End Sub
```

The C language syntax is as follows:

```
ODio1 LED = New oDio1;

Sub void main(void)
{
 LED.IOLine = 31;
 LED.Direction = cvOutput;
 While (True)
 {
 LED.Invert;
 OOPic.Wait = 100;
 }
}
```

All code in this book will be written in Basic, which will reach the widest audience. The accompanying CD-ROM will have all the code samples written in C as well, but the C code will not appear in the book.

What Is In This Book . . . and What Isn't

This book was written to help you to get the most from your OOPic microcontroller. I feel that it will be most useful for beginners, but I suspect that even you old-timers will find something you didn't know here. I have included the syntax guides for all the current OOPic objects, functions, and programming options, but I have not included examples for all known objects and all known cases. That is a combinatorial explosion due to the extreme flexibility of the OOPic system and is just not possible to contain in a single book. A great deal of explanation is given to the core capabilities of the OOPic, but only a fraction of the OOPic objects will be given specific attention here. Fear not; plenty of material is included to help you along with your own projects.

In a nutshell, here is what this book has to offer:

- Chapter 1, "OOPic Family Values," contains full hardware, firmware, and configuration data for all OOPic variants. This includes connector pinouts, power supply options, and object lists.
- Chapter 2, "The OOPic IDE and Compiler," consists of installation details and help for installing and configuring the OOPic IDE software, which is also included on the CD-ROM.
- Chapter 3, "OOPic Object Standard Properties," discusses the basic OOPic concepts of the VC and how all objects are designed to work together. This includes complex linking and user-defined class files.
- Chapter 4, previously mentioned, walks you through your first VC and shows you how to use OOPic integer math, the standard programming syntax, and all the built-in OOPic functions.
- Chapter 5, "Analog-to-Digital and Hobby Servos (Project: Push My Finger)," builds on prior information by introducing *analog-to-digital* (A2D) objects, hobby servos, and more software and hardware interfacing techniques.

- Chapter 6, "OOPic Timers, Clocks, LCDs and SONAR (Project: SONAR Ping)," gets more complex with the addition of timers, clocks, LCDs, and another toy—a SONAR distance ranger.
- Chapter 7, "OOPic Events, Keypads, and Serial I/O (Project: A Mini-terminal)," delves into the sticky issues revolving around serial I/O with a PC and introduces many new objects, including a keypad encoder.
- Chapter 8, "OOPic Interfacing and Electronics (and Steppers and Seven-Segment LEDs)," drops back and discusses electronic interfacing to the OOPic, including independent power supplies, I/O interface precautions, and common circuits used with embedded processors. Seven-segment displays and stepper motors are topics here as well.
- Chapter 9, "OOPic I2C and Distributed DDE Programming (Project: Remote Control)," demonstrates how to interface two OOPic devices together using DDELink objects and introduces a form of remote motor control and *electrically erasable programmable read-only memory* (EEPROM) programming.
- Chapter 10, "OOPic Robotics and URCP (Project: A Robot That Toes the Line)," deals specifically with robots and, as such, is devoted entirely to OOPic II and later. It discusses object-oriented motors, bumpers, and the design of an inexpensive line-following robot.
- Chapter 11, "OOPic R Serial Control Protocol (SCP)," is an introduction to the new OOPic language, SCP, and gives an example interface between a Palm *personal digital assistant* (PDA) and an OOPic R robot.

Many first-time embedded hobbyists are drawn to the OOPic because of its simplicity and the large number of supported hardware objects that are easy to connect and use. Because of this, I've tried to keep the "techspeak" to a minimum to help you understand how to use the OOPic controller. I will introduce many concepts that might be confusing to those of you inexperienced with electronics or programming. Use these concepts as a starting point toward expanding your understanding of not only the OOPic controller, but of the big, wide magical world of electronics and programming, in general. This book won't get you all the way there, but it will help you find those topics you'll want to study further. There *is* magic in the world, and you can be part of what creates it.

I have tried to include solutions to all the known OOPic problems. I'm a long-time participant in the OOPic discussion groups, and I have read and helped with many of these issues. I have tried to include as many complex projects as I could to help illustrate some of the more advanced capabilities of the OOPic. I have also included some easier projects to help those new to the OOPic understand its capabilities. So if you are looking for a way to get a smiley face to display in that troublesome LCD module you have, it may not be here. If you are looking for how to get I2C communications working between two OOPic controllers, you're in luck.

With that said, this book still has more to offer. The appendices include a full listing of OOPic objects broken down by the *flavor* of the OOPic you're looking for. OOPic I, OOPic II, OOPic R, and OOPic C are fully covered in all their glory. At the time of this book's writing, the OOPic firmware revisions A.1.X and B.1.X have been declared obsolete. Because

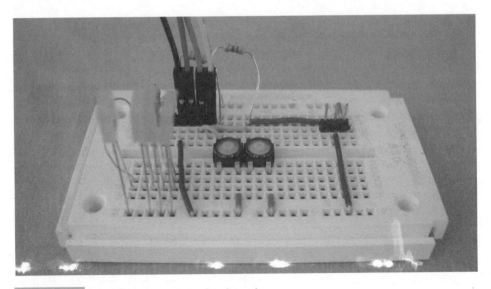

Figure I-1 A solderless prototyping board

of that, I will not be including any A.1 or B.1 information and will concentrate on the A.2, B.2, and higher firmware revisions. I'd rather start out up-to-date than start out behind, so I am leaving the obsolete OOPic variants out of this discussion. The Savage Innovations web site continues to have information on these obsolete firmware revisions, however.

What You Will Need to Build the Projects in This Book

To start out, you're going to need an OOPic.[1] I will introduce you to some very nifty add-on boards that other companies provide for the OOPic, but in most cases you'll be building projects initially on solderless prototyping board (see Figure I-1).

This wonderful invention enables you to use 22-gauge, solid connect wire to make connections between components that are not permanent and don't require a soldering iron. You can find it at Radio Shack™ as well as any number of online electronics shops.

For many of your projects, you can use the Magevation *Logic Status Module* shown in Figure I-2. This handy device connects to your OOPic board and enables you to instantly see the status of all I/O lines as they are changing. It doesn't even get in the way of your projects because it is a *passthrough* board and no signals are blocked or interfered with. I'll use this board later on in the book.

---

1. Well, duh. But I suppose you should know it doesn't come with the purchase price of the book!

Figure I-2 Magnevation® Logic Status Module

I recommend that you get an OOPic starter kit because that will give you a battery connector and a programming cable so you don't have to build one. If you prefer to build one yourself, I'll show you how in Chapter 1, so you won't have to wait long to get started.

As an embedded controller hobbyist, you will find that you will spend your time evenly between hardware (those things you can touch) and software (those things you can program). This is why so many of us do this; we get to play with the magic at both ends of the computer! So not only are your compilers important to you, so too are your tools. Here are the tools I recommend:

- **Soldering iron and solder**   Although you have the prototyping board, eventually you'll need to solder *something*. Use 60/40 resin core electronics solder *only*.
- **Needle-nosed pliers**   For bending and grabbing stuff.
- **Diagonal cutters or wire cutters**   The purpose of these is obvious, but you'll abuse them with nonintended uses as well I'll bet.
- **Wire stripper**   For dealing with the wire you just cut.
- **A selection of slotted and Phillips screwdrivers, from big to small**   I'd tell you what size to get, but you'd ignore me anyway because it depends on your project.
- **Digital voltmeter**   This is the single most useful tool in your toolbox, other than your brain. They're cheap too and you can find them anywhere.
- **A bunch of other stuff and a great junkbox**   It's hard to predict what you will need and use for these projects. It depends upon what you have, and as a hobbyist, it is very important to have a well-stuffed junkbox!
- **Eye protection**   Solder splashes and snipped wires fly at surprisingly high velocities.

Perhaps the most obvious tool you will need is a computer running Microsoft Windows® 95, 98, NT, 2000, or XP. A plain DOS shell won't work, so you will need a Windows operating system to run the OOPic *Integrated Development Environment* (IDE); it sounds cool either way. As you move up the list of these operating systems, problems will increase when getting the IDE to work. I have an old Pentium® 133 laptop running Windows 95 that serves me quite well. The problems that you will encounter when installing and configuring the IDE have to do with the ever-changing interface to the PC Parallel port. If you are using the OOPic R or C, you can breath easier; those problems don't exist for you because you can program your device via the serial port.

Standards and Conventions Used in This Book

First off, I would like to acknowledge all the copyrights, registered names, and trademarks of Microsoft Windows 95, 98, NT, 2000, and XP. Also to be acknowledged is Parallax Inc. and the Basic Stamp®. I would also like to thank Savage Innovations for creating the OOPic and letting me write about it. Magnevation must also be thanked for the use of their excellent OOPic products.

When referring to an attribute or method of an OOPic object, I will use an *italic* font. For example, the OOPic object has a *wait* method, and it will be referred to thus: OOPic.*Wait*. This does not count when I am writing a code sample, which will look like this:

```
Dim LED as New oDio1
```

When an OOPic object is described along with its attributes and methods (called *properties* in the OOPic manuals), it will be displayed in a table that clearly shows what is what. The table will be preceded by the Savage Innovations OOPic object icon.

What Is Happening in the OOPic World

As I am writing this book, Savage Innovations has released some OOPic devices that can be programmed in new ways. The OOPic I, with the A.1.X and A.2.X firmware versions, is programmed using a parallel port cable. The IDE programs it using I2C protocols over this port and talks directly to the EEPROM chip on the OOPic carrier board. The OOPic II, with the B.1.X and B.2.X firmware, is also programmed in this manner. The OOPic R and C, with the B.2.X+ firmware, can be programmed over the serial port and utilize SCP. All these programming options will be described in subsequent chapters.

The OOPic I and II have been released on the same carrier board, and the OOPic R is a different carrier with a built-in serial port and with many new features on the board. The OOPic C is a 24-pin carrier board that can be plugged into any carrier that supports the Parallax Stamp or any other company's Stamp variant. These are new, and I have included all the information known at this time, but it may not be complete.

Even more exciting is the *brand-new* OOPic B.2+ firmware that enables communication with the OOPic R and C via serial communications. More than just communicating, you can take control through this SCP via the serial channel. This book will also explore what is known about communication between a Palm M100 and an OOPic R. The robotics applications for SCP are endless. In fact, as Savage Innovations releases more variants of the OOPic platform, you'll quickly find that they are being tuned more and more for robotics and automation. Since robots are, of course, at the tip of the coolness pyramid, the OOPic is increasingly better for the robot hobbyist.

As I write this book, more third-party companies are building OOPic enhancements, OOPic-based robots, OOPic add-ons, embedded controllers, and robot boards. By the time you read these words, you'll have a huge selection of products to entice and entertain you. What lucky times these are!

ABOUT THE AUTHOR

Dennis Clark is the author of a series of articles on behavioral robots for the European hobbyist magazine *Elecktor Electronics*, and is co-author of McGraw-Hill's *Building Robot Drive Trains*. A resident of Fort Collins, CO, he is also the designer of the TraCY robot controller for The Robotics Club of Yahoo!, as well as several other useful robotics function add-ons and boards.

OOPIC FAMILY VALUES

CONTENTS AT A GLANCE

For a while, we had just the *Object-Oriented Programmable integrated circuit* (OOPic), and with the release of the OOPic II, we had the OOPic I and OOPic II. Now several varieties of OOPic are available, each one more capable than the last (and more complex). It's gotten so that the dedicated hobbyist just can't figure out which feature is in which device. This chapter comes to the rescue by detailing all the available OOPic variants, what their feature sets are, and what objects are supported by their firmware (at the time of this writing, of course).

Common Features in All OOPic Devices

There are features in the hardware and software that are common amongst the various OOPic variants. Here I list these and what they mean to you.

HARDWARE FEATURES

Despite all the diversity between the OOPic variants, many common features exist. They all work together and can all network together, which demands some uniformity of design and function. This is a list of common features dealing with the OOPic hardware.

- All OOPic devices use the same *Integrated Development Environment* (IDE) and compiler. You just select the firmware target from a pulldown list in the Tools menu.
- All use *Inter-IC* (I2C) *electrically erasable programmable read-only memory* (EEPROM). In fact, you can *remove* them and move them between OOPic boards. You can also upgrade all OOPic boards with larger EEPROM if you desire, but let's face it; no one has seen an OOPic program get bigger than about 4KB without using a large amount of EEPROM space for string storage or constant tables.
- All use the same syntax. This is a big deal. It means you can upgrade your OOPic and just recompile the code for the newer firmware revision with no need to port or translate.
- As I have already implied , all OOPic source code is forward compatible. When you move to a newer OOPic platform, your legacy source is still fine. You may find that newer firmware has added objects that cause some of your code to become obsolete, but you can move it unchanged and know it will still run as expected.
- All OOPic boards are fully built and tested, with no kits.
- All OOPic boards can be programmed using the PC parallel port I2C protocol. The OOPic C platform will require some additional parts to fall under this umbrella, but it can be done.

All PICMicro microcontrollers that are used in the OOPic are clocked at 20MHz. Because the *PICMicro* (PIC) divides this clock by 4, the effective clock speed is 5MHz.

This information might be useful to you if you are entering specific numbers into PIC registers using the oRAM or PIC objects to get specific baud rates or *pulse width modulation* (PWM) divisors. I'm not recommending that you actually do any of these things, mind you. It's just that you should be aware of this information.

SOFTWARE FEATURES

All OOPic objects, except the oDIOXX objects, have an *Operate* property that is used to activate them. In general, the object will then update once every pass through the *virtual circuit* (VC) list. Approximately 80 percent of the OOPic processing power is dedicated to the VCs mentioned in the introduction; the other 20 percent is allocated to OOPic code. This suggests that a VC operates four times faster than code written outside a VC. In actuality, because instructions need to be fetched from the serial EEPROM and the VC code is internal to the chip, the actual speedup is more like 80 times. So, yes, this means you really should write VC-based code whenever possible so it will run as quickly as possible.

Hardware objects require that you select *input/output* (I/O) lines for them to use. Some hardware objects like oPWM and oSerial require that you select from specific I/O lines where the PICMicro has special hardware to support those functions.

Processing objects have a variety of I/O functionality. Two types of input and output capabilities are available: the *flag* and the *value*. A flag is a boolean (0 or 1) signal that is used to enable or disable an object, or signal completion. A value is a number that is either needed as an input or is the result of an operation for an output. Almost every processing object with an output pointer for a Boolean or a value also has an output result flag that can be linked to. The debug window functions with all OOPic versions and objects.

The OOPic Programming and Networking Cables

Before I dive into the descriptions of the OOPic varieties, it would be wise to describe the programming and networking cables that the OOPic devices use. If you didn't get an OOPic Starter Kit that includes cables, you'll be most interested right now in building your cables or at least finding them. Read on and I'll show you what you need to get or build to program your OOPic.

THE I2C PROGRAMMING AND DEBUGGING CABLES

When OOPic devices are programmed over the PC parallel port, the computer writes directly to the EEPROM on the OOPic board. The programmer holds the OOPic controller chip in RESET mode until the program has been written into the EEPROM. Figure 1-1 shows how this programming cable is built. Table 1-1 details the parts required and suggests where you can get them. Figure 1-2 shows the pin orientation of the cable's connectors.

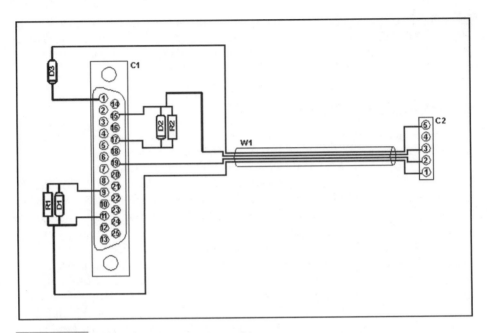

Figure 1-1 OOPic I2C programming cable

TABLE 1-1 PROGRAMMING CABLE PARTS LIST		
SYMBOL	**DESCRIPTION**	**PART NUMBER**
J1	Male 25-pin D connector	Radio Shack #276-1547
J2	Female 5-pin *Session Initiation Protocol* (SIP) connector	Digikey #WM2003-ND
D1, D2, D3	1N914 or 1N4148 diode	Radio Shack #276-1620
R1, R2	10K resistor	Radio Shack #271-1335
W1	Four-conductor cable	Radio Shack #278-777

OOPIC I2C PROGRAMMING PORT

The OOPic programming port is also called the I2C *local network* for the OOPic. This port is used to program the I2C EEPROM labeled as E0 on all OOPic controller boards. This local network is also the network that any I2C coprocessor, such as I2C thermometers, SONAR, or other EEPROM devices, must be connected to. The LSDA and LSCL pins are also available on the I/O connector of the *OOPic S* boards, and this connector is described later in this chapter. The oI2C object *communicates* on this network. Table 1-2 shows the

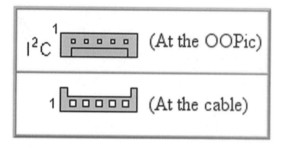

Figure 1-2 Five-pin connector orientation

TABLE 1-2 OOPIC PROGRAMMING PORT PIN DEFINITIONS

PIN	NAME	DIR.	DESCRIPTION	FUNCTION
1	LSDA	In	Local I2C Data	I2C serial data generated by OOPic programmer (PC) when reading or writing the EEPROM.
2	GND		Ground	Common
3	LSCL	Both	Local I2C Clock	I2C serial clock generated by OOPic programmer (PC) when reading or writing the EEPROM.
4	+5		Regulated +5V	Usually not used, but it can be.
5	RESET	Both	Reset (active low)	Pulled low to disable the OOPic when programming the EEPROM.

pin definitions of the OOPic programming port. In the table, "Dir." refers to the direction the TTL signal is going in reference to the OOPic board only during programming. Except during EEPROM programming, the OOPic controller is the master on this bus. The programming port does not use the PIC hardware I2C device; it is a software I2C communications port.

OOPIC I2C NETWORKING PORT

The OOPic network ports are used to network other OOPic devices and *only* other OOPic devices. You will use the oDDELink object to communicate between OOPic boards on this bus. This includes using the I2C debugging features of the OOPic IDE. Refer to Figure 1-2 to see the pin orientation of the connectors used, and Table 1-3 shows the pin descriptions of the OOPic networking port.

Your OOPic boards will all daisy chain using the networking connectors on their respective boards. The *OOPic R* board only has one networking connector so it will be at the end of the chain. The OOPic networking bus uses the PIC internal hardware I2C serial controller.

TABLE 1-3   OOPIC NETWORKING PORT PIN DEFINITIONS

PIN	NAME	DIR.	DESCRIPTION	FUNCTION
1	SDA	Both	Local I2C data	I2C serial data generated the current I2C master or slave.
2	GND		Ground	Common
3	SCL	Both	Local I2C Clock	I2C serial clock generated by the I2C master on the bus.
4	+5		Regulated +5V	Can be used to power another OOPic board.
5	RESET	Both	Reset (active low)	Pulled low to reset all OOPic devices.

OOPIC R AND OOPIC C SERIAL PROGRAMMING CABLE

The OOPic R and *OOPic C* boards can use a standard RS-232 serial cable to program their EEPROM. The standard nine-pin D RS-232 cable is the cable of choice here. This cable has a male nine-pin D connector on one end, and a female nine-pin D connector on the other. It is important to make sure that the cable is a *straight-through* cable, which means that, for instance, pin 1 must connect to pin 1 on the other end of the cable, and so on for all the other pins. All the wires must be there as well, because some lines are configured on the board such that the computer can determine the existence of the OOPic R or, in the case of the OOPic C, the existence of a Stamp-like carrier board.

This cable is available at a number of surplus shops, PC supply houses, and junk drawers. It can also be found described as an *EGA extension cable* in some surplus catalogs.

OOPic I (A.2 Firmware) Description

The OOPic I is the classic OOPic, the first to be made available. The current firmware revision is A.2.X, and the A.1.X version has been declared obsolete. With the release of the OOPic (OOPic I), Savage Innovations developed what is now called the S board. This is board is commonly seen, with the 40 pin I/O connector on one edge and the prototyping area in the center between the 40-pin PIC chip and the I/O connector.

HARDWARE DETAILS

Figure 1-3 shows a breakdown of the hardware features on the OOPic I standard board, and Table 1-4 details the pin descriptions of the I/O connector. Sometimes it's hard to tell which

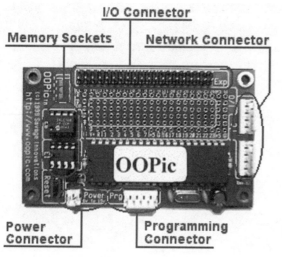

Figure 1-3 OOPic I
hardware layout

TABLE 1-4   OOPIC I/O CONNECTOR PIN DEFINITIONS

PIN	NAME	FUNCTION	PIN	NAME	FUNCTION
1	LSDA	Local I2C serial data	2	GND	Ground
3	LSCL	Local I2C serial clock	4	Power	+6–18V power supply
5	Reset	Active low OOPic reset	6	I/O 15	I/O Group 1, bit 7
7	I/O 1	Analog-to-digital (A2D) 1	8	I/O 14	I/O Group 1, bit 6
9	I/O 2	A2D 2	10	I/O 13	I/O Group 1, bit 5
11	I/O 3	A2D 3	12	I/O 12	I/O Group 1, bit 4
13	I/O 4	A2D 4	14	I/O 11	I/O Group 1, bit 3
15	I/O 5	I/O line 5	16	I/O 10	I/O Group 1, bit 2
17	I/O 6	I/O line 6	18	I/O 9	I/O Group 1, bit 1
19	I/O 7	I/O line 7	20	I/O 8	I/O Group 1, bit 0
21	+5V	Regulated +5V out	22	+5V	Regulated +5V out
23	GND	Ground	24	GND	Ground
25	I/O 16	I/O Group 2, bit 0			
		Timer 1 crystal	26	I/O 31	I/O Group 3, bit 7
27	I/O 17	I/O Group 2, bit 1 (PWM2)			

(continued)

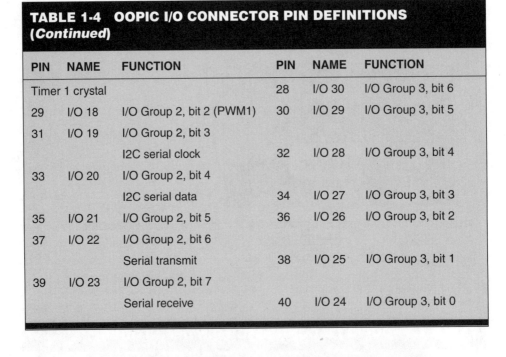

TABLE 1-4   OOPIC I/O CONNECTOR PIN DEFINITIONS (Continued)

PIN	NAME	FUNCTION	PIN	NAME	FUNCTION
Timer 1 crystal			28	I/O 30	I/O Group 3, bit 6
29	I/O 18	I/O Group 2, bit 2 (PWM1)	30	I/O 29	I/O Group 3, bit 5
31	I/O 19	I/O Group 2, bit 3			
		I2C serial clock	32	I/O 28	I/O Group 3, bit 4
33	I/O 20	I/O Group 2, bit 4			
		I2C serial data	34	I/O 27	I/O Group 3, bit 3
35	I/O 21	I/O Group 2, bit 5	36	I/O 26	I/O Group 3, bit 2
37	I/O 22	I/O Group 2, bit 6			
		Serial transmit	38	I/O 25	I/O Group 3, bit 1
39	I/O 23	I/O Group 2, bit 7			
		Serial receive	40	I/O 24	I/O Group 3, bit 0

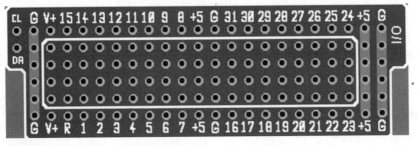

Figure 1-4 OOPic I prototyping area

I/O line goes to which pin on the connector, as you can tell in Figure 1-4, which shows the OOPic I prototyping area. The I/O pins and power labels that you see on the upper part of the prototyping area are for the pins on the outside (upper) row of the I/O connector. The I/O pins and power options that are written on the lower side of the prototyping area are for the pins on the inner (lower) row of the I/O connector. See? Now it's simple to tell.

OOPIC I FEATURES

This section looks at the specifications for the OOPic I, which uses the same *random access memory* (RAM) space for variables as it does for objects. You can still save memory, however, by using Byte and Word variables instead of oByte and oWord variable objects. Code is stored in a removable EEPROM, and object data and variables are stored in the PICmicro internal RAM. The following features are what you would compare when deciding which OOPic fits your needs:

- PIC16C74b microcontroller clocked at 20MHz
- Four 8-bit A2D channels
- 86 bytes of object memory
- Variable RAM that uses object memory space
- 4KB of program code space (this is what is shipped; you can install more)
- Parallel port (I2C) that is programmable only

OOPIC I OBJECT LISTING

Table 1-5 is the full list of the OOPic I programming objects. For a more detailed description, see the appendices or the extremely detailed OOPic site at www.oopic.com. This table outlines the most important points so you can figure out which object is in which version. The OOPic I concentrates on providing a basic set of generic hardware and pro-

TABLE 1-5   OOPIC I OBJECT LIST

HARDWARE OBJECTS

OBJECT	DESCRIPTION
oA2D	Provides a numerical measurement of a voltage.
oDIO1	Provides a 1-bit digital I/O.
oDIO16	Provides a 16-bit digital I/O with I/O 8–15 and I/O 24–31 *Least Significant Byte* (LSB)-*Most Sigificant Byte* (0MSB+).
oDIO16x	Provides a 32-bit digital I/O using a 1-of-3 decoder.
oDIO4	Provides a 4-bit digital I/O (7 possible choices).
oDIO8	Provides an 8-bit digital I/O (3 possible choices).
oI2C	Provides access to an I2C device on the local (programming) bus.
oKeypad	Reads a 4 × 4 keypad matrix.
oPWM	Provides a PWM output (two available).
oSerial	Provides an asynchronous serial I/O port using the *Universal Asynchronous Receiver Transmitter* (UART) hardware.
oSerialPort	Provides a buffered asynchronous serial I/O port with flow control.
oServo	Controls a hobby servo.
oTimer	Provides a 16-bit, high-speed counter.

(continued)

TABLE 1-5   OOPIC I OBJECT LIST (*Continued*)

PROCESSING OBJECTS

OBJECT	DESCRIPTION
oCounter	Provides counting functions, normal and quadrature encoded.
oDataStrobe	Provides a data strobe in response to a value being written to it.
oDDELink	Provides a *Dynamic Data Exchange* (DDE) link over the I2C network.
oDebounce	Provides logic-state debounce functions.
oEvent	Runs program code in response to an event.
oFanOut	Copies an input *flag* value to other multiple objects.
oGate	Provides logic-gate and *flag* linking functions.
oIndex	Provides indexing functions to an oBuffer object.
oMath	Provides mathematical functions and links values between objects.
oOneShot	Produces a one-pulse output in response to logic transition.
oSrvSync	Provides a method of synchronizing multiple hobby servos.
oRandomizer	Provides a random number.
oRTC	Maintains a (nonpersistent) real-time clock.
oWire	Copies a Boolean value or *flag* to another object.

VARIABLE OBJECTS

OBJECT	DESCRIPTION
oBit	Maintains a 1-bit variable.
oBuffer	Maintains a variable-size data buffer/string variable (bytes only).
oByte	Maintains an 8-bit (1-byte) variable.
oEEProm	Provides access to the nonvolatile EEPROM memory of E0 only.
oNibble	Maintains a 4-bit variable.
oRAM	Provides access to the OOPic RAM and registers.
oWord	Maintains a 16-bit (2-byte) variable.

USER-DEFINABLE OBJECTS

OBJECT	DESCRIPTION
oUserClass	Uses an OOPic program to provide a new object definition (not VC).

SYSTEM OBJECTS	
OBJECT	**DESCRIPTION**
OOPic	Provides control of and maintains information about the OOPic.
Pic	Provides access to the Microchip PIC16C74b register memory.

cessing objects. Subsequent OOPic releases have based their more complex objects on this original set of objects.

The oRAM object has been superceded by the easier-to-use oPIC object. I do not recommend using the oRAM object.

OOPic II (B.2 Firmware) Description

The OOPic II looks exactly the same as the OOPic I. The connectors, prototyping area, and PC board are identical. The similarity ends there because the OOPic II is expnded to include a large number of robotic-specific and complex objects. The OOPic B.2 firmware has made the older B.1 firmware obsolete. As with the older OOPic I code, information on the OOPic web site still provides details on the B.1 firmware.

HARDWARE DETAILS

The OOPic II uses the standard S board form factor, and its hardware layout is exactly the same as the OOPic I, as shown in Figure 1-3. The OOPic II has the exact same I/O connector pinout as the OOPic I as well. Refer to Table 1-4 and Figure 1-4 for the details.

OOPIC II FEATURES

This section outlines the specifications for the OOPic II, which uses a separate RAM space. Code is stored in a removable EEPROM, and object data and variables are stored in the PICmicro internal RAM. The following features are what you would compare when deciding which OOPic best fits your needs:

- PIC16F77 microcontroller clocked at 20MHz
- Four 8-bit A2D channels
- 86 bytes of object memory
- 72 bytes of variable memory space
- 8KB of program code space (this is what is shipped; you can install more)
- Parallel port (I2C) that is programmable only

OOPIC II OBJECT LISTING

Table 1-6 is the full list of the OOPic II programming objects. For a more detailed description, see the appendices or the extremely detailed OOPic site. This list contains some fairly complex objects, so you may want to use it as a quick reference and skip to the appendices right away.

TABLE 1-6   OOPIC II OBJECT LIST

HARDWARE OBJECTS

OBJECT	DESCRIPTION
oA2D	Provides a numerical measurement of a voltage.
oA2DX	An expanded version of the oA2D object that reads an analog voltage and detects when it has exceeded a threshold.
oBumper4	Reads a four-switch bumper and converts the input to a URCP heading value.
oBumper8	Reads an eight-switch bumper and converts the input to a URCP heading value.
oButton	Reads the state of a switch and controls the state of a *light-emitting diode* (LED).
oCompassDN	Reads a Dinsmore 1490 compass and converts the data to a URCP heading value.
oCompassDV	Reads a Devantech compass and converts the data to a URCP heading value.
oCompassVX	Reads a Devantech compass and converts the data to a URCP heading value.
oDCMotor	Uses URCP speed values to control a DC motor using a LMD18200 H-bridge (a direction bit, a brake bit, and a PWM output).
oDCMotor2	Uses URCP speed values to control a DC motor using an L293-style H-bridge (two direction bits and a PWM output).
oDCMotorMT	Uses URCP speed values to control a DC motor using a Mondo-Tronics-style H-bridge (a direction bit and a PWM output).
oDCMotorWZ	Uses URCP speed values to control a DC motor using a Wirz 203 Motor H-bridge (a direction bit, a brake bit, and a PWM output).
oDIO1	Provides a 1-bit digital I/O.
oDIO16	Provides a 16-bit digital I/O with I/O 8–15 and I/O 24–31 (LSB-MSB).
oDIO16x	Provides a 32-bit digital I/O using a 1-of-3 decoder.
oDIO4	Provides a 4-bit digital I/O (7 possible choices).

oDIO8	Provides an 8-bit digital I/O (3 possible choices).
oFreq	Outputs a frequency to a sound transducer on I/O 21 only.
oI2C	Provides access to an I2C device on the local (programming) bus.
oIRPD1	Reads a single *infrared* (IR) proximity detector (one input).
oIRPD2	Reads a dual IR proximity detector (two inputs and one output).
oIRRange	Reads a Sharp GP2D12 analog IR ranging module.
oJoystick	Reads an Atari-style digital joystick.
oKeypad	Reads a 4 × 4 keypad matrix.
oKeypadX	Scans up to an 8 × 8 matrix of switches and determines which switch is pressed.
oLCD	Controls a LCD display that uses the 44780 chip set.
oLCDSE(T)	Control Scott Edwards Serial LCD Display (T stands for terminal command set).
oLCDWZ	Controls a Wirz Electronics SLI-OEM LCD display controller.
oMotorMind	Controls a Solutions^3 Motor Mind module.
oPWM	Provides a PWM output (two available).
oPWMX	Version of the oPWM that outputs a 4-bit resolution PWM pulse on any I/O line.
oSerial	Provides an asynchronous serial I/O port using the hardware UART.
oSerialPort	Provides a buffered asynchronous serial I/O port with flow control.
oSerialX	Provides serial inputs or outputs on any I/O line with flow control.
oServo	Controls a hobby servo.
oServoSE	Controls a Scott Edwards Serial Servo controller.
oServoSP1	Uses URCP speed values to control a modified servo. 0 sends no pulses out; otherwise, this is identical to oServoX.
oServoSP2	Uses URCP speed values to control a modified servo (experimental).
oServoX	Uses URCP heading values to control an *unmodified* hobby servo.
oSonarDV	Reads a Devantech SRF04 ultrasonic rangefinder.
oSonarPL	Reads a Polaroid 6500 sonar ranging module.
oSPI	Provides SPI serial I/O.
oSPO256	Controls an SP0256 voice synthesizer chip.
oStepper	Controls a stepper motor.
oStepperL	Controls a stepper motor with multiplexed and latched drivers.
oStepperSP	URCP speed values to control a stepper motor.
oStepperSPL	URCP speed values to control a stepper motor with multiplexed and latched drivers.

(continued)

TABLE 1-6   OOPIC II OBJECT LIST (*Continued*)

HARDWARE OBJECTS

OBJECT	DESCRIPTION
oTimer(X)	Provides a 16-bit high-speed counter (X enables MSB to be linked).
oTone	Outputs one or more low-frequency tones on an I/O line at the same time.
oTracker	Implements a line tracker that translates direction in URCP headings.
oUVTronHM	Reads a Hamamatsu UVTron flame detector (A2D line used).
oVideoIC	Overlays text on a video signal using Intuitive Circuits' OSD232 Onscreen Display Character Overlay Board.

PROCESSING OBJECTS

OBJECT	DESCRIPTION
oBus(O, I, C)	Copies a value from one object to another.
oChanged(O, C)	Detects when an object's value has changed.
oClock	Provides a programmable low-speed logic clock.
oCompare(0, 2, C)	Compares object values with one reference, two references, or zero.
oConverter	Provides conversion functions for seven-segment displays, stepper motors, and so on.
oCountDown(O)	Decreases an object's value until it is 0.
oCounter	Provides counting functions, normal and quadrature encoded.
oDataStrobe	Provides a data strobe in response to a value being written to it.
oDDDELink	Provides a DDE link over the I2C network.
oDebounce	Provides logic-state debounce functions.
oDivider	Provides a clock divider.
oEvent(X, C)	Runs program code in response to an event.
oFanOut(C)	Copies an input flag value to other multiple objects.
oFlipFlop(C)	Implements a set-reset FlipFlop.
oGate(C)	Provides logic-gate and flag-linking functions.
oIndex(C)	Provides indexing functions to an oBuffer object.
oMath(O, I, C)	Provides mathematical functions and links a value between objects.

oNavCon(I, C)	Provides differential steering calculations for two motor-drive robots.
oNavConEI(C)	Provides differential steering calculations with encoder inputs.
oOneShot	Produces a one-pulse output in response to logic transition.
oRamp(I, C)	Calculates a URCP speed ramp up or down for DC motor control.
oRandomizer	Provides a random number.
oRepeat	Provides a key-repeat function.
oRTC	Maintains a (nonpersistent) real-time clock.
oSrvSync	Provides a method of synchronizing multiple hobby servos.
oWire(C)	Copies a Boolean value or flag to another object (one-input oGate).

VARIABLE OBJECTS

OBJECT	DESCRIPTION
oBit	Maintains a 1-bit variable.
oBuffer	Maintains a variable-size data buffer/string variable (bytes only).
oByte	Maintains an 8-bit (1-byte) variable.
oEEProm	Provides access to the nonvolatile EEPROM memory of E0 only.
oNibble	Maintains a 4-bit variable.
oRAM	Provides access to the OOPic RAM and registers.
oWord	Maintains a 16-bit (2-byte) variable.

USER-DEFINABLE OBJECTS

OBJECT	DESCRIPTION
oUserClass	Uses an OOPic program to provide a new object definition (not VC).

SYSTEM OBJECTS

OBJECT	DESCRIPTION
OOPic	Provides control of and maintains information about the OOPic.
Pic	Provides access to the Microchip PIC16F77 register memory.

You'll see a lot of (C) or (O, I, C) suffixes in the processing objects. These objects have had *clocking* inputs added to their VC objects, and many have input (I) and output (O) values in the object instead of a link to another object as well. Finally, you will note a lot of hardware objects that use the *Uniform Robotic Control Protocol* (URCP), which enables direction or speed to be encoded into a single byte of data. A signed data byte simplifies a lot of robotics' direction giving, which is discussed further in subsequent chapters.

OOPic II+ (B.2+ Firmware) Description

The OOPic II+ adds three new capabilities to the OOPic series: an internal fast EEPROM that acts like persistent variable storage (*sByte*), a new 10-bit A2D with a total of seven channels, and the *Serial Control Protocol* (SCP), which requires an RS-232 serial connection. Because the OOPic II+ uses the S Board form factor like the OOPic I and II, RS-232-level translation hardware will need to be added by the user. The SCP can be used to program and control an OOPic II+ and liberates the OOPic II+ from the PC parallel port.

HARDWARE DETAILS

The OOPic II+ uses the standard S board form factor, and its hardware layout is exactly the same as the OOPic I, as shown in Figure 1-3. Adding three new A2D channels changes the I/O connector pinout in a minor way. I/O lines 22 and 23, Serial transmit and Serial receive, may not be used as digital I/O, only as serial I/O. These differences are detailed in Table 1-7.

TABLE 1-7   OOPIC II+ I/O CONNECTOR PIN DEFINITIONS

PIN	NAME	FUNCTION	PIN	NAME	FUNCTION
1	LSDA	Local I2C serial data	2	GND	Ground
3	LSCL	Local I2C serial clock	4	Power	+6V–18V power supply
5	Reset	Active low OOPic reset	6	I/O 15	I/O Group 1, bit 7
7	I/O 1	A2D 1	8	I/O 14	I/O Group 1, bit 6
9	I/O 2	A2D 2	10	I/O 13	I/O Group 1, bit 5
11	I/O 3	A2D 3	12	I/O 12	I/O Group 1, bit 4
13	I/O 4	A2D 4	14	I/O 11	I/O Group 1, bit 3
15	I/O 5	A2D 5	16	I/O 10	I/O Group 1, bit 2
17	I/O 6	A2D 6	18	I/O 9	I/O Group 1, bit 1
19	I/O 7	A2D 7	20	I/O 8	I/O Group 1, bit 0
21	+5V	Regulated +5V out	22	+5V	Regulated +5V out
23	GND	Ground	24	GND	Ground

PIN	NAME	FUNCTION	PIN	NAME	FUNCTION
25	I/O 16	I/O Group 2, bit 0			
		Timer 1 crystal	26	I/O 31	I/O Group 3, bit 7
27	I/O 17	I/O Group 2, bit 1 (PWM2)			
		Timer 1 crystal	28	I/O 30	I/O Group 3, bit 6
29	I/O 18	I/O Group 2, bit 2 (PWM1)	30	I/O 29	I/O Group 3, bit 5
31	I/O 19	I/O Group 2, bit 3			
		I2C serial clock	32	I/O 28	I/O Group 3, bit 4
33	I/O 20	I/O Group 2, bit 4			
		I2C serial data	34	I/O 27	I/O Group 3, bit 3
35	I/O 21	I/O Group 2, bit 5	36	I/O 26	I/O Group 3, bit 2
37	I/O 22	Serial transmit	38	I/O 25	I/O Group 3, bit 1
39	I/O 23	Serial receive	40	I/O 24	I/O Group 3, bit 0

OOPIC II+ FEATURES

This section outlines the specifications for the OOPic II+, which uses a separate RAM space. Code is stored in a removable EEPROM, and object data and variables are stored in the PICmicro internal RAM. A new data type is being used in addition to the Byte and Word data types known as the sByte. The sByte variables are stored in the PICMicro internal EEPROM space and will retain their values even after the OOPic II+ is turned off. No sWord exists, and only bytes can be stored in sByte variables. The following features are what you would compare when deciding which OOPic best fits your needs:

- PIC16F877 microcontroller clocked at 20MHz
- Seven 10-bit A2D channels
- 86 bytes of object memory
- 72 bytes of variable memory space
- 8KB of program code space (this is what is shipped; you can install more)
- 256 bytes of fast internal EEPROM variable space (sByte)
- Parallel port (I2C) and serial port that are programmable (with additional RS-232 hardware)

The OOPic II+ serial lines I/O line 22 and I/O line 23 are monopolized by SCP when it is functional, so oSerial, oSerialPort, and oSerialX must be used with care. Regardless of which object controls the serial port, you may not define I/O 22 and 23 as digital lines. When SCP is disabled, those other serial objects can be used normally. More details about this will be provided in Chapter 11.

OOPIC II+ OBJECT LISTING

Only two new objects have been added with the OOPic II+, oA2D10 and oSequencer (see Table 1-8). The oA2D10 object is a 10-bit A2D object that can have either four 10-bit A2D channels or seven 10-bit A2D channels active, depending upon how it's used. The oSequencer object is a complex addition that is almost an entirely new language embedded into the OOPic's main language. It uses the internal fast EEPROM to store a sequence of bytes to be sent to an object, such as sending a series of notes for a song to the oFreq object. The entire OOPic II object list will not be included here only to add a single object, so I'll just note the new one.

OOPic R (B.2+ Firmware) Description

The OOPic R uses the same firmware as the OOPic II+, the B.2+ revision, but it has a radically different PC board form factor. The OOPic R has a number of interfaces built onto the board that will simplify its use in a robot or even as an experimenter's board. This board is a special-case board and in no way shows *Things To Come*. The OOPic S-Board form factor is still the standard OOPic release base and will remain so as long as 40-pin *dual in-line package* (DIP) PICMicro processors exist. Those OOPic users that loathe the use of a soldering iron will love the OOPic R.

The board contains a speaker, three push buttons, three LEDs, a built-in RS-232-level converter, and an associated DB9 serial cable connector. The OOPic R board also has special hobby servo connectors, a consolidated DC motor connector with direction and PWM pins going to it already, and two on-board 5V regulators (one for the OOPic onboard chip set and one for the I/O connector). The board also offers space for an optional LM7805-pinout compatible regulator for the I/O port connectors.

Also selectable is the ability to use raw battery power from a special connector instead of any kind of regulator. This is especially useful for powering your DC motors or hobby

TABLE 1-8 OOPIC II+ ADDITIONAL OBJECTS	
HARDWARE OBJECTS	
OBJECT	**DESCRIPTION**
oA2D10	Provides a 10-bit numerical measurement of a voltage.
PROCESSING OBJECTS	
oSequencer	Sequences a series of numbers from fast EEPROM to an object.

servos. Half the OOPic processor's I/O lines are taken up by these onboard functions, so only 16 I/O ports appear on the 40-pin I/O connector. The OOPic R I/O connector is broken into four separate power bus options; that is, every four I/O ports have their own selectable power supply.

HARDWARE DETAILS

The OOPic R is a far more complex board than the OOPic S series boards, so it will be broken down by sections with the more interesting bits detailed separately for clarity. Figure 1-5 shows the OOPic R board in all its glory and notes the various subsystems that are built onto the board. Table 1-9 shows the pinout details for the 56-pin I/O connector on the OOPic R board. Just like the other OOPic variants, a 40-pin cable can be used to hook up with the OOPic R board, but this connector also enables all 16 available I/O ports to use a standard Futaba® or Hitec® hobby servo connector as well.

Another difference worth noting is that the OOPic R I/O connector does not include the local I2C bus LSDA and LSCL lines anymore. You will need to use the I2C programming connector on the board to gain access to that bus. Also, only one I2C networking port connector is located on the OOPic R board, which means it must be at the end of any daisy chain of I2C linked OOPic devices. Note that all the pins closest to the outside edge of the board (pins 41 to 56) are ground.

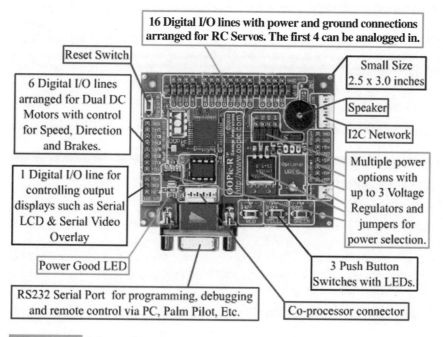

16 Digital I/O lines with power and ground connections arranged for RC Servos. The first 4 can be analogged in.

Reset Switch

6 Digital I/O lines arranged for Dual DC Motors with control for Speed, Direction and Brakes.

1 Digital I/O line for controlling output displays such as Serial LCD & Serial Video Overlay

Power Good LED

RS232 Serial Port for programming, debugging and remote control via PC, Palm Pilot, Etc.

Small Size 2.5 x 3.0 inches

Speaker

I2C Network

Multiple power options with up to 3 Voltage Regulators and jumpers for power selection.

3 Push Button Switches with LEDs.

Co-processor connector

Figure 1-5 OOPic R PC board layout

TABLE 1-9 OOPIC R I/O CONNECTOR PIN DEFINITIONS

PIN	NAME	FUNCTION	PIN	NAME	PIN	NAME
1	+5V	Regulated +5V out	2	GND		
3	I/O 1	A2D 1	4	S1 Power	41	GND
5	I/O 2	A2D 2	6	S1 Power	42	GND
7	I/O 3	A2D 3	8	S1 Power	43	GND
9	I/O 4	A2D 4	10	S1 Power	44	GND
11	+5V	Regulated +5V out	12	GND		
13	I/O 8	I/O Group 1, bit 0	14	S2 Power	45	GND
15	I/O 9	I/O Group 1, bit 1	16	S2 Power	46	GND
17	I/O 10	I/O Group 1, bit 2	18	S2 Power	47	GND
19	I/O 11	I/O Group 1, bit 3	20	S2 Power	48	GND
21	I/O 12	I/O Group 1, bit 4	22	S3 Power	49	GND
23	I/O 13	I/O Group 1, bit 5	24	S3 Power	50	GND
25	I/O 14	I/O Group 1, bit 6	26	S3 Power	51	GND
27	I/O 15	I/O Group 1, bit 7	28	S3 Power	52	GND
29	+5V	Regulated +5V out	30			GND
31	I/O 28	I/O Group 3, bit 4	32	S4 Power	53	GND
33	I/O 29	I/O Group 3, bit 5	34	S4 Power	54	GND
35	I/O 30	I/O Group 3, bit 6	36	S4 Power	55	GND
37	I/O 31	I/O Group 3, bit 7	38	S4 Power	56	GND
39	+5V	Regulated +5V out	40			GND

OOPIC R FEATURES

Here we'll examine the specifications for the OOPic R. Similar to OOPic II+, the OOPic R uses a separate RAM space and code is stored in a removable EEPROM. Also like OOPic II+, object data and variables are stored in the PICmicro internal RAM. The sByte variables are also used and will retain their values even after the OOPic R is turned off. No sWord exists, so only bytes can be stored in sByte variables. The following features are what you would compare when deciding on an OOPic:

- PIC16F877 microcontroller clocked at 20MHz
- Four 10-bit A2D channels
- 86 bytes of object memory
- 72 bytes of variable memory space

- 8KB of program code space (this is what is shipped; you can install more)
- 256 bytes of fast internal EEPROM variable space
- Parallel port (I2C) and serial port that are programmable
- SCP enabled with onboard RS-232 hardware and a DB9 serial port connector
- Three momentary contact push button switches
- Three LEDs: red, yellow, and green
- Multiple power selectivity by banks of I/O ports
- Special connectors for servos, DC motor controllers, and serial LCD
- Onboard speaker
- Two 5V regulators and space for an optional third regulator

As stated earlier, here the OOPic R serial lines I/O line 22 and I/O line 23 are monopolized by SCP when it is functional, so oSerial, oSerialPort, and oSerialX must be used with care. When SCP is disabled, those other serial objects can be used normally. All the dedicated OOPic R I/O lines come out to pads that are labeled on the top of the board except for two: I/O 22 (Tx) and I/O 23 (Rx). These *time-to-live* (TTL) signal lines run to pads under the DB9 serial connector. Turn your board over so that you are looking at the back of the board with the DB9 connector facing toward you. You will see three pads near the edge of the PC board that have no pins sticking through. The pad on the left is Rx, I/O 23, and the pad on the right is Tx, I/O 22. The pad in the middle isn't connected (see Figure 1-6).

Power Selection Jumpers

The OOPic R has a rich variety of options for powering the I/O connectors' power pins, labeled +P on the I/O connector. Figure 1-7 shows how the jumpers map to the I/O banks

Figure 1-6 OOPic R I/O 22 and I/O 23

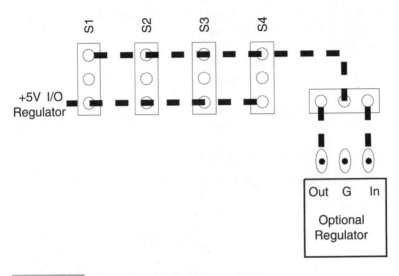

Figure 1-7 OOPic R power jumper mapping

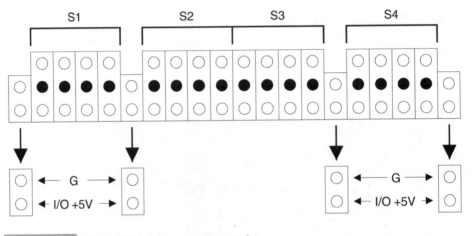

Figure 1-8 OOPic R I/O bank power mapping

and how the *PSA* jumper selects the +SP power source. Figure 1-8 shows which banks of I/O pins are affected by the power jumpers.

As an example of how to program these pins, let's say you want to power your four hacked hobby servos on your four-wheel-drive robot with a separate 7.2V battery. All other I/O devices are powered by the onboard 5V high-current regulator. Because we know that regulating the supply to our motors is just an excuse to heat our robot, we aren't going to regulate it. I/O lines 28 to 31 are chosen as our servo I/O lines, so we set the *PSL* jumper to the right, which selects the +SP power line directly. We then set jumpers *S1*, *S2*, and *S3* to the lower position, which selects an onboard 5V high-current regulator for them. Finally, we set jumper *S4* to the upper position, which selects the power source chosen by the PSL jumper for that bank. Figure 1-9 shows our final power selection jumpers in their proper positions, and Figure 1-10 shows where we'll connect all our batteries. Table 1-10 lists the

Figure 1-9 Power jumper setting example

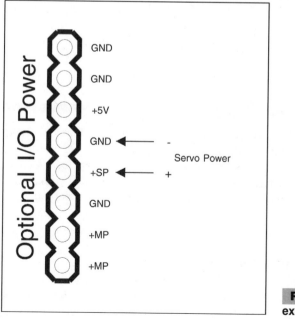

Figure 1-10 Battery hookup example

TABLE 1-10   OPTIONAL POWER CONNECTOR

PIN	NAME	DESCRIPTION
1	GND	Ground reference
2	GND	Ground reference
3	+5V	Regulated +5V out
4	GND	Ground reference
5	+SP	Separate hobby servo power input
6	GND	Ground reference
7	+MP	Main unregulated power input
8	+MP	Main unregulated power input

pin definitions for the *optional power connector*. Remember, this is the *optional* power connector. The main power connector is just below this one, where we've attached the main battery for OOPic power.

Pushbutton Switches and LEDs

The OOPic R board uses the three additional A2D ports as digital I/O to read the three pushbutton switches and write the three LEDs. Each of these I/O lines handles one switch and one LED through a clever arrangement of resistors. The new B.2 firmware object oButton is especially designed to handle this kind of switch and LED arrangement in a number of ways. To be even more useful, the OOPic R board's LEDs are three different colors: red, yellow, and green. Figure 1-11 details how the switches and LEDs are configured on the board.

Dual DC Motor Control Connector

This connector is laid out in such a way that both hardware PWM lines and four digital control lines are routed conveniently to plug into a variety of H-bridge motor driver boards. The B.2 firmware includes several objects that can use this connector: *oDCMotor*, *oDCMotor2*, *oDCMotorMT*, and *oDCMotorWZ*. These objects cover any kind of DC motor you can find or build that isn't controlled by a serial data link. Table 1-11 shows the pin definitions of the connector.

Networking and Programming Connectors

The OOPic R has an I2C programming connector that is also the only connection point for the local I2C bus used for offboard I2C coprocessors. It also has an I2C networking port that is used to link to other OOPic devices, but unlike the S-Boards, only one networking

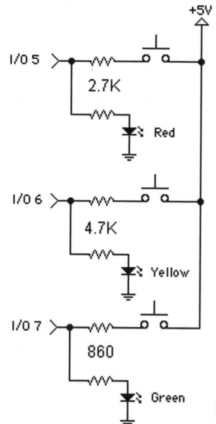

Figure 1-11 OOPic R push button and LED circuit

TABLE 1-11   DUAL H-BRIDGE CONNECTOR

PIN	NAME	DESCRIPTION
1	GND	Ground
2	Channel-A control-1	I/O 24 as either a direction or brake pin.
3	Channel-A PWM	I/O 17 PWM output.
4	Channel-A control-2	I/O 25 as either a direction or brake pin.
5	Channel-B control-1	I/O 26 as either a direction or brake pin.
6	Channel-B PWM	I/O 18 PWM output.
7	Channel-B control-2	I/O 27 as either a direction or brake pin.
8	+5V	+5V from the onboard high-current 5V regulator.

connector is used. Thus, the OOPic R must be at the end of the I2C chain. These I2C ports function exactly as their S-Board counterparts.

The OOPic R also has a DB9 serial connector port and the associated RS-232-level translation hardware required to connect it to any computer serial port. The board can be programmed through this port just as it can be programmed through the I2C programming port. The OOPic R may also be programmed via the new serial port using SCP commands and it can be controlled via this serial port in the same manner by the external computer using SCP commands to access the VC components currently programmed and running. When SCP is enabled, the serial port is unavailable to any other object, such as oSerial and oSerialPort. The oSerialX object uses any I/O port, so it is unaffected by SCP activity. More details about SCP programming are covered in Chapter ??. The onboard DB9 connector only connects the Tx and Rx lines, so no flow control can be done using this connector unless you wire the signal lines to it manually and add additional RS-232-level conversion hardware.

The Rest of the Interesting Features

Certain other features are simple, useful, and interesting (and a few are boring but useful, such as the power-on LED), but they don't warrant all that much description. Here are two such features:

- A speaker is attached to I/O line 21 that can be used with the oFreq and oTone objects.
- A three-pin connector that uses I/O line 16 for use with TTL serial connected devices, such as a serial LCD controller or a serial video overlay device, all of which have associated OOPic object definitions, of course.

OOPIC R OBJECT LISTING

The OOPic R has the same object list as the OOPic II+. See Table 1-6 and Table 1-8 for those lists.

OOPic C (B.2+ Firmware) Description

The OOPic C is an even greater diversion from the classic OOPic S-Board form factor. It is a 24-pin carrier lead chip that is pin-compatible with all the Stamp carrier boards out there. This means that anyone who wants to try something a little different can use an OOPic C and preserve their investment in their current carrier boards. This flavor of OOPic has been created to allow a gentle introduction into the world of the VC and OOP programming à la OOPic.

HARDWARE DETAILS

The OOPic C has 16 digital I/O lines, which make up 16 of the 24 DIP pins on the carrier. These are the same 16 I/O lines used by the oDIO16 OOPic object. In addition to those I/O lines, support exists for an I2C programming port with pads on the far right side of the chip. These ports have the exact same pinout and pin location as the programming (local I2C bus) and I2C networking ports discussed earlier in the chapter. Support also exists for an I2C networking port with pads on the far left side of the chip, four A2D lines on the pads, the hardware UART Tx and Rx I/O lines on the pads, both PWM lines out to the pads, and a couple more I/O lines. Thus, it's not really hampered by only having 24 pins on the carrier board. Figure 1-12 shows the pinout of the OOPic C carrier chip. Table 1-12 lists the pins' descriptions, and Table 1-13 lists descriptions of the pads on the carrier chip.

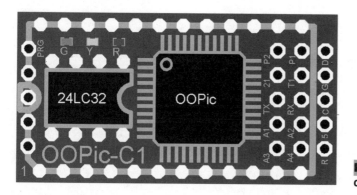

Figure 1-12 OOPic C chip pinout

TABLE 1-12 OOPIC C PIN DESCRIPTION

PIN	NAME	DESCRIPTION	PIN	NAME	DESCRIPTION
1	Tx	RS-232 out from OOPic C	2	Rx	RS-232 into OOPic C
3	ATN	Active high OOPic reset	4	GND	Ground
5	I/O 8	I/O Group 1, bit 0	6	I/O 9	I/O Group 1, bit 1
7	I/O 10	I/O Group 1, bit 2	8	I/O 11	I/O Group 1, bit 3
9	I/O 12	I/O Group 1, bit 4	10	I/O 13	I/O Group 1, bit 5
11	I/O 14	I/O Group 1, bit 6	12	I/O 15	I/O Group 1, bit 7
13	I/O 24	I/O Group 3, bit 0	14	I/O 25	I/O Group 3, bit 1
15	I/O 26	I/O Group 3, bit 2	16	I/O 27	I/O Group 3, bit 3
17	I/O 28	I/O Group 3, bit 4	18	I/O 29	I/O Group 3, bit 5
19	I/O 30	I/O Group 3, bit 6	20	I/O 31	I/O Group 3, bit 7
21	+5V	Regulated +5 in/out	22	Reset	Active low reset in/out
23	GND	Ground	24	Vin	Unregulated +9−18V

TABLE 1-13   OOPIC C SOLDER PAD DEFINITIONS

PRG PADS ON FAR LEFT (STARTING WITH PIN 1 ON THE BOTTOM LEFT)

PAD	NAME	DESCRIPTION	PAD	NAME	DESCRIPTION
1	LSDL	Local I2C serial data	2	GND	Ground
3	LSCA	Local I2C serial clock	4	+5V	Regulated +5V reference
5	RESET	Active Low Reset			

PADS ON RIGHT

PAD	NAME	DESCRIPTION	PAD	NAME	DESCRIPTION
A1	I/O 1	10-bit A2D1	A2	I/O 2	10-bit A2D2
A3	I/O 3	10-bit A2D3	A4	I/O 4	10-bit A2D4
TX	I/O 22	UART serial transmit	RX	I/O 23	UART serial receive
21	I/O 21	oFreq output line	T1	I/O 16	16-bit timer input
P1	I/O 18	PWM-1	P2	I/O 17	PWM-2

PADS ON FAR RIGHT

PAD	NAME	DESCRIPTION	PAD	NAME	DESCRIPTION
R	RESET	Active low Reset	5	+5V	Regulated +5V reference
C	I/O 19	I2C networking serial clock	G	GND	Ground
D	I/O 20	I2C networking serial data			

The OOPic C carrier board pins are through-hole posts that extend above the chip carrier high enough to be used as connections to other chips or even *daughter board* piggy-backed chips. The OOPic C is also offered in a *no lead* option so that the chip carrier can be placed into very tight spaces or mounted directly to other devices by soldering directly to the holes where the chip carrier leads would otherwise have been. Figure 1-13 shows the OOPic C with and without its chip carrier leads.

OOPIC C FEATURES

The specifications for the OOPic C are outlined here. The OOPic C uses a separate RAM space for variables so you don't take up space that could be used by objects. Code is stored in a removable EEPROM, and object data and variables are stored in the PICmicro internal RAM. The *sByte* is utilized here, and its variables are stored in the PICMicro internal EEP-ROM space and will retain their values even after the OOPic C is turned off. No sWord is

Figure 1-13 OOPic C options

used, so only bytes can be stored in sByte variables. The following features are what you would compare when deciding which OOPic best fits your needs:

- PIC16F877 microcontroller clocked at 20MHz
- Four 10-bit A2D channels
- 86 bytes of object memory
- 72 bytes of variable memory space
- 8KB of program code space (this is what is shipped; you can install more)
- 256 bytes of fast internal EEPROM variable space (sByte)
- Parallel port (I2C) and a serial port that are programmable
- Three LEDs: red, yellow, and green
- Availability on a 24-pin Stamp-compatible carrier chip
- 18 I/O lines available on chip pins, and 10 more I/O lines available on connector pads
- Pins 1 and 2 are RS232 serial lines that can be directly connected to an RS232 cable

The OOPic C serial lines (I/O Lines 22 and 23) are monopolized by SCP when it is operating, so oSerial and oSerialPort can't be used. When SCP is disabled, those I/O Lines can be used with oSerial and oSerialPort.

OOPIC C OBJECT LISTING

The OOPic C objects are exactly the same as those in the OOPic R and OOPic II+. See Tables 1-6 and 1-8 for the listings.

THE OOPIC IDE AND COMPILER

CONTENTS AT A GLANCE

The OOPic hardware comes with its associated *Integrated Development Environment* (IDE) and compiler for free. Neither component will function without the other, so you are paying for both of them when you get an OOPic controller. The IDE handles the basics of file editing, connecting to the OOPic via a programming cable, setting language and firmware options, and downloading the compiled file to the OOPic. The compiler is separate from the IDE and is called by the IDE to convert your source code to OOPic execution code for downloading to an OOPic board.

This chapter goes through the process of installing the IDE and compiler, debugging OOPic objects, and configuring and troubleshooting an OOPic programming cable installation. This book assumes you are using the IDE that comes bundled with the included CD-ROM. If you are having trouble with your OOPic programs, read the "CurrentlBuglist.pdf" file in the root directory of the CD-ROM. This file gives the currently known bugs with the compiler included on the CD-ROM.

Installing the OOPic Environment

This process can be anything from painless (95 percent of the time) to very painful, depending on your operating system and PC hardware. I'll deal with them in order from easiest to most complex. You can get the latest installer package from the OOPic web site located at www.oopic.com. The included CD-ROM has all you'll need at the time of this writing. It will work with all examples and current OOPic firmware and boards. Follow these tips to minimize problems with the installation:

- Turn off all virus checking software before installing the OOPic software.
- Follow ALL installer directions.
- Reboot your computer after you have installed the OOPic software and BEFORE you try to use the software.

WINDOWS 95 AND WINDOWS 98

This is the simplest way to go. In fact, it is recommended that you just go out and buy a cheap laptop at your favorite auction house with Windows 95 on it; use that as your dedicated embedded controller development toy machine. It doesn't have to be super fast; I have a Pentium™ 133 laptop that works great with the OOPic IDE.

Installation Procedure

First, make sure you have at least 10MB of hard disk space available, preferably 20MB. You should have at least 32MB of RAM for Windows 95 or Windows 98; my laptop has 40MB and does fine. You also need an IBM-compatible DB25 parallel port connector or a DB9 serial port connector (depending on which OOPic you are using). To install the IDE, follow these steps:

1. Put the included OOPic Development CD-ROM into your CD-ROM drive.
2. Select *Environment\OOPicFV501.exe* and double-click it to start the installer.
3. Follow the instructions and use the default settings for everything.

That's it; the OOPic IDE is now installed. On occasion, you will get an alert window whose text begins, "The OLE system files are in use and cannot be updated." Simply fol-

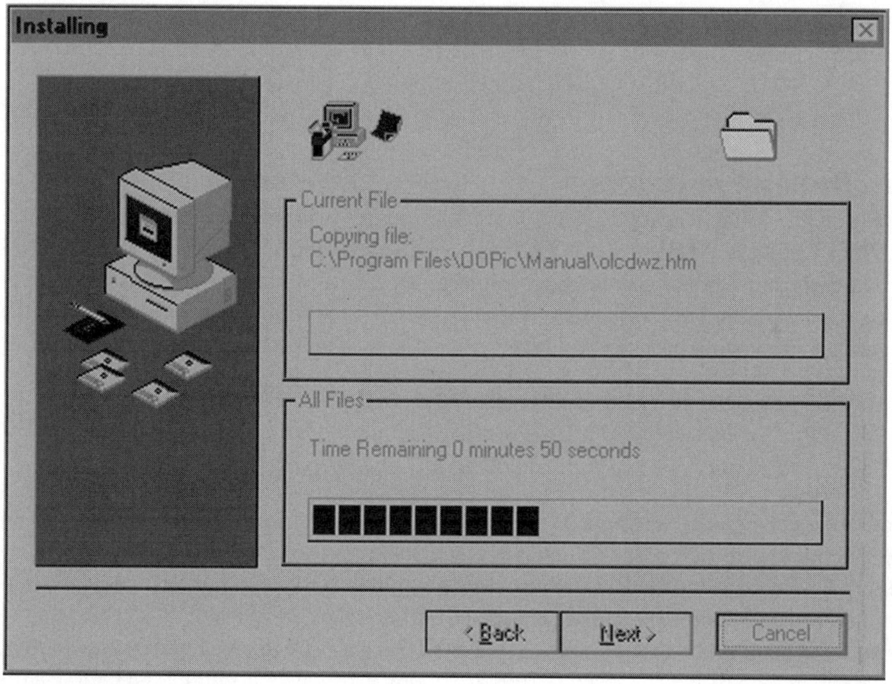

Figure 2-1 Installer progress window

low the instructions, hit *Yes*, and continue letting the installer reboot your system and restart the installation. Just keep clicking *Next* and the install will continue on its merry way and work fine. As long as you keep seeing the window displayed in Figure 2-1, your installation will succeed.

When completed, the installer will have done its magic and in the C:\Program Files\OOPic directory you will find oopic.exe (the IDE executable), oopicMK.exe (the compiler) and the following directories:

- **Manual**   This is the manual files directory. The files are all in HTML and are accessed through the Help menu in the OOPic IDE.
- **Samples**   Dozens of example programs are provided here, showing the use of OOPic objects.
- **Source**   This is the Visual Basic™ source code for the IDE. This is great fun for all you hackers out there that want a shot at tweaking the OOPic IDE to suit yourself. Note that you will need your own Visual Basic development environment to use these files.

WINDOWS NT, WINDOWS 2000, AND WINDOWS XP

These versions of the Microsoft operating system are more resistant to lockups, freezes, and crashing because they restrict what a program can access. They also force proper "be-good" rules on unruly code. Unfortunately, these new operating systems prevent the real-time access to the hardware parallel port that the OOPic (and all other parallel port-based programmers) require. To provide access to the parallel port, Savage Innovations has provided the *Port95nt.exe* installer that installs utilities, allowing access to this hardware.

Installation Procedure

Make sure you have at least 10MB of hard disk space available, preferably 20MB. You should have at least 64MB of RAM for these operating systems to run well. You also need an IBM-compatible DB25 parallel port connector or a DB9 serial port connector (depending on which OOPic you are using). You *must* install files in the following order for the OOPic IDE to function properly:

1. Select *\Patches\Port95NT.exe* and double click on it, follow all directions.
2. Select *Environment\OOPicFV501.exe* and double-click it to start the installer.
3. Follow the instructions and use the default settings for everything.

You may get a similar error to the OLE error discussed earlier. Just let the installer do what it wants and you'll be fine. When completed, in the C:\Program Files\OOPic directory you will find oopic.exe (the IDE executable); oopicMK.exe (the compiler); and the Manual, Samples, and Source directories described earlier.

There are other patch installs in the CD-ROM \Patches directory named "setup_<file>. exe. Use them only if their respective files are giving errors when running the OOPic software. They are rarely needed.

WINDOWS XP ADDITIONAL PROCEDURE

Windows XP is a completely new operating system that is similar to Windows NT and Windows 2000. However, some folks have had difficulties getting the OOPic IDE to work with Windows XP. If this happens to you and your OOPic IDE complains about not being able to find MSCOMDLG or COMDLG32 files, then you will have to untangle this not-unheard-of snag. This problem has a solution, however. After following the previous installation instructions correctly, execute the following steps:

1. Put the included OOPic Development CD-ROM into your CD-ROM drive.
2. Select one of the files named *\Patches\setup_<file>.exe* if the OOpic software is having problems with the system file named <file>. Double click on the file to start the installer.
3. Follow the instructions and use the default settings for everything.

Only use this fix as a last resort. Do not assume you need it.

The OOPic installer includes these files and they should not need to be installed using this patch, but sometimes it is required. If you start having problems with the OOPic IDE after you install an operating system update or another software package, simply reinstall the OOPic IDE. Don't try to mess with any of the other files; just do another full install. None of your files will be lost.

Programming Options for OOPic Variants

You can use many possible combinations of firmware, cabling, and languages to program your OOPic controller. This section will show you how to set them properly. No matter which firmware version or board layout you use (and this includes the third-party OOPic controller board), the processes for setting them up are similar.

SETTING OOPIC OPTIONS

You need to set certain options to ensure the OOPic IDE generates the correct code for your board. Here are some additional options you'll set to make your coding easier and more comfortable.

Setting OOPic IDE Views

A few different views can be used in the IDE to amuse and educate you. They are found in the Views menu pulldown window shown in Figure 2-2:

- **Toolbar**   This is a default setting. Clicking this selection will enable the toolbar, the one with all the useful buttons on it, to be displayed (see Figure 2-3). Leave this set; it's just too useful.
- **Opp Codes**   Clicking this selection will cause the compiled OOPic tokens to be displayed in the right window pane (in my opinion, this is about as interesting as

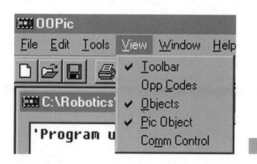

Figure 2-2 OOPic Views menu

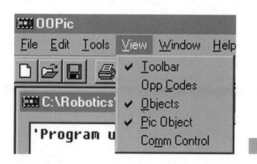

Figure 2-3 OOPic IDE Toolbar

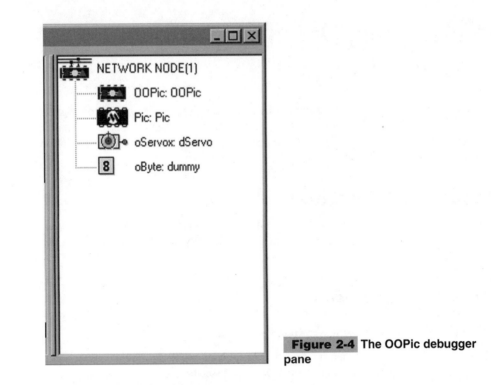

Figure 2-4 The OOPic debugger pane

watching paint dry). If something is not working properly, a true OOPic hacker can use this option to track down the offending code.

- **Objects** This is a default selection. This view shows the graphical OOPic object list after a program has been compiled. When enabled, the OOPic debug functionality is accessed through this pane by clicking on the objects you are interested in. See Figure 2-4 for a short example. Debugging will be discussed later in this chapter.
- **Pic Object** Clicking this selection will include the PIC object (shown in Figure 2-4) in the debugger pane. If it is not clicked, then this object will not be present. It's really not essential to anyone but the *most* hardcore OOPic hacker.
- **Comm Control** Clicking this selection will enable the serial port command-line window. This window enables you to send *Serial Control Protocol* (SCP) commands to an OOPic R, OOPic C, or OOPic II+ device. If you use the oSerial object in your OOPic program, it will enable you to also interact with your OOPic over the serial port.

Setting the OOPic Language Option

All the examples presented here use the Basic language (C examples appear on the CD-ROM, just not in print), but the OOPic supports the C and Java™ syntax as well. A language can be selected in two ways:

- Select your language, in this case Basic, from the Tools pulldown menu, as shown in Figure 2-5.

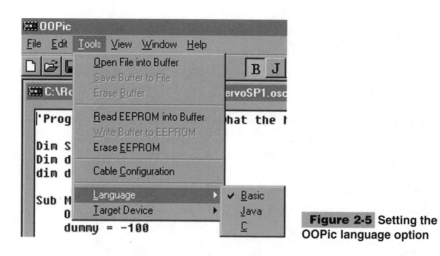

Figure 2-5 Setting the OOPic language option

- Select your language, in this case Basic, but press the B button on the Toolbar, which is also shown selected in the upper right of Figure 2-5.

Setting the OOPic Firmware Version

This setting is essential, and no default setting exists because the OOPic IDE has no way of knowing which OOPic variant you are going to be programming for. The compiler generates different code for each of the OOPic variant firmware options. The options currently supported by the OOPic IDE are shown in Figure 2-6. Table 2-1 shows how to set the firmware option.

The OOPic I and II devices can have multiple firmware revisions, as you can see. Check their labels carefully before selecting your firmware. If you choose incorrectly, your program won't run.

PARALLEL PORT PROGRAMMING

All OOPic variants can be programmed via the parallel port cable using I2C bus logic. The OOPic C requires soldering a programming port connector on the chip carrier, to program the chip using the parallel port. The parallel port programming process communicates over the cable directly to the OOPic EEPROM while holding the OOPic PICMicro in a reset state so that it is disabled. Before you can use your programming cable, the OOPic IDE either has to discover the cable or you need to tell it where this cable is. You then calibrate the speed of your programming cable and you're ready to roll.

OOPic Parallel Port Cabling

Obviously, before you can program your OOPic using your parallel port cable, you need to connect it to your PC and to the five-pin programming port connector on your board.

THE OOPIC IDE AND COMPILER

2

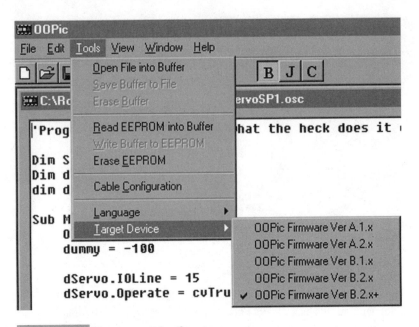

Figure 2-6 Firmware selection menu

TABLE 2-1	OOPIC DEVICES AND FIRMWARE
NAME	**FIRMWARE**
OOPic (OOPic I)	A.1.X
	A.2.X
OOPic II	B.1.X
	B.2.X
OOPic II+	
OOPic R	
OOPic C	B.2.X+

Figure 2-7 shows you what the parallel port looks like on the back of, in this case, a laptop computer. If you didn't get a parallel port programming cable with your OOPic (you only purchased a board and not the complete starter kit), refer to "The I2C Programming and Debugging Cable" section in Chapter 1, "OOPic Family Values," for instructions on building one of your own.

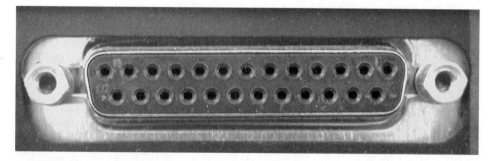

Figure 2-7 The parallel port connector

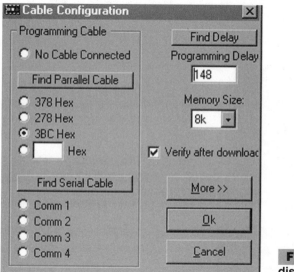

Figure 2-8 Parallel port discovery

OOPic Parallel Cable Calibration

Before you connect and program your OOPic, you first need to get the OOPic IDE ready to program. The first steps in this process are telling the IDE how to find the parallel port and cable, and then to calibrate the computer system. To discover and calibrate your parallel port cable, click the Tools menu and go to Cable Configuration. Figure 2-8 shows what this tool looks like. Note that you can discover either a parallel port cable or a serial port cable, but only one of these will be active at a time.

If you already know which parallel port you will be connecting to, you can check that selection, if it is one of the three most common ones, or you can enter it into the Hex box. I prefer to just let the OOPic IDE find and select the correct port. It's easier this way, and you get a free system check that guarantees everything will work.

To get the OOPic IDE to discover your cable automatically, simply plug the cable into the parallel port connector, *not* into the OOPic board, and press the *Find Parallel Cable* button. When you have the IDE discover the cable, it displays a dialog box that tells you how to connect the cable. If the IDE finds your parallel cable, you will get a dialog box that looks like Figure 2-9. In this case, the IDE found the cable at address Hex 3BC. Press OK and the IDE automatically selects the entry, as shown in Figure 2-8.

If you have plugged your cable into a parallel port connector and pressed the Find Parallel Cable button, and you get the error shown in Figure 2-10, proceed to "Troubleshooting OOPic Programming Problems" later in this chapter to address the problem.

The next step in the I2C cable calibration process is to find the system delay value. This value can range from 20 to 2,000, so don't be alarmed by what you see. The value you get depends on the speed of your computer and its parts. To calibrate your cable, press the Find Delay button shown in Figure 2-8. You then get a dialog box telling you what to do to perform the test, as shown in Figure 2-11. Follow those instructions and press OK. The OOPic IDE then shows a dialog box telling you that the test will take 10 seconds, so press OK and wait for the test to complete.

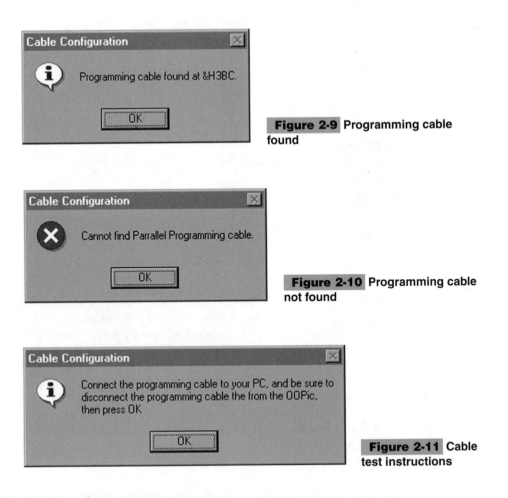

Figure 2-9 Programming cable found

Figure 2-10 Programming cable not found

Figure 2-11 Cable test instructions

This test sends bits of data down the wires and times them to get a proper synchronous operation between the data and the clock lines of the I2C programming interface. When the OOPic IDE is satisfied with its results, the dialog box shown in Figure 2-12 is displayed, showing the newly discovered timing value. Unless you suspect the value it finds is wrong, press Yes and the value will be placed into the Programming Delay window and be used by the OOPic IDE when programming.

Now leave the *Verify after download* box checked. It's a quick and easy validation of the EEPROM on your OOPic board.

Because you can change the EEPROM on your OOPic boards, and because the OOPic variants ship with either 4KB or 8KB of EEPROM code, you must now enter the size of your EEPROM in the Memory Size window. See Figure 2-8 for the location of that entry. The OOPic IDE doesn't know and it can't query the OOPic board to find out for itself either; you have to tell it.

You are now finished; the OOPic IDE has been properly configured and calibrated to program your OOPic boards. Click OK to exit this dialog.

SERIAL PORT PROGRAMMING

The OOPic R and C are easy to program via the serial port, which eliminates any possible problems you might have with your PC parallel port when running in the newest Microsoft operating systems. If you add RS-232-level translation hardware to an OOPic II+, it too can be serial port programmed. The OOPic R is designed with serial port programming in mind, so nothing more needs to be added to what is already on the board. The OOPic C has the necessary RS-232 translation hardware built into it, and it is expected to be loaded onto a Parallax Inc. Stamp-compatible carrier board that already has the DB9 serial connector installed.

If you are using the OOPic II+, you need to do more than provide level translation for the transmit and receive serial lines. Pins 8 and 9 need to be shorted together on the DB9 connector so that the OOPic IDE recognizes a proper serial connection to program. Figure 2-13 shows a schematic example of how to build a proper RS-232-level translator circuit and create the needed signals that the OOPic will require for serial port programming.

OOPic Serial Port Cabling

To connect your PC to an OOPic for serial port programming, you only need a straight-through DB9 male to DB9 female serial cable. Often this is labeled as either an *enhanced*

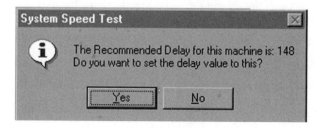

Figure 2-12 Program delay value found

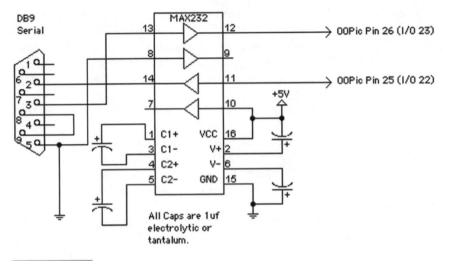

Figure 2-13 OOPic II+ RS232 circuit

graphics adapter (EGA) extender cable or a monitor extender cable in surplus catalogs. Figure 2-14 shows the port you are looking. In Chapter 1, Figure 1-5 illustrates the OOPic R board and shows the DB9 connector to use when programming using the serial port. For the OOPic R, OOPic C, and OOPic II+, just look for that DB9 connector.

OOPic Serial Port Discovery

Unlike the parallel port programming requirements, the OOPic serial port programming does not need calibration. It uses the standard serial communications protocol whose timing considerations are handled properly by the system hardware and software. The discovery of the OOPic serial programming cable consists of the OOPic IDE polling the serial ports looking for a port that responds with a "v" when a "V" is transmitted on it at 9600 baud. Only the OOPic firmware B.2.X+ versions will respond correctly and tell the IDE that there is a serially programmable OOPic out there on the cable.

To set your serial programming cable, click the Tools menu and select Cable Configuration. Referring to Figure 2-8, if you know which serial port you've connected your programming cable to, you can simply select it from the four options listed under the Find Serial Cable button. If you are like me, clueless or perhaps simply lazy, press the Find Serial Cable button. The OOPic IDE then displays a dialog window like Figure 2-14, giving you connection instructions for hooking up the cable. Follow those instructions and press OK.

If you've done everything properly, you will get a window telling you that the programming cable was found and which com port it is connected to. Press OK and the OOPic IDE selects the proper port. Leave *Verify after download* checked and enter your OOPic board EEPROM size. Chapter 1 details all the possible values that ship with the various OOPic variants. If you've installed your own EEPROM in your OOPic board, enter the correct size.

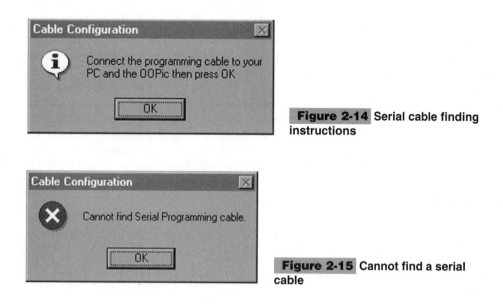

Figure 2-14 Serial cable finding instructions

Figure 2-15 Cannot find a serial cable

The OOPic IDE won't know what you have installed and it can't query the OOPic board to find out; you have to tell it. Press OK, and you're done here.

If you get the window shown in Figure 2-15 displaying the error "Cannot find Serial Programming cable," something is incorrect in your cable or your OOPic board.

If the OOPic IDE can't find the serial cable, only three reasons are possible:

■ Pins are bent on either the connector on your PC or on the serial cable. Be careful when bending the pins back; they will usually break off and you'll have to get another cable.

■ The serial cable does not connect all nine wires between the two ends or the cable is damaged and one or more wires are no longer connected. Use your trusty *Digital Volt Meter* (DVM) to do a continuity test on the cable. Replace the cable if it shows an open wire.

■ Your home brew, *do-it-yourself* (DIY) RS-232-level converter and DB9 circuit is faulty. Refer to Figure 2-13 for how to build a proper one for the OOPic II+ or C. Remember that the OOPic C has the RS-232-level translation built in; all you need to do is configure the DB9 cable properly.

TROUBLESHOOTING OOPIC PROGRAMMING PROBLEMS

By this time, you have the OOPic IDE installed and running, but you're looking at this section of the book because you either can't get the OOPic IDE to find your parallel cable or you have a problem downloading code to your OOPic EEPROM. It's most likely that the problem is the parallel cable.

Savage Innovations has tried to design the OOPic parallel port programming hardware to enable the OOPic programming cable and a printer to reside together on the parallel port.

This is done by using pins that usually don't have any other function on the parallel port. Unfortunately, some computer manufacturers have also noticed that some pins are usually not used and left them out of their parallel port designs. If this is the case with your computer, nothing can be done for your problem. You just can't use the OOPic parallel port programming capability with this computer. This is most often seen with laptops. If you have found this problem with a desktop or desk-side computer, all you need to do is add another parallel port card and use that instead. Sometimes the problem is something else altogether. Let's continue on to see if your problem can be corrected and your parallel port cable can work properly. Don't panic yet!

Parallel Port Cable Not Found

If, while trying to discover your parallel cable, you get an error telling you one cannot be found (refer to Figure 2-10), often very little can be done to correct the problem. However, you can check for a few obvious faults:

1. Check the programming cable for bent pins. Be careful when straightening them; they break off easily.
2. Check the programming cable for good connections and a proper circuit if you built the cable yourself. Refer to Figure 1-1 for the proper way to build an OOPic parallel port programming cable.

Remember that the parallel port programming cable does not need to be plugged into the OOPic board to be found by the OOPic IDE.

If the cable checks out fine, you need to find out where the parallel port is. To do this, access the System Properties dialog box that provides information about the hardware in your PC. Every Microsoft operating system variant has a way to do this that is either a little or a lot different from the others. In Windows 95, the process is as follows:

1. Right-click My Computer.
2. Click Properties.
3. Click Device Manager.
4. Click Ports (com and LPT). Figure 2-16 shows what the System Properties dialog box looks like. From here, we can find either the COM port (serial) or the LPT port (Parallel). To find the parallel port address, continue your quest.
5. Click Printer.
6. Click Resources.

Figure 2-17 shows that the address you need for your printer port is listed as the Input/Output Range setting. In this case, 03BC, or 3BC for short, is the address of the parallel port. This value is used to select the address value in the Cable Configuration dialog box shown in Figure 2-8. If this value isn't one of the values with radio buttons to the left of it, enter it into the empty box at the bottom of that list labeled Hex. We're going to be using this dialog box, so don't close it yet.

Plug the OOPic parallel port programming cable into the PC LPT port and connect it to the programming connector on your OOPic board. Since you left the Cable Configuration dialog box open (you *did,* didn't you?), press the More button. This opens the cable

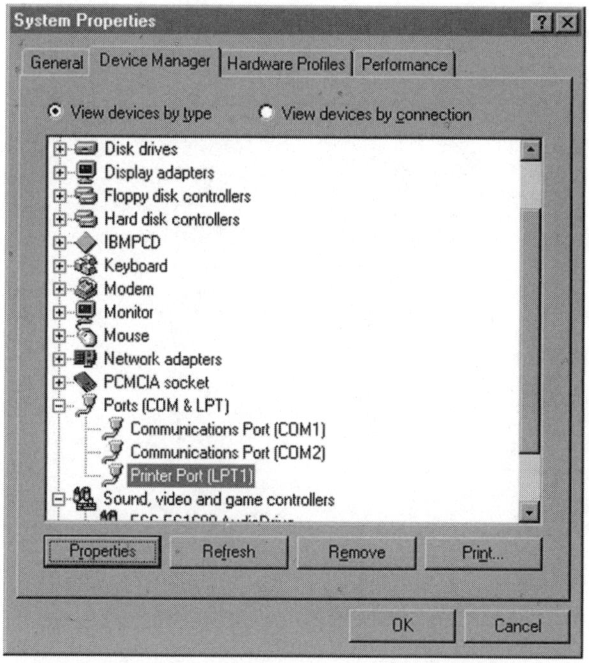

Figure 2-16 System Properties dialog box

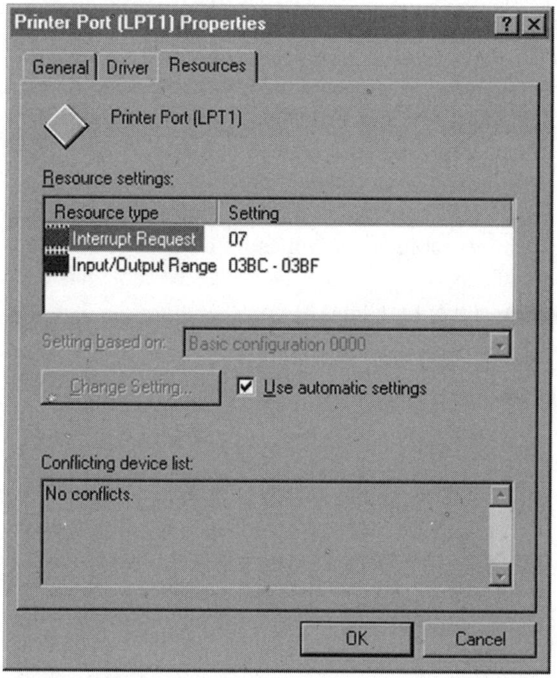

Figure 2-17 Printer port properties dialog box

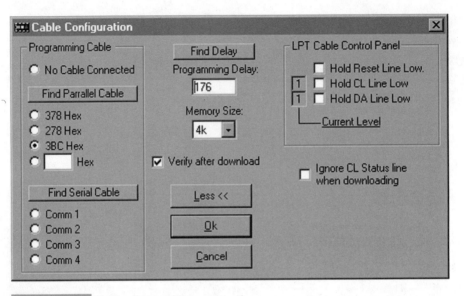

Figure 2-18 Cable diagnostics

diagnostic section shown in Figure 2-18. Apply power to your OOPic board and make sure that the Hold Reset Line Low box is *not* checked.

Observe the Current Level boxes to the left of the Hold CL Line Low and Hold DA Line Low check boxes. If they are flickering between 0 and 1, your cable is fine and should work. These are the lines that would not have connected if you were using a nonstandard parallel port card. If these lines are not flickering and you've confirmed that your cable and battery are good on your OOPic board, you have a parallel port that cannot be used with the OOPic parallel port programming cable.

Jump to the "Build a Buffered Parallel Port Cable" section at the end of this chapter to learn how to construct a buffered parallel port cable or see Appendix D, "OOPic Products, Accessories and Resources," for sellers of this type of cable. The buffered cable has been found to correct all known problems with OOPic parallel port programming.

Serial Port Cable Not Found

If, when you are doing a serial cable discovery from the Cable Configuration dialog box (refer to Figure 2-8), you get an error like that shown in Figure 2-15, a problem exists with your serial cable connection to your OOPic board. If you are using an OOPic R, the fault lies in your cable; check it for bent pins or missing signal wires. You want a complete cable with all wires present that is straight through and with no transmit/receive line swapping.

If you are using an OOPic C or II+ with RS-232-level translation hardware added, check that pins 8 and 9 of the DB9 connector are shorted together. If they are not, the cable may *not* be discovered as a programming interface for the OOPic board.

Error Writing to EEPROM

If you get an error message similar to Figure 2-19, the problem lies with writing or reading the EEPROM on the OOPic board. If you are using the parallel port cable, check the items listed in the error dialog box (in Figure 2-19) first. If this error is consistently given, try these solutions:

- Check your power to the OOPic board. Use your DVM and place the black lead on any G (ground) pad and your red lead on any +5 pin on the board it should read 5V. If it does not, change the batteries.
- Recalibrate your programming cable, as discussed earlier in this chapter.
- Check pin 19 on your parallel port cable to make sure it's grounded. It's possible that the parallel port (typically on your laptop) is not grounded. It may be another level as a result of some test circuitry on your computer. Place the DVM's negative (black) lead on the metal case of the parallel connector. Place a thin wire (in a pinch a metal paperclip works) into the hole for pin 19 (see Figure 2-20). It should read 0V. If it reads 5V or near that value, you have a nonstandard connector. Check pins 18 and 20. If they are 0V (grounded), then you can modify your programming cable by moving the wire in the DB25 connector housing to either pin 20 or pin 18. Refer to Figure 1-1 to see how this cable is wired.

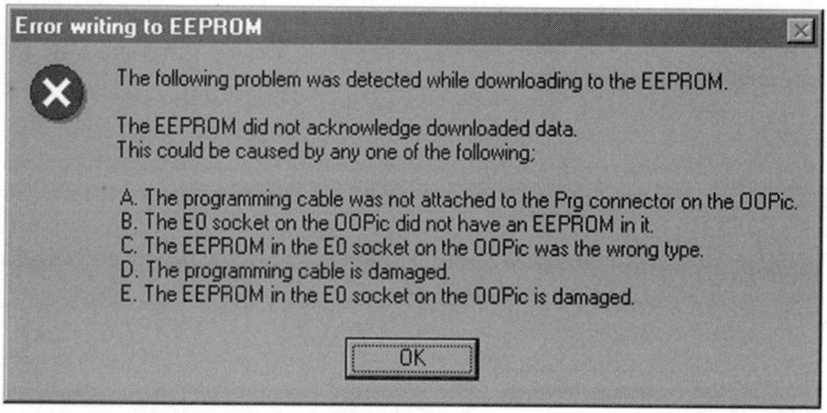

2

THE OOPIC IDE AND COMPILER

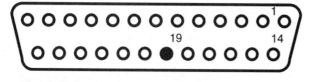

Error writing to EEPROM

⊗ The following problem was detected while downloading to the EEPROM.

The EEPROM did not acknowledge downloaded data.
This could be caused by any one of the following:

A. The programming cable was not attached to the Prg connector on the OOPic.
B. The E0 socket on the OOPic did not have an EEPROM in it.
C. The EEPROM in the E0 socket on the OOPic was the wrong type.
D. The programming cable is damaged.
E. The EEPROM in the E0 socket on the OOPic is damaged.

[OK]

Figure 2-19 EEPROM write error

◯◯◯◯◯◯◯◯◯◯◯◯◯
                    1
        19   14
◯◯◯◯◯◯●◯◯◯◯◯

Figure 2-20 Looking for pin 19

- Check the EEPROM on the OOPic board to make sure it has no bent pins.

If you are using the parallel port cable and your error occurs only intermittently, you have some sleuthing to do:

- Make sure that no other device drivers are using your parallel port, other than a printer. External disk drives or backup devices' drivers will poll[1] the parallel port and this can wreak havoc on an OOPic session. Another symptom of device polling is if the OOPic resets constantly while the programming or debugging cable is plugged in. Also, make sure that you don't have any sessions of parallel-port-based oscilloscopes or logic analyzers active when you are using OOPic IDE.
- It is possible you have an earlier OOPic I board that uses SC and *destination address* (DA) pull-up resistors that are too large for a low-quality cable. If the resistors to the left and right of the OOPic chip are 4.7KB (yellow, purple, red, gold) or 10KB (brown, black, orange, gold), you need to alter or replace them. You can't miss these resistors; they are the only two on the OOPic board. To improve the transition speed of the clock and data lines you have two choices. Either carefully solder a 4.7KB resistor on top of each current resistor (piggybacking) or, if you are skilled with a soldering iron and solder sucker, remove those resistors and replace them with 2.7KB (red, purple, red, and gold) resistors.
- Finally, check the OOPic and EEPROM for bent pins or pins that are loose in their sockets.

If you are using a serial port cable, the problems should be easier to find and solve. In fact, problems rarely occur, but if they do, refer to this checklist:

- If you are using a serial cable to program an OOPic R, C, or II+ (with level translation hardware), your OOPic may have reached a bad state and needs a hard reset. To perform a hard reset for the downloader on any OOPic B.21 firmware device, follow these steps:

  1. Power off the OOPic board.
  2. Carefully remove the EEPROM.
  3. Power the Oopic board on and start a download session.
  4. When the IDE stops and shows the dialog box in Figure 2-21, insert the EEPROM while the board is still on.
  5. Press *OK*. The download will now work properly.

Figure 2-21 Insert EEPROM message

---

1. Polling is when the computer has a program that occasionally reads or writes an input/output (I/O) pin looking for a specific response.

■ It's possible that another program may be using the serial port. Turn off any other program that could be using the serial port, such as modem or networking connections, terminal emulation windows, or other embedded controller IDE programs.

■ Finally, if you are using an OOPic C or II+ with your own DIY RS-232-level translation circuit (in the fine tradition of hackers everywhere), look for any of the following problems:

■ Check your RS-232-level translation circuit for proper operation.

■ The OOPic C pin 1 must connect to pin 2 on the DB9 female connector or the RS-232 transmit line must connect to pin 2 on the DB9 connector. Pin 25 of the PIC chip connects to the input of the transmit buffer on your level translator chip.

■ The OOPic C pin 2 must connect to pin 3 on the DB9 connector or the RS-232 receive line must connect to pin 3 on the DB9 connector. Pin 26 of the PIC chip connects to the output of the receive buffer on your level translator chip.

■ Pins 8 and 9 must be connected together at the DB9 connector. This is how the OOPic IDE knows a cable exists.

■ Pin 5 on the DB9 connector must connect to ground.

Other Troubleshooting Tips

Here are a few more tricks and tips to make sure your OOPic is healthy when you suspect a problem:

■ The OOPic R board has a tiny *light-emitting diode* (LED) next to the parallel port programming connector labeled *EAC*. If this LED is lit, your OOPic R board is running and fetching instructions from the EEPROM. Thus, your PIC is working.

■ You can troubleshoot the wiring in your DIY programming cable by using the Cable Configuration dialog box. Click the Tools menu and select the Cable Configuration function. Then press More to get the display shown in Figure 2-18 and perform these checks:

■ To check the reset line, click the check box labeled *Hold Reset Line Low*. Using your (by now very familiar) DVM, check that pin 5 of the OOPic side cable connector is near 0V. If this value is 5V or so, or it is wandering around, you've goofed up your wiring of the reset line and it must be fixed. Now uncheck the Hold Reset Line Low box and make sure that the reset line goes back to or near 5V. If it doesn't, you goofed again. Refer to Figure 1-1 for how you should have built your parallel port cable.

■ To check the clock line, click the checkbox labeled *Hold CL Line Low*. Pin 3 of the OOPic side cable connector should measure 0V. If it doesn't, your cable is not wired properly. Uncheck the Hold CL Line Low box to make sure the clock line goes back to or near 5V. If it doesn't return to 5V or so, your cable is not wired properly. Figure 1-1, again, shows the proper way to wire the cable.

■ To check the data line, click the checkbox labeled *Hold DA Line Low*. Pin 1 of the OOPic side cable connector should measure 0V. If it doesn't, you wired your cable improperly. Uncheck that box and verify that the DA line goes back to or near 5V.

2

THE OOPIC IDE AND COMPILER

- In the case of the DA and CL lines, the OOPic diagnostic window will show a *1* if the line is at or near 5V and a *0* if the line is at or near 0V. You may be able to debug your cable with that information and may not need to use the DVM.

- If you have completed the previous diagnostic and found that your cable is not operating properly, but it is wired correctly, you probably have a laptop with a nonstandard parallel port, which will never work with an OOPic parallel port programming cable—sorry. Jump to "Build a Buffered Parallel Port Cable" at the end of this chapter to learn how to construct a buffered parallel port cable or see Appendix D for sellers of this type of cable. The buffered cable has been found to correct all known problems with OOPic parallel port programming.

If nothing so far has helped, check out the OOPic group on www.YahooGroups.com. This is a *very* active user group and they may have discovered new information that can help you solve your problem.

Debugging Virtual Circuits

The OOPic has a friendly built-in debugging facility that enables you to directly view the current state of any object in your OOPic program. To use it, select Objects from the View menu, as shown in Figure 2-2. Next, you must write your program, compile it, and download it to your OOPic board. You will then see your objects listed in graphical format in the right window pane of your editor window, as shown in Figure 2-4.

If you are using the parallel port programming cable, you *must* unplug it from the programming connector and plug it into one of the networking connectors on the board. For the OOPic S and R boards, the networking connectors are on the right side. On the OOPic C, the networking connector is the set of five pads on the far outside right of the chip carrier. If you are using a serial port to program your OOPic, you don't need to do anything else but leave the cable connected.

To use the debug facility, you must first assign a node address to your OOPic board. This is done by adding the following line to your **main** subroutine:

OOPic.Node = 1

For the OOPic R and OOPic C you need to add the following line to your main subroutine:

OOPic.SNode = 1

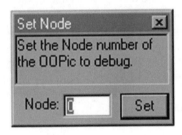

Figure 2-22 Set OOPic node.

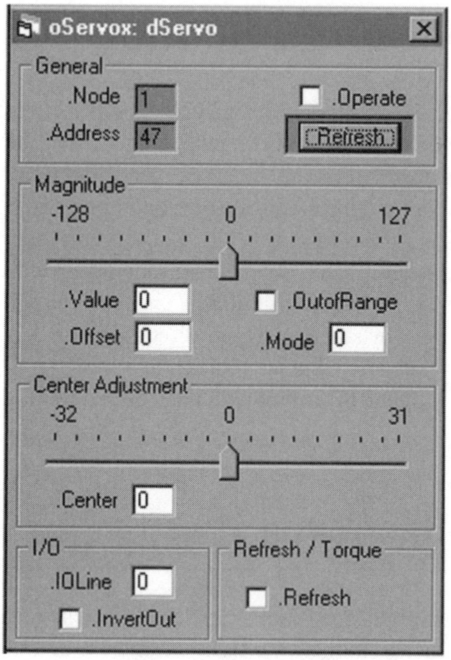

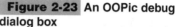

Figure 2-23 An OOPic debug dialog box

In this case, we've chosen address 1. You can choose any address up to 254, but don't use address 0 or 255.

Now you need to tell the IDE your OOPic's address. To choose an OOPic address, click the NETWORK NODE at the top of the debug window, and the dialog box shown in Figure 2-22 will display. Enter the node ID of the OOPic you are debugging and press OK. In this case, node 1 has been chosen for all of the examples.

To look at any object's value, flag, or setting, all you need to do is click that object in the debug window and that object's dialog box will display, as shown in Figure 2-23, where all the properties are now displayed. This is how you find the address of an object if you want to communicate between OOPics using DDELink. The node and address of the object are displayed in the upper left of the dialog window. You can use the debug dialog window to start events, enable and disable objects, and change object values or read object values that have been set by the program. Use the Refresh button to get the latest information from the object. If you change a value, hitting Return will cause a refresh to occur.

If the OOPic is responding to requests from the IDE for debug information, the .Node box in the upper left will be green. If the OOPic under test does *not* respond properly or times out, the .Node box will be red. Similarly, if the object desired returns bad or incorrect header information, the .Address box will be red. If the OOPic IDE is getting properly formatted information for that object, the .Address box will be green.

Note: Make sure you close all object debug windows before you change and recompile a project. If you don't, the IDE will get the OOPic object addresses confused, and the debug dialog boxes will no longer work properly.

2

THE OOPIC IDE AND COMPILER

OOPic B.2.X+ SCP Communication

The latest OOPic firmware used in the OOPic R, C, and II+ is designed to enable queries and control any OOPic object in a program by using a serial port connection. In fact, the OOPic debug dialog boxes use SCP to get their information and change objects in the target OOPic. This serial port control makes it possible to have a higher level computer control the OOPic programmatically without requiring users to program in a custom protocol themselves. This means that another computer can query the OOPic for sensor readings and set servo and PWM values directly. This exciting new OOPic technology enables your PC or Palm Pilot® to be a higher-level robotic mind, for instance, in a very direct manner.

I'm not going to go into any details about SCP because Chapter 11 lists the full SCP protocol and offers a detailed example using a Palm M105 to control an OOPic-based robot.

Build a Buffered Parallel Port Cable

Late in the writing of this book, I helped design a buffered parallel port cable whose idea and working prototype were proposed by Tran Duong Dung on the OOPic YahooGroup discussion list. Folks in the United States and Canada have tested it on several systems that refused to work with the standard OOPic programming cable, and it appears to solve all known and reported parallel port programming problems. It even works just fine as an OOPic debugging network cable. This is exciting news for many OOPic users and deserves a place here as the reigning "silver bullet" of parallel port solutions for the OOPic system. Others on the OOPic YahooGroup have expressed interest in building a buffered cable solution for sale to those who don't want to do the work of building one at home. Check Appendix D out for those vendors that offer a kit or product. For those that love to work with a soldering iron, Figure 2-24 shows the schematic for building an OOPic buffered programming cable.

Make sure you connect *all* the grounds shown on the DB25 cable, which is a fix for one of the parallel port programming problems. Be careful with your soldering, especially if you are attempting to build the buffer into the cable housing of the DB25 connector. The 74HCT244 has been chosen because both inputs and outputs appear on both sides of the chip and are lined up with the DB25 connector pins that they connect to. You can build this circuit onto another PC board if you want and use an intermediate cable, but you'll still need to make your own DB25 connector or modify the OOPic standard cable to connect all the DB25 printer port grounds together.

The buffered programming cable is slightly different from the standard OOPic programming cable. When discovering the cable, the buffered programming cable should be plugged into a board that is already powered on. This will properly power the 74HCT244 chip so that signals will be routed correctly. Sometimes it works even when not plugged in, but don't count on it. This cable also works for both programming and debugging on all OOPic variants.

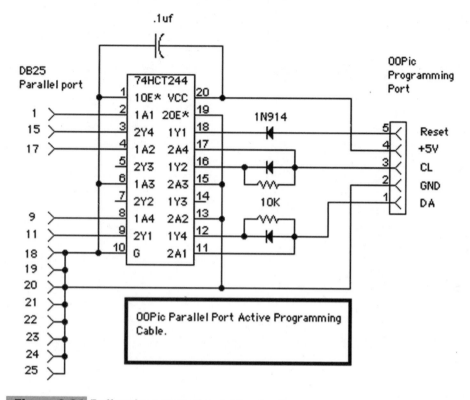

Figure 2-24 Buffered programming cable circuit

    This buffered programming cable has been designed under the assumption that most pro-
gramming problems stem from parallel ports not having enough drive[2] to deal with the
OOPic circuits on the boards. The cable gives the port a single digital device to drive that
is very low power and high speed. I do not guarantee that this solution will work for all problems, but it has worked for everyone who has tested it.

---

2. Drive is the ability to supply the current necessary for signals to travel between two points.

OOPIC OBJECT STANDARD PROPERTIES

Many common features and capabilities are shared among OOPic objects and the revisions of the OOPic flavors. This chapter defines and describes these commonalties. Some objects deviate slightly from the standard OOPic object pattern, and when one of these deviants (I've always wanted to use that word in a book) is used, the difference will be pointed out.

This chapter doesn't deal with the specifics of any single object, but rather with what is the same or similar among all the objects. Here, too, you will see how to link objects together to create *virtual circuits* (VCs). Sometimes you'll want to link flags for logical operations or to signal when to start or stop a process. Other times you will want to link actual values to control the specific action of a device. The OOPic has the capability to do both, with restrictions, which will be defined later.

This chapter starts out simple and gets progressively more complex. Feel free to stop reading at any time and come back later. This chapter also contains topics that won't appear again later in any other chapters.

How OOPic Objects Work

Understanding how the OOPic works will help you design better OOPic code and it will perhaps make some objects easier to understand and use. Knowing how the OOPic schedules when an object will run can help you write programs that will need less debugging as well. (Debugging is the act of preventing small "creatures" from crawling out of your project and biting you.) Knowing how the OOPic works will also make you sound smarter and will give you better geek chic at your next robotics meeting.

The two pieces to this puzzle are the OOPic *object list loop* and the *servo list loop*. These lists of objects operate independently from each other with minimal interaction.

THE OOPIC OBJECT LIST LOOP

When an OOPic object is created (a *New* is used in your code), its object memory is allocated in object memory space and is placed into the OOPic object list. Remember that the OOPic has 86 bytes of object memory space and each object takes up some of that space. This OOPic object list consists of pointers[1] to program code that is stored in the PICMicro permanent program space inside the PICMicro chip. Unlike the serial *electrically erasable programmable read-only memory* (EEPROM) memory that stores your program code that you write, this memory is high-speed, and the code that is stored there runs as native machine code in the PICMicro controller. This object code runs at about 100 times the speed of your program code, which must be interpreted by the program running in the background of the PICMicro chip. This is why you want as much of your program running as a VC as possible. It will make your program operate much faster that way.

Basically, creating and running a program on a computer or a microcontroller can be done in two ways: The program can be compiled to native machine code or it can be interpreted. *Native* code means that the program is running in the machine language that the computer, in this case the PICMicro, understands. The native machine language is binary,

---

1. A pointer is like a sign that shows where something is. In this case, it shows where program code is located, like an address.

all ones and zeros, and it is specific to that computer and won't run on any other type of machine.

Interpreted languages, like the OOPic and many other easy-to-use embedded controller languages, do not run in the computer directly. The program that is being executed is actually a list of *tokens* (objects that represent an action) that are read in and decoded. Then the native code, whose purpose that token represents, is executed before moving on to the next token in the list. All this activity that goes on before the actual native code runs takes time, time that isn't spent on your program, but rather is spent on figuring out what your program wants to do. So, unlike a compiled program, the interpreted program is actually not running. Instead a program is running that reads your program and then initiates actions that your program tokens tell it to run. Is that clear? Figure 3-1 shows the steps that take place when a compiled-to-native program is run, and Figure 3-2 shows what takes place when an interpreted program is run.

The OOPic places a higher priority on the operation of its VCs over "regular" code. Approximately, 80 percent of the PICMicro's run time is spent on VC code; 20 percent is then spent on code logic that is not VC. "What?" you say. "That's not 100 times more!" Indeed, it isn't. In that 20 percent of the time that the OOPic spends on your non-VC code logic, the OOPic program must fetch a token from the serial EEPROM, which is *very* slow compared to the OOPic internal memory. The OOPic must then figure out what that token is, find out where the native instructions are for that token, and finally execute them. That takes a *lot* of time. By comparison, the native compiled code (the OOPic objects) only needs to fetch the next instruction and the code runs. The hardware knows where everything is, and nothing needs to be looked up or interpreted. *That* is why you want your program to use as many VCs as possible, because you want it to run quickly.

Let's get back to the discussion about the OOPic object list loop. As soon as an object's *Operate* property has been enabled, the OOPic sequencing program can execute its functionality. Enabling and disabling an object by turning on or off its Operate property does

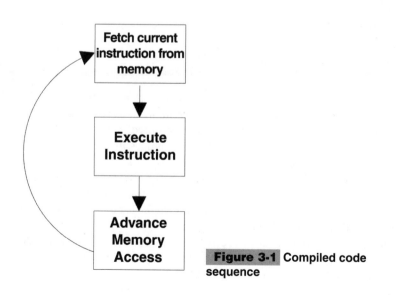

Figure 3-1 Compiled code sequence

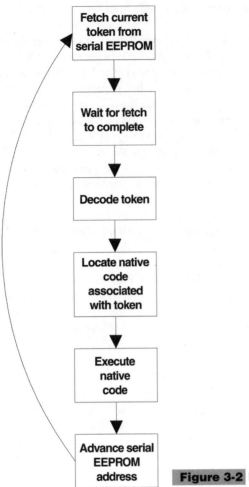

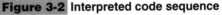
Figure 3-2 Interpreted code sequence

nothing to its place in the object list. When the OOPic sequencing program reads an object that is not operating, it just skips it and goes to the next one in the list. Once that object is created, it resides forever in the object list, whether it's being used or not. So don't add objects you aren't using; they will take up object memory and processing time and accomplish nothing.

The OOPic operating system continually steps through the OOPic object list and executes the defined properties of each object in the order in which it was created by your program. When the OOPic gets to the end of the list, it starts over at the beginning. If you want a VC to go as fast as possible, you should keep the object list as short as possible. If you have functionality that doesn't have to be fast, you can keep it in your programmed logic area and not use a VC to implement it. If you want two objects to operate closely together in time, they should be created (with a *New*) in adjacent lines.

Because each object is executed once during each pass through the list, any change that occurs to an object will be seen by every object that is linked to it during that pass. The same thing occurs if another change occurs during the next pass through the loop. Any change is guaranteed to be seen by every object in the list. If you have a VC whose signals are timing critical, make sure that you know which object needs a linked signal in what order. If you don't, then your VC may not operate like you think it should. The one exception to this rule is the oTimer object (more on this in later chapters). It takes time to get through the list, and during that time, one object may see one timer value and an object farther down the object list may see another one.

Some OOPic objects refer to the object list loop in their definitions. For instance, the oOneShot object states that when its input flag changes from 0 to 1, the Result flag is set to 1 for an amount of time equal to one object loop. At the end of the object list loop, the OneShot output is set to 0. This object is interesting because if you link the output of the OneShot object to an oDio1 object and connect an oscilloscope to that output, the pulse width will be the actual length of time it takes to run through your entire VC OOPic object loop. Cool, huh?

OSERVO, OSERIALX, OSONAR: THE EXCEPTIONS TO THE RULE

The VC object list loop has one set of exceptions that includes oServo, oSerialX, oSonar, and a few other objects. These defined objects are placed in a separate special queue as well as the OOPic object list. This queue allows these special objects to run with very consistent timing specifications. These queues are discussed in their respecitve chapters. The oServo object in particular has restrictions on its use and is discussed in Chapter 5, "Analog-to-Digital and Hobby Servos (Project: Push My Finger)."

Object Attributes and How to Use Them

All OOPic objects have properties known in OO parlance as *attributes* and *methods*. Attributes are values or flags that can be changed or read outside of an object. OOPic object *Result*, *Value*, and *Flag* properties are all attributes. Attributes modify the operation of an OOPic object in some manner. Each object has its own unique attributes that help define it. The shared ones will be discussed here, and the unique ones will be discussed when an object is introduced in subsequent chapters.

Methods are subroutines inside an object that can be called or *signaled* from outside an object. In both cases, changing an attribute or running a method inside a particular object has no affect on any other object (except oTimer and OOPic, which are covered later). Usually, the only objects that have methods are the oDio hardware objects and the data objects. These typically shift bits or perform set and clear operations on the data contained in the objects.

Most of the code samples in this chapter are compatible with the OOPic A.1.X firmware, so that everyone, can type them in to see how they work. They also compile and run fine on all later firmware, as well. Some code samples are for later firmware versions only and this is noted in their text.

COMMON PROPERTIES SHARED BY OBJECTS

A few properties reside in all objects of a certain type. Hardware objects differ from processing objects, which differ from data objects. They all have a few things in common, however, and the subgroups have some common properties as well. All objects have a default input and output. You don't need to specify a property when you read, write, or link to these default inputs or outputs. The OOPic manual page entries specify which output or input is the default property for each object.

The default value is the input or output value or the result of an object that is accessed when you reference just the object with no specified property (it has the dot [.] connection). As an example, Listing 3-1 shows two equivalent ways to set the value of an oByte object.

```
Dim Dummy as New oByte

Sub Main()

    'Set the default value of the variable
    Dummy = 255

    'Another way to set this value
    Dummy.Value = 255

End Sub
```

Listing 3.1

Every OOPic Object Has This Property

You'll find the Address property in every object, and occasionally you'll need to use it. It is the memory address of the object in the OOPic object list where all the properties of the object are kept. You would use this if you were tinkering with the object internals like a hacker, or if you were using oDDELink to get two OOPic micros to talk to each other and share information. Because the Address property is omniprresent, I don't include it in the properties list of the objects I discuss in the subsequent chapters. Not many properties are shared by all objects.

Common Hardware Object Properties

Hardware objects have many properties that work the same in every object. The common properties are as follows:

- **Operate**   This turns the hardware object on.
- **Value**   This is the value written to or read from the hardware object.
- **String**   Where it makes sense (and in a few cases it doesn't), any hardware object that has a Value property also has a String property that allows its value to be represented

as a string. This is useful when you wish to display the value on an LCD or send it to a computer for display.

Setting the Operate property to 0 doesn't mean that the object stops functioning, that would create unstable conditions for other parts of a VC. It means that the object is not responding to input changes, it holds the state of its last change until Operate has been set back to 1.

Common Processing Object Properties

Processing objects are quite diverse, but they do share a few common properties:

- **Result**   This value is the result of whatever processing needs to be done by the object.
- **Output**   Optionally, you may not want to read the result of an operation but would rather link it directly to another object. This property accomplishes that.
- **Operate**   I'm sure you know what this property does by now.

Common Data Object Properties

The data objects haven't been discussed much, because they really are only used in special circumstances. When you need a great deal of control and information about data, and you need to link them to a VC, you'll need to use a data object. The common data object properties are as follows:

- **Value**   This is what is read or written to as the default property of a data object.
- **String**   Any value that can be represented by a number can also be represented by a string.
- **Signed**   This property is part of the oWord and oByte objects and tells the OOPic to make sure the sign is saved when the value is moved to another object or used in a math operation.
- **Methods**   A method on a data object will do things such as clearing all bits, setting all bits, shifting them, and so on. The methods of an oBit object differ from the methods of an oWord object, but all have the Set, Clear, and Invert methods.

Certain data objects contain data that are unique to that object and whose use is complex. The oBuffer, oEEPROM, and oRAM objects are very complex and have methods and attributes you won't find anywhere else.

Linking OOPic Objects Together

The capability to link objects together into a VC is what makes the OOPic so unique. This linking capability is shared by all OOPic objects no matter what they do. Two types of links can be made among OOPic objects:

- **Flag links**   These links signal an object to begin or end processing. The flag signals that something has occurred or that something needs to occur. Some objects have

3

OOPIC OBJECT STANDARD PROPERTIES

pointers that link to another object's flag; some have flags that can be linked via processing objects to another object.

■ **Value links**   These links carry values between objects. Bytes, words, and bits can all be values that are linked, primarily between hardware objects and processing objects. All processing objects have a choice of either an Output pointer or a Result value, with a couple exceptions noted later.

LINKING FLAG PROPERTIES

Any flag output from any object can be linked to any flag input of another object. This can be done using one of five commonly used processing objects:

■ **oGate**   This object has multiple inputs that can be combined to create a number of Boolean logic structures. oGate is commonly used to link activation flags together to logically select the next OOPic object that will process information in some manner. oGate is a logical *OR* function, but this can be changed to any logical gate by setting *InvertIn*, *InvertOut*, or *Exclusive* properties. You need to understand Boolean logic and Boolean algebra to fully utilize these interesting properties. It's easy to learn, so go for it.

■ **oWire**   This object is simply an oGate object with only one input and one output. It doesn't appear in the original OOPic I object list, but it is rather a simplified construct built into the newer compilers.

■ **oFanout**   Because linking a single flag to multiple other objects takes up memory space for each link, the oFanout object has been created. oFanout enables you to link one output flag to up to four different input flags, which saves memory and enables an inversion of the flag's logical value for any of the outputs as well.

■ **oOneShot**   This object outputs a single pulse whose length is the time it takes the OOPic object list loop to complete when its input changes. No more pulses will be sent until the input resets to zero.

■ **oFlipFlop (OOPic II and newer)**   This is a logical structure that emulates a sort of *Set-Reset* flip-flop. This object that decodes a two-bit input has a few other uses as well.

A few other processing objects such as oCountdown and oDebounce also link flags, but only after some processing has been done on them. Note that each time you link a flag to another object it takes up memory space, so don't make links you aren't going to use; you'll be wasting time and memory.

Example of Flag Linking

At this point, you could use an example of how a flag is linked between two objects so that one can send a message to the other. Listing 3-2 shows how a bumper input on a robot is used to activate a backup event. The oDebounce object is used to condition the switch input and link it to the event object. This code sample also shows you how to create an event that will execute when a flag is set. You'll be doing a lot more of this in subsequent chapters, so this section does not dwell on this example and the objects in it.

```
 'Robot bumper VC setup
Dim BMPR as New oDio1                      'bumper input
Dim bmpLook as New oEvent                  'event started by the bumper
Dim bmp as New oDebounce                   'debounce the bumper switch

Sub Main()
'This sets up the bumper event

    BMPR.IoLine = 26                       'This is the I/O line to use
    BMPR.Direction = cvInput               'I/O line is an input
    bmp.InvertIn = cvTrue                  'Invert the I/O line input
    bmp.Input.Link(BMPR)                   'link I/O line to debounce object
    bmp.Operate = cvTrue                   'And enable it
    bmp.Period = 2                         'Debounce period
    bmp.Output.Link(bmpLook.Operate)       'link debounce to event Sub

End Sub

Sub bmpLook_CODE()
'Bumper event handler

End Sub
```

Listing 3-2

MORE ON OGATE AND FLAG LINKING

 The oGate processing object is actually very versatile, and it is the object to use when you need to look at more than one input flag before enabling or triggering another object in a VC.

oGate and oGateC (All OOPic Firmware Versions, oGateC B.X.X. and Later)

The oGate object is actually a number of objects, depending on how you configure it. The oGateC object is the same as oGate, but with the optional ability to clock it, which means you can issue a virtual pulse to the object that will last one loop through the object list loop. The oGate memory size is 2 plus the number of input bytes (oGateC is 4 plus the number of input bytes). Table 3-1 outlines the oGate Properties.

When you read through the properties list of oGate, you see a lot of *InvertInNs* (where N = 1 to 8), *InvertOuts,* and *Exclusives* listed. These properties enable you to customize oGate to the logic you want. As a default, oGate is an OR gate, but by using these other properties, you can create any logical gate you want. Figure 3-3 shows the various (and by no means all-inclusive) logical gates that you can create with oGate and their truth tables. A truth table shows the output from a logical function given the logical inputs. If you know Boolean algebra and deMorgan's theorem, you don't even need to look at these tables, but you can extend the logic to include all the configurable inputs of oGate. The circles at the inputs or outputs of a gate denote a logical inversion. The standard logical representations have been included alongside the ones created by using the inverting inputs or outputs of the oGate object. To get the XOR (Exclusive OR) and XNOR (Exclusive NOR) gates, you use the *Exclusive* property and in the case of XNOR, the *InvertOut* property. Many more

3

OOPIC OBJECT STANDARD PROPERTIES

TABLE 3-1   OGATE PROPERTIES

ClockIn (flag pointer)	Links to the flag bit that will be used to clock (signal) this object to perform an operation.
Exclusive (bit)	When set to 1, this makes the inputs of the OR gate exclusive; the logic table is below in Figure 3.3.
Input(1–8) (flag pointer)	Links to flag properties of an object that will be used as an input to the object.
InvertC (bit, flag)	When set to 1, this inverts the logical sense of the *ClockIn* property; that is, a 0 will clock the object instead of a 1.
InvertIn(1–8) (bit, flag)	When a selected input's *InvertIn* flag is set to 1, the logical sense of that input is inverted; a 1 becomes a 0 and vice versa.
InvertOut (bit, flag)	When this property is set to 1, the output of the object is logically inverted.
Operate (bit, flag)	This must be set to 1 (cvTrue) in order for the object to function (default property).
Output (flag pointer)	This links to a Flag property in an object that will be affected by the output of this object's logical operation.
Result (bit, flag)	This is the result of the logical operations defined by the input of the oGate object. This value reflects the result *before* the output is inverted if that property is set.
Width (nibble)	Returns the number of inputs defined when this object is dimensioned.

possibilities exist, including creating flip-flops and other simple logic gates, but you can explore those on your own.

LINKING VALUE PROPERTIES

All hardware and data objects have a Value property that contains the results of a reading or setting. This is the default property that is referenced when you link output or input to a hardware object. All processing objects have both an Output pointer that is used to link to a Value property of a hardware or data object and a Result value that can be linked to a processing object's pointer. The Value and Result properties are read or written when your program code reads from an object without specifying a property.

Value and Result properties can also be linked with the following most commonly used processing objects:

■ **oMath** This processing object is used to handle simple math and conversion functions. It is also commonly used to link Result or Value properties between objects

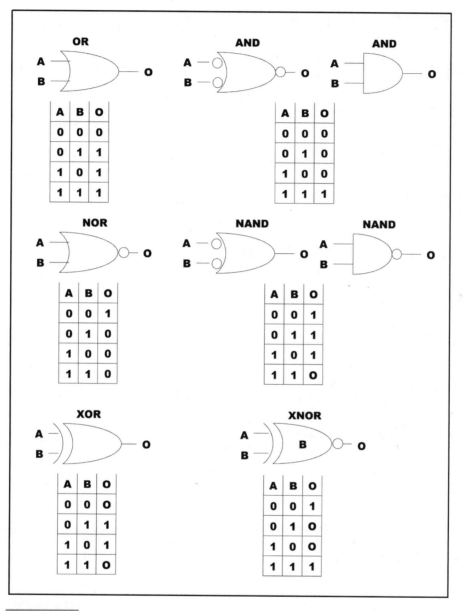

Figure 3-3 Truth tables for logic gates

and does no math or conversions at all. An unlinked input of oMath will be read as all zeros.

- **oBus (OOPic II and newer)**   This processing object is strictly used to link the default Value or Result properties between objects. You can also invert the inputs before linking them to the next object.

The oConvert and oCompare processing objects, as well as a few others, link values between objects, but only after processing their data in some way.

Value Linking Example

You probably could use an example of linking value, output, or result properties in a VC right now. Listing 3-3 shows how you would link the value property read from an oA2D hardware object to an oDio8 hardware object for display on eight *light-emitting diodes* (LEDs).

```
Dim Pot as New oA2D          'reading a pot as input
Dim Display as New oDio8     'LED output display
Dim Connect as New oMath     'Here's how we link them

Sub Main()

    OOPic.node = 1           'So we can use debug

    'Set up the hardware objects first
    Pot.IOLine = 1           'Put on I/O line 1
    Pot.Operate = cvTrue     'And start it

    Display.IOGroup = 1      'Use I/O lines 8-15
    Display.Direction = cvOutput 'Kind of obvious
    Display = 0              'Clear them out

    'Now set up the oMath object
    Connect.Mode = cvOr      'Does nothing, but needs set
    Connect.Negative = cvFalse 'As above, just needs set
    Connect.Input1.Link(Pot) 'Gets the A2D value
    Connect.Output.Link(Display) 'Connect to output port
    Connect.Operate = cvTrue 'Turn it on

End Sub
```

Listing 3-3

You'll note that this program is 100 percent VC. No code logic is running; it's all in the VC. Notice that nothing is linked to the *Input2* property of the oMath object. This was done knowing that it would cause Input2 to be left as zeros and enable the oMath object to be nothing more than a link.

Please also note that an infinite loop is not included at the end of the Main() subroutine. Many people put loops there because they have an irrational fear of their code falling off the edge of the world or something. If you start your VCs, they will run even when no logical code is left. Once the VC is started, it runs until the OOPic is reprogrammed or turned off. In fact, the OOPic object reads the serial EEPROM and executes the program, which is run completely by the OOPic object and clearly doesn't need another program to run. A pure VC program is a thing of beauty.

The Magnevation OOPic Logic Status Module is used to debug this program. It's cool to see all the LEDs cycle through the pattern as the value of the *potentiometer, variable resis-*

Figure 3-4 Logic status module and demo

tor (Pot) attached to *input/output* (I/O) line 1 is changed. Figure 3-4 shows a picture of the setup. This board is simple to use and comes in handy when you've goofed something up and need a quick readout of your I/O lines. If you want to run this example code, all you need to do is connect the center pin (wiper) of a trimmer pot to pin 7 of the OOPic I/O connector and then connect +5V and Ground to the other two pins on the trimmer pot. You can get those signals on pins 21 and 23 of the OOPic I/O connector. Which side of the pot you connect the voltage to doesn't matter. The logic status module gets its power from the OOPic board, so no other wiring is necessary. It's very neat and simple.

NOTE: These processing objects will only link the default values; no other byte or word properties will work. Sorry, you'll have to find another way to link the Rate property of a stepper motor object to some other processing object.

LINKING PROPERTIES USING THE ORAM OBJECT

One tool can be used to link almost any property of one object to any property of another object: the oRam object. This is a very complex object and you need to really know the internal workings of the OOPic object to do it. This is for true hackers only, so if you are fearless and understand microcontrollers then read on!

One of the oRam object's uses is to link properties that otherwise can't be linked. Finding those properties within the object is perhaps confusing but really not very complicated. This is a topic that can only be explained by example.

oRam Linking Example

Each OOPic object uses a specified amount of object memory space, depending upon its configuration. The OOPic object lists in Appendix A, "OOPic A.2.X Objects," Appendix B, "OOPic B.2.X Objects," and Appendix C, " OOPic B.2.X+ Objects," provide this information. All the OOPic object properties are contained in these RAM locations. The oRam object enables you to access these properties and link them into a VC.

This capability has some practical limitations, however. Byte-sized properties are good candidates for oRam linking, whereas Flag-, Nibble-, and Word-sized properties within other objects are not. Why? Because the OOPic objects have their properties *packed* into their RAM locations. This means that if you have four Flags and a Nibble property, they could all be located in the same byte. The Value property of the oRam object is a byte, so if you want to link to a Flag or Nibble property that has other properties in that same byte, you would need to look into that byte first. Then you would save that value and use combinations of oMath objects to mask bits you don't want to change. This would require logical ANDing and ORing of bits to only change the Flag or Nibble property you wanted to change. The Value property is only one byte wide, so a Word-sized property can't be linked in a single pass. This means that two oRam objects will be needed to assemble a Word-sized property and link it to a VC. Again, it would take more objects to assemble two bytes into a word.

Although these tasks are possible, they will use up a lot of object memory space. You will have to decide if it is worth the effort and space. Sometimes it is, and sometimes you get lucky and a property of less than the byte size is located in a RAM location all by itself. You can then link that property directly.

This example is one of those fortunate ones. Let's say you want to link the Rate property of the oStepper object to a VC so that changes in the readings of an oA2D object will affect the step rate of a stepper motor. The oStepper object's Value property is the number of steps the stepper is to take, not its speed. This means that an oBus object cannot link to Rate. The oStepper object is included in Appendix B; look at its property list and note that it uses 8 bytes of object memory. Which byte holds the Rate property? Listing 3-4 shows how you'll

```
Dim stepA as New oStepper

sub main()

    OOPic.Node = 1          'For debugging

    stepA.Rate = 100        'Speed of step
    stepA.Direction = 0     'Direction of step
    stepA.Free = 0          'locked or free rotor
    stepA.InvertOut = 0     'reverse step pattern
    stepA.Unsigned = 0      'allow negative steps
    stepA.IOgroup = 1       'I/O lines 8-15
    stepA.nibble = 0        'I/O lines 8-11 chosen
    stepA.Mode = 1          'Turn continuously
    stepA.Phasing = 1       'Two phase pattern
    stepA.Operate = 1       'enable the object

End Sub 'End Main()
```

Listing 3-4

find it. This program shows only a stepper program, but you must set the Rate property so that the compiler specifies where the Rate property resides.

Compile this program with a target firmware, either B.2.X or B.2.X+. At the time of this book's writing, the B.1.X firmware had a bug that wouldn't let oRam properly link. After compiling, select Opp_Codes under the View menu. You'll see that the right side window of the OOPic IDE now shows a bunch of text that looks like Figure 3-5.

Now you get to the hard part. Where is the Rate property? Scroll down the right panel until you find the line that reads:

```
----:<main-Begin>:
```

This denotes the beginning of the section of operand codes for the main() subroutine. Scroll farther down this list until you find these lines:

```
0035:100 'Number
0036:045 '->stepA+Offset4
0037:186 'Let.RATE
0038:100 'Number
0039:045 '->stepA+Offset4
0040:186 'Let.DIRECTION
```

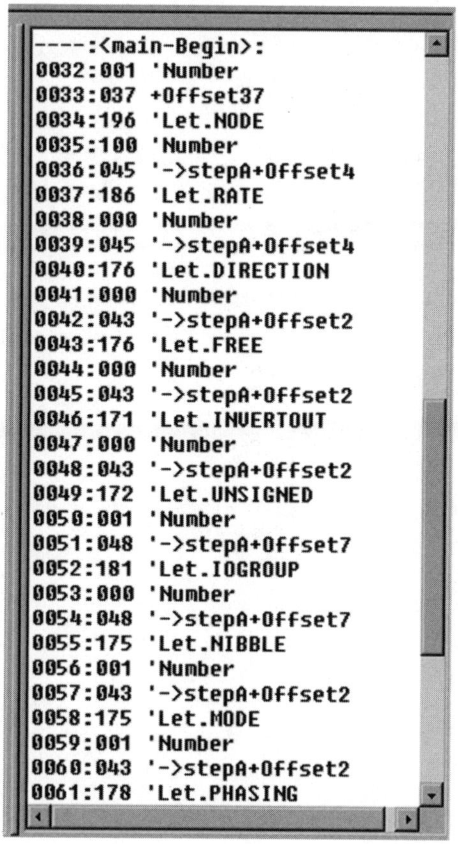

```
----:<main-Begin>:
0032:001 'Number
0033:037 +Offset37
0034:196 'Let.NODE
0035:100 'Number
0036:045 '->stepA+Offset4
0037:186 'Let.RATE
0038:000 'Number
0039:045 '->stepA+Offset4
0040:176 'Let.DIRECTION
0041:000 'Number
0042:043 '->stepA+Offset2
0043:176 'Let.FREE
0044:000 'Number
0045:043 '->stepA+Offset2
0046:171 'Let.INVERTOUT
0047:000 'Number
0048:043 '->stepA+Offset2
0049:172 'Let.UNSIGNED
0050:001 'Number
0051:048 '->stepA+Offset7
0052:181 'Let.IOGROUP
0053:000 'Number
0054:048 '->stepA+Offset7
0055:175 'Let.NIBBLE
0056:001 'Number
0057:043 '->stepA+Offset2
0058:175 'Let.MODE
0059:001 'Number
0060:043 '->stepA+Offset2
0061:178 'Let.PHASING
```

Figure 3-5 Opp_code listing

Three columns are used here. The first is the line number of the file; these numbers may be different on your display if you wrote your code a little differently than what is listed here. The second column, the one that starts after the colon (:), is the actual opp code, or Token, as it is more commonly known.

The third column, the one after the tick (') comment, is the text explanation of the opp code. The *stepA*+Offset4 followed by the *Let.RATE* means that the Rate property is the fifth byte in the oStepper object, or the base address plus 4. If you look through the rest of the oStepper object's properties, you will see another flag that shares that address, *Let.Direction,* which must use bit 7 of byte 5. This is not good news; it means that two properties are using that byte, and this complicates your link to the Rate property. Every time your link to Rate is written to, you will overwrite the Direction property. This problem can be fixed with an oMathI object to some extent. It means you will lose some of the oStepper object's functionality, namely the capability to use signed numbers to define which direction to turn, but that is not a huge problem.

Listing 3-5 shows how you can create a VC that will enable you to correct for the direction bit when you are writing to the oRam object. The listing also shows you which pins to connect to which devices.

```
'RamStep.osc (BASIC language syntax)
'OOPIC II version B.2.X, Compiler version 5.01
'
' Stepper controller for either unipolar or bipolar stepper
' motor - It just depends how you connect the dots...
' From bits 0 to 3 the phases are A-B-C-D, very simple.
' For bipolar steppers coil1 is 0 & 2, coil2 is 1 & 3.
' Put a pot on A2D line 1 (pin 7 of connector) and make
' sure that you can't short to ground or +5 and you can
' use this pot to adjust the stepping speed.  I used a 100K
' pot with 330 ohm resistors to ground and +5V at each end
' and the wiper to A2D port.  Steps <value> steps in one
' direction then steps <value> steps in the other direction.
'
' Stepper ouputs
'   Bit 0 = I/O 8 = Pin 20 = Phase A (bipolar winding A1)
'   Bit 1 = I/O 9 = Pin 18 = Phase B (bipolar winding B1)
'   Bit 2 = I/O 10= Pin 16 = Phase C (bipolar winding A2)
'   Bit 3 = I/O 11= Pin 14 = Phase D (bipolar winding B2)
' A2D 1 = I/O 1 = Pin 7
'
' Copyright Dennis Clark 2003
' Permission is granted to use this code anyway you like,
' as long as you say I told you ;-).

Dim stepA as New oStepper
Dim rate as New oA2D
Dim look as New oRam
Dim conn as New oMathI

'Workaround additions for B.2.2 firmware
Dim dummy as oByte
Dim dodge as oBus

sub main()
```

```
        OOPic.Node = 1                      'For debugging
    stepA.IOgroup = 1                    'I/O lines 8-15
        stepA.nibble = 0                    'I/O lines 8-11 chosen
        stepA.Mode = 0                      'Turn specified steps
        stepA.Unsigned = 1                  'Allow direction change in mode 0
        stepA.Phasing = 1                   'Two phase pattern
        stepA.Operate = 1                   'enable the object

        rate.IOLine = 1                     'A2D1
        rate.Operate = 1                    'Turn it on

    '***** Workaround code *****
    dodge.input.link(rate)
    dodge.output.link(dummy)
    dodge.operate = 1
    '*************************

    look.Location = stepA.Address + 4   'Where Rate is

    '***** Workaround code *****
    'conn.input1.link(rate)              'A2D link
    conn.input1.link(dummy)             'A2D link
    '*************************

    conn.input1.link(rate)              'A2D link
    conn.output.link(look)              'oStepper.Rate link
    conn.mode = cvOR                    'OR the two inputs
    conn = 0                            'set direction here
    conn.Operate = 1                    'turn it on

    do
        stepA.Value = 100               '100 steps this direction
        conn = 128                      'use this to set direction
        while(stepA.NonZero = 1)        'Look for all done
wend

        stepA.Value = 100               '100 steps this direction
        conn = 0                        'go other direction
        while (stepA.NonZero = 1)       'wait until all stepped
        wend
    loop
```

Listing 3-5

A bug in the B.2.2 firmware prevents oA2D from being linked directly to an oMath object. Listing 3-5 includes a workaround with an intermediate oByte object that solves the problem. The bug is fixed in Firmware revision B.2.3.

An oMathI object has been used so that the *Value* input, which is *Input2* to the oMath object, will be ORed with *Input1*, which is the oA2D output. By using oMathI, you don't have to use any other objects to set the direction bit, and because oMathI uses Value for Input2, you can link with a VC or simply assign that value in code, as shown previously. Using a single object to both hold the direction bit and do the math needed to combine them saves memory by using fewer objects. Listing 3-5 runs a stepper in one direction for 100 steps and then switches and runs the other direction. The Pot connected to the oA2D port sets the speed of the stepper. That wasn't too painful, was it?

I checked my discovery of where the Rate property was in the oStepper object by defining an oStepper object and eight oByte objects. I then created the oStepper and set Rate to 127 (all 1's in the variable). Using debug, I looked at all the fields in oRam to see what was set to 1's. Byte number 5 (address + 4 bytes offset) hit the jackpot, confirming where the property was.

All the objects you want to use in this manner can be discovered and modified in a similar way. Because all objects are different, the exact process will be different for each object. This section has shown you a process that can be applied to any property you need to have a unique link to.

The Oddball Objects: Nonstandard Properties

A few OOPic objects don't exactly match the commonly seen property lists previously discussed. The following sections list these nonstandard properties and what makes them different from other objects.

ODDBALL HARDWARE OBJECTS

These are the hardware objects that differ from the majority of their brethren. Usually, a hardware object has an *operate* property. Except for oDio16, the following hardware objects do not have this property:

- **oDio1**   A single bit input or output.
- **oDio4**   A four-bit I/O port.
- **oDio8**   An eight-bit I/O port.
- **oDio16**   A 16-bit I/O port. (This object is just a bit different; it has the Operate property.)
- **oDio16X**   A multiplexed port designed to work with a 1-of-4 demultiplexer that enables 32 bits of I/O to be done with only 11 I/O lines (8 to 15 and 5 to 7).

ODDBALL PROCESSING OBJECTS

Processing objects can either have both a Result value and an Output link or just one of them for the results of the processing. oMath is the exception here; it has an Output link and a Value result.

The oIndex object enables VC and program access to the oBuffer data object. Thus, its results don't really work with the Output and Result paradigm either. The oIndex object requires links to an oBuffer, an Index value somewhere, and an output to some other Value property.

OBJECTS THAT ARE DIFFERENT FROM ANY OTHER TYPE

A few more objects defy categorization and are outlined in the following sections.

oUserClass (All OOPic Firmware Versions)

This object enables you to break your code up into objects like a normal OO-based language. You can't link your object to a VC, but it will enable you to reuse code easily between projects. The steps are simple:

1. Write a normal OOPic program, debug it, and save it. You don't need to use the .osc suffix if you don't want to; whatever the file is called is how you need to define it in your main project program.

2. Using the Dim statement, declare an object of type <filename>. From that point on, you can use object notation to access subroutines as methods and data as attributes.

oUserClass has two restrictions:

■ You can't reference the OOPic object from within the user class. This is because the OOPic object is intrinsic and isn't created with Dim within the class. Thus, no reference exists for it within the namespace of the class when it is created in a user program.

■ You can't reference any method of the user class from within the class. The reason for this is similar to the previous restriction. Because this class is given a name by the Dim statement that defines the class, and the class does not know what this name is, you can't call any method from within the class because the namespace isn't known ahead of time. The OOPic OS does not have the concept of "this," as Java does, nor of "me," as Visual Basic does.

Normally, the first restriction would create trouble if you needed to delay for a specific amount of time. However, an undocumented command in the OOPic is called:

```
Do N times
Loop
```

where N is the number of times through the loop. This code will ideally generate a 5-millisecond delay for the count through the loop. It may well do that, but after testing it with a scope and some trial and error, I came up with an equation for the delay that is approximately:

$$T = 7N + 12 \text{ milliseconds (range of N = 1–65,535)}.$$

This means that the smallest delay you can have is about 19 milliseconds, which is 5 times smaller than an *OOPic.delay* = 1 delay.

The really nice effect of using oUserClass is that the namespaces of the main program and the user class are distinct. You don't need to worry about making sure all the global variables have unique names; the oUserClass handles that for you. An example is in order.

Throughout this book I use my own LCD routines to write to a nonserial LCD. Because I always use these routines, they are a logical set of functions I can place into a user class and reference them when I need to. Listing 3-6 shows my LCD routines, all encapsulated together and called dlclcd.osc.

```
'dlclcd.osc
'Chapter 3 oUserClass example via LCD routines.
'Also adds the undocumented do N times: loop command.
'
'For OOPic I A.2.X or later firmware
'
'Copyright Dennis Clark 2003
'Permission is granted to use this code anyway you like
'as long as you say you got it from me.

'LCD objects
Dim LCD as New oDataStrobe
Dim nibs as New oDio4
Dim RS as New oDio1
Dim E as New oDio1

Sub Main()

     nibs.IOGroup = 3              'I/O lines 28-31
     nibs.nibble = 1
     nibs.Direction = cvOutput

     RS.IOLine = 27               'LCD RS line
     RS.Direction = cvOutput
     RS = 1                        'set to data (no instructions!)

     E.IOLine = 26                 'LCD E line
     E.Direction = cvOutput
     E = 0                         'enable off

     'The oDataStrobe object set up in Nibble mode will
     'output the upper nibble first and then the lower
     'nibble.  Keep this in mind when reading LCD init
     'routines.
     LCD.Output.Link(nibs)         'Data to LCD
     LCD.Strobe.Link(E)            'LCD enable line
     LCD.Mode = cv4Bit             'doing 4 bit LCD
     LCD.Nibble = cvLow            'when doing single transfers
     LCD.OnChange = cvFalse        'transfer when value written
     LCD.Operate = cvTrue

End Sub

Sub Init()
'This initializes the LCD.
'Because we can't use the OOPic object in a User Class, we
'have to have some other way to waste time.  Welcome the
'undocumented command:
'                    do N times: loop
'
'Where N is any number from 1 to 65535.  The approximate
'delay generated is 7N + 12 milliseconds.  Counting code
'bytes says that it should be 5N milliseconds, but my ad hoc
'measurements give my formula as closer to reality.
```

```
    do 3 times: loop              'wait for LCD to come up
    LCD.Mode = cv8Bit             'only one strobe here
    RS = 0                        'instruction mode
    LCD = &H33
    do 1 times: loop
    LCD = &H33
    do 1 times: loop
    LCD = &H33
    do 1 times: loop
    LCD = &H22
    do 1 times: loop
    LCD.Mode = cv4Bit             'now in 4 bit mode
    LCD = &H28
    do 1 times: loop
    LCD = &H08                    '2 lines, font 0
    do 1 times: loop
    LCD = &H01                    'screen off
    do 1 times: loop
    LCD = &H06                    'screen on
    do 1 times: loop
    LCD = 12
    do 1 times: loop
    LCD = 6
    do 1 times: loop
    RS = 1                        'data mode

End Sub

Sub Clear()

    RS = 0                        'instruction mode
    LCD = 1                       'clear display
    do 1 times: loop
    LCD = 2                       'home cursor
    do 1 times: loop
    RS = 1                        'data mode

End Sub

Sub Locate(row as byte ,colm as byte)
'This locates the cursor in a specific location
'This works with 2 line displays and 4x20 displays.
'If you have a 4x16 then change the "+ 20" to be
'"+ 16" below.

    RS = 0                        'instruction mode
    if row < 2 then
        LCD = (128 + (64*row) + colm)
    End If
    If row = 2 then
        LCD = 128 + 20 + colm
    End If
    If row = 3 then
        LCD = 128 + 64 + 20 + colm
    End If
    do 3 times: loop
    RS = 1

End Sub
```

Listing 3-6

Listing 3-7 is the simple main program that defines and uses the previous program like the self-contained class that it is. When using this user class within a program, make sure you note the I/O lines used by the LCD object.

```
'userclass.osc
'Chapter 3 oUserClass example via LCD routines
'
'This program will define a user class consisting of the LCD
'routines that you'll get to see often throughout the rest of
'the book.
'
'For OOPic I A.2.X or later firmware
'
'Copyright Dennis Clark 2003
'Permission is granted to use this code anyway you like
'as long as you say you got it from me.

'If this next line won't compile, saying "file not found"
'then give the explicit path to the file, as in:
'"c:\robotics\OOPic\book\dlclcd.osc" or wherever you have
'the file located.
Dim LCD as New oUserClass("dlclcd.osc") 'That's all there is.

Sub Main()

    LCD.Init                    'initialize the display
    LCD.Clear                   'clear screen
    LCD.Locate(0,0)
    LCD.LCD.String = "OOPic oUserClass"
    LCD.Locate(1,0)
    LCD.LCD.String="All done."   'Looks odd, but...

End Sub
```

Listing 3-7

OOPic (All OOPic Firmware Versions)

The OOPic object controls the activity of the OOPic interpreter object. It has some very useful methods that will be shown later. Anything you do with the OOPic object affects the entire program. Some system-level activities only the OOPic can do.

OOPic is actually the object that reads the EEPROM and executes the program stored there. In essence, your code is running inside another object. The OOPic memory size is 10 bytes (an OOPic has 96 bytes of object space, but the OOPic object itself takes 10 of those bytes). Table 3-2 lists the OOPic properties.

The OOPIC object is always present; you can't turn it off. It is the only object that can have systemwide effects when you use it. The most common use for the OOPic object is for code delays and its clock bits (Hz1 and Hz60). It is often useful to know at what point the OOPic comes up, and the *StartStat* property can tell you that. StartStat may be useful, for instance, if you are using an EEPROM to store data and program status information to know where to start after a reset condition has so that your program doesn't just start from the beginning causing you to lose all the status or data that you have stored in the EEPROM..

TABLE 3-2 OOPIC PROPERTIES

Delay (word)	A write to this property will delay that number (0 to 65,535) times 1/100 of a second before executing the next instruction.
ExtVRef (bit)	This is discussed in a later chapter on *analog-to-digital* (A2D) operation. When set to 1, this sets the internal voltage reference to the voltage appearing on I/O line 4 and should *not* to exceed 5V.
Hz1 (bit, flag)	This read-only bit has a duty cycle of once per second; actually, it's closer to 0.99957Hz if you want to be picky.
Hz60 (bit, flag)	This read-only bit has a duty cycle of $\frac{1}{60}$ of a second. For a 60Hz clock, this is useful for clocking your RTC objects.
Node (byte)	Zero means no I2C node is assigned; 1 through 127 is the DDELink I2C node number of this OOPic. This is used for IDE debugging and for DDELink communications between OOPic controllers.
Operate (bit, flag)	If you set this bit to 1, the OOPic goes into low-power halt mode, effectively turning it off (default property).
Pause (bit, flag)	Similar to Operate, setting this bit to 1 will pause the OOPic, and it stops executing instructions.
PullUp (bit)	Setting this bit to 1 enables the internal pull-up resistors on I/O lines 8 through 15. This is useful if you want buttons on those inputs and don't want to use your own pull-up resistors. This mode is used by oKeypad and oKeypadX.
Reset (bit, flag)	When set to 1, this causes an OOPic software reset.
Snode (byte)	Sets the SCP (see Chapter 11) address of the OOPic for serial port debug and other communications uses.
StartStat (nibble)	This read-only property returns the reason for the last OOPic reset. 0 is standard power on; the OOPic just turned on. 1 is a reset by the hardware reset line. 2 is a reset caused by a power brownout (low voltage on Vcc). 3 is a reset caused by the Watch-Dog timer or the Reset property being set.

3

OOPIC OBJECT STANDARD PROPERTIES

PIC (All OOPic Firmware Versions)

 The PIC object gives you access to all the registers and memory of the PICMicro controller being used. This allows you to look at and modify settings used in the PICMicro. It also means you can modify settings that the OOPic operating system has set and depends on, and *that* means the PIC object is very dangerous if fiddled with, even if you know what you are doing. Table 3-3 lists the PIC systems used by the OOPic operating system. Fooling with them will surely cause something to go wrong.

If you don't know what those registers in Table 3-3 do, you shouldn't be using the PIC object. It is only visible in the debug pane if you have enabled a view of it in the View menu item Pic Object.

The PIC object is not examined in detail here. To do that would require a datasheet about the PICMicro your version of OOPic is running on, and that clearly isn't the focus of this book. A full listing of the registers you can access via the PIC object can be found in Appendix A. You can find out all you need to know about the underlying hardware of the OOPic you're using by looking up the hardware specification in Chapter 1, "OOPic Family Values," and by going to www.microchip.com to download the data PDF file that concerns you. It's a lot of fun to work directly with the hardware, but be very careful. Although you won't hurt the chip, you may mess up your running program.

TABLE 3-3   DON'T TOUCH THESE

FSR	PIE1
INDF	PIE2
INTCON	PIR1
OPTION	PIR2
PCON	STATUS
PCL	TMR0
PCLATH	TRISA
PORTA	

Now What? Where to Go from Here

The purpose behind this chapter is to explain some of the inner workings of the OOPic microcontroller. A *lot* is going on behind the scenes to make the OOPic simple to program, and a lot more is taking place here than just a PICMicro controller. Where do you go from here? Continue reading about the OOPic and how to solve a host of common embedded programming projects—that's where!

YOUR FIRST OOPIC PROGRAM, OOPIC I/O (PROJECT #1: DAS BLINKEN LIGHT)

CONTENTS AT A GLANCE

The OOPic is an embedded controller that is remarkably easy to program, but, like all programming environments, good ways and less-than-good ways can be followed to accomplish a task. This chapter shows you a couple of good programming techniques that will help you create reliable projects quickly and with less debugging. Also in this chapter, you'll begin your OOPic programming adventures by examining *input/output* (I/O) concepts with the oDIO1 hardware object, the oWire processing object, and the oDebounce processing object. Along the way you learn how to link *flag* properties.

Anything can be accomplished with the OOPic microcontroller in one of two ways. The first is to use a *virtual circuit* (VC); the other is to use standard logic programming. You will see an example of each and compare these programs to another popular and easy-to-use microcontroller as a way of examining what OO programming means and how VCs make your programming life easier. Because you have to use procedural code instead of VCs sometimes, the intrinsic functions of the OOPic procedural code and how the OOPic handles math and looping constructs are discussed as well. Writing embedded programs involves some quirks, which are discussed along with some of the not-so-obvious issues when dealing with the OOPic specifically.

The oDio Hardware Objects

The oDio objects are the simplest of the OOPic hardware objects. They don't do any fancy sensor operations; they just move bits between the PICMicro controller and the outside world. Although that sounds boring, in reality this can accomplish a lot, such as timing SONAR signals, outputting frequencies, and controlling other devices with the oDio lines. oDio objects are at the core of all the other more interesting OOPic hardware objects. It is fitting that you should learn about them first.

COMMON FEATURES OF ODIO HARDWARE OBJECTS

First off, the oDio16x object is not going to be discussed. This object uses 11 bits to create a 16-bit digital I/O port by multiplexing I/O lines 8 through 15 by using I/O lines 5 through 7. If you really need a few more bits of digital I/O, this is the object to use. The oDio16x object is completely different from the other oDio hardware objects and that's all I'm going to say about it.

The rest of the oDio hardware objects all have the same methods and many of the same attribute properties. The methods that all oDio hardware objects have are as follows:

- **Clear**   Clears the Value property to zero; in effect, all I/O lines are cleared to zero.
- **Dec**   Decrements the Value property by one. The I/O lines are likewise affected.
- **Inc**   Increments the Value property by one. The I/O lines are likewise affected.
- **Invert**   Inverts each bit of the Value property. This is a one's complement of all the I/O lines.

- **Lshift**   Shifts the bits in the Value property left one bit. The I/O lines are likewise affected.
- **Rshift**   Shifts the bits in the Value property right one bit. Again, the I/O lines are likewise affected.
- **Set**   Sets the Value property to its maximum positive integer value. The I/O lines are all thereby set to 1.

The oDio1 object does not use the LShift, RShift, Inc, or Dec methods. All oDio hardware objects have the properties outlined in Table 4-1.

TABLE 4-1   COMMON PROPERTIES

Address (address pointer)	The address of the OOPic object, from 0 to 127.
Direction (bit, flag)	Sets the Bit, Nibble, Byte, or Word I/O direction, either input or output.
String (string)	The string representation of the Value property.
Value(bit, nibble, byte, word)	The number that is written to or read from the I/O lines specified. This is the default value of the object (as discussed in Chapter 3, " OOPic Object Standard Properties").

All bits and flags may be specified or compared using a variety of *aliases,* or alternate names, which make more sense when used in a program in certain places. Those aliases are displayed in Table 4-2.

From this point on, the oDio hardware objects differ in their properties and actions.

The rest of this section lists the oDio objects and what they do. Again, oDio16x has been left out because it differs so strongly from the rest of the oDio hardware objects.

TABLE 4-2   BIT AND FLAG ALIAS OPTIONS

ALIASES FOR 0	ALIASES FOR 1
cvOff	cvOn
cvLow	cvHigh
cvFalse	cvTrue
cvOutput	cvInput

ODIO1

Memory Size: 1 Byte

UNIQUE PROPERTIES

I/O line (byte)	OOPic I/O line 1–31; 0 is disabled.
NonZero (bit, flag)	Returns 0 if Value is not equal to zero; it returns 1 otherwise

This is the most basic hardware I/O object. It is the definition of a single I/O line. oDio1 can be changed from an input to an output and vice versa dynamically via program statements, but not by a VC. Only a flag may be linked in a VC. The NonZero flag enables the Value property to be linked into a VC as a Flag property.

ODIO4

Memory Size: 1 byte

UNIQUE PROPERTIES

IOGroup (nibble)	Selects one of three I/O groups: 8–15, 16–23, or 23–31.
Nibble (bit, flag)	Selects the upper or lower nibble of the selected I/O group.

IOGROUP + NIBBLE SELECTION MATRIX

IOGroup	Nibble	Description
0	X	Object is disabled.
0	X	Object is disabled.
1	0	I/O lines 8–11
1	1	I/O lines 12–15
2	0	I/O lines 16–19
2	1	I/O lines 20–23
3	0	I/O lines 24–27
3	1	I/O lines 28–31

The Direction property sets the direction for the entire nibble; individual bits are not specified in an oDio4 hardware object. The Direction bit can, however, be changed dynamically in the program so the object can be either all inputs or all outputs at any chosen time.

ODIO8

Memory size: 1 byte

UNIQUE PROPERTIES

IOGroup (nibble)   Selects one of three 8-bit I/O groups:

0: Object is disabled

1: I/O lines 8–15

2: I/O lines 16–23

3: I/O lines 24–31

The Direction property sets the direction for the entire byte; individual bits are not specified in an oDio8 hardware object. The Direction bit can, however, be changed dynamically in the program so the object can be either all inputs or all outputs at any chosen time.

ODIO16

Memory size: 1 byte

UNIQUE PROPERTIES

IOGroup (nibble)   Enables the I/O lines 8–15 and 24–31.

0: The object is disabled.

1: I/O lines 8–15 and 24–31 are enabled as the *least significant byte* (LSB) and *most significant byte* (MSB) of the 16-bit Value property.

Only one set of I/O lines is supported for the oDio16 hardware object. The I/O lines are arranged as shown in Table 4-3.

TABLE 4-3   BIT ORDER FOR ODIO16 OBJECT

MSB	LSB
I/O lines 31–24	I/O lines 15–8

oDebounce Processing Object

When you use a switch or a push button to tell a microprocessor to start or stop something, an unusual phenomenon occurs called *switch bounce*. When you flip a mechanical switch, the contacts do not simply touch each other, and that's the end of the story. What actually happens is that the metal contacts bounce many times before they actually settle down in contact with each other. Because these bounces may occur a hundred or more times within a few milliseconds, and our microcontrollers can sample that button hundreds of times within that same few milliseconds, the controller can first read a 0, then a 1, and then a 0. This bouncing causes erratic behavior in our programs.

To eliminate the effects of switch bounce, we use an object called oDebounce. Usually, a good-quality switch will settle down within about 20 milliseconds. I like to use the OOPic oDebounce object with a Period property of 2, or about 32 milliseconds, for my switch debounce.

ODEBOUNCE

Memory size: 5 bytes

UNIQUE PROPERTIES

Input (flag pointer)	A link (pointer) to a flag denoting the switch connection input.
InvertIn (bit, flag)	0: When the Input property is 1, the Result property is 1 after the debounce.
	1: When the Input property is 0, the Result property is 1 after the debounce.

InvertOut (bit, flag)	0: The Result property is copied to the flag linked to the Output property.
	1: The Result property is inverted and then copied to the flag linked to the Output property.
Operate (bit, flag)	When set to 1, the object is enabled; otherwise, it is disabled (default property).
Output (flag pointer)	A link (pointer) to a flag that will be updated with the debounced value.
Period (byte)	The number of 1/60 second time increments that must pass before accepting the switch change.
Result (bit, flag)	The result of the debounce operation.

The difference between the InvertIn and InvertOut properties is not immediately obvious. When you invert the input, both the Result property and the flag linked to the Output property will have the same value. If you invert the output, then the Result will be of the same polarity as the input, and the flag linked to the Output property will be the inverted polarity. You might have some need for that; I've always just inverted the input. Remember, oDebounce only works with a logical 0 to a logical 1 transition. This should suggest a reason why you can invert both the input and the output to this processing object.

Obviously, with delays of over two seconds, more than just switch debounce can be done with this object. Let's say you have a user interface and you don't want the user to be able to change his or her mind faster than every two seconds. In this case, you can set the Period property to be 120 and only one logical 0 to logical 1 transition will be recognized within any two-second time period.

oWire Processing Object

To connect a Bit and a Flag property between two OOPic objects, you can use a simple oWire processing object. Not much can be said about the oWire object; it exists for the sole purpose of linking a Bit and a Flag property together. A clocked version of this object exists, but you won't use it in this example.

OWIRE

Memory size: 3 bytes

UNIQUE PROPERTIES (UNCLOCKED VERSION)

Input (flag pointer)	A link (pointer) to the signal input.
InvertIn (bit, flag)	0: When the Input property is 1, the Result property is 1.
	1: When the Input property is 0, the Result property is 1.
Operate (bit, flag)	When set to 1, the object is enabled; otherwise, it is disabled (default property).
Output (flag pointer)	A link (pointer) to a flag that will be updated with the input value.
Result (bit, flag)	The result of the input bit evaluation.

Blinking an LED, the Embedded World's Version of "Hello, World"

I'm sure you remember the very first program you ever wrote on your computer. All it did was print out the line, "Hello, World." Because most microcontrollers don't have a display to print anything out on, the newcomer to embedded programming must write his or her first program to blink a *light-emitting diode* (LED). I've blinked LEDs on just about every embedded platform I've ever used as my first program to see how the *Integrated Development*

Environment (IDE) and compiler operate. With the OOPic, I'm not going to be any different. This chapter's project program will be blinking an LED when you push a button.

This will be explained in a somewhat backward fashion in order to make a point. You will first write a procedural program in the OOPic language, similar to what you would write if you were using a Parallax Basic Stamp II microcontroller or any similarly programmed embedded controller. Then you'll see a program that takes full advantage of the OOPic microcontroller's *Object Oriented* (OO) VC paradigm, so you can see how much simpler the OOPic is to program than a strictly procedural programming language. After that, you'll learn how I designed the program and how to create the electronic circuit you need to see your LED blinking.

PROCEDURAL PROGRAMMING THE OOPIC

This OOPic program is built using procedural programming techniques. This is how you would program, for instance, a Parallax Basic Stamp. The code can only do one thing at a time, and while it is doing that one thing, nothing else is done. Even though this program runs perfectly fine, it is not using the OOPic to its full capacity. In fact, most of the OOPic's time is spent twiddling its fingers, so to speak, waiting to be told what to do next (see Listing 4-1).

```
'procblink.osc
'Chapter 4 Example "Das Blinken Light"
'Demonstrates procedural programming in the OOPic.
'
'Copyright Dennis Clark 2003
'Permission is granted to use this code anyway you like
'as long as you say you got it from me.

Dim button as New oDio1
Dim LED as New oDio1

Sub Main()

    button.IOLine = 8          'Choose button I/O line
    button.Direction = cvInput 'Make it an input

    LED.IOLine = 15            'Choose LED I/O line
    LED.Direction = cvOutput   'Make it an output

    do
        while button = cvHigh  'Wait for button press
        wend

        OOPic.delay = 2        'Debounce wait

        while button = cvLow   'While button is down
            LED = NOT LED       'Blink the LED
            OOPic.delay = 50
        wend
    loop

End Sub
```

Listing 4-1

Notice all the loops and waits being used to look for button presses and releases. The code works by first waiting for the button to be pressed. It then waits for a 20-millisecond debounce period and loops through the LED blink routine. Once each pass, it looks for the button to be released. The LED is off until the button is pressed, and then it blinks at a 1Hz rate until the button is released.

If you wanted the program to do something else while the LED is blinking, you would have to code the new activities within the overall do loop, and something might get missed. Eventually, this do loop could get very complex and timing problems would arise. By comparison, let's look at an OO event-based program that does the same thing but uses no traditional code constructs.

OBJECT-ORIENTED EVENT-BASED PROGRAMMING

The basic OO concept is that you don't program based on a list of actions that need to be taken or data that needs to be moved around. You program by considering discrete objects that are to be manipulated in some manner. This is a subtle and, at the same time, grand viewpoint change. It takes time to understand it and to change the way you think as well. It took me the better part of a year to finally get the point.

In this program, you have two important objects, the button and the LED, and you want them to work together so that when you push the button, the LED blinks, and when you let off the button, the LED goes out. The objective is to send messages through your program with one object telling it when to do something, and the other object actually doing it. This next program shows how to do just that, and all that happens is that messages move around among objects telling them what to do (see Listing 4-2).

```
'OOBlink.osc
'Chapter 4 Example "Das Blinken Light"
'Demonstrates oDio1, oWire and oDebounce objects and
'the use of OO design in the OOPic.
'
'Copyright Dennis Clark 2003
'Permission is granted to use this code anyway you like
'as long as you say you got it from me.

Dim button as New oDio1
Dim LED as New oDio1
Dim wire as New oWire
Dim smooth as New oDebounce

Sub Main()

    OOPic.node = 1                      'For debugging

    button.IOLine = 8                   'Choose button I/O line
    button.Direction = cvInput          'Make it an input

    LED.IOLine = 15                     'Choose LED I/O line
    LED.Direction = cvOutput            'Make it an output
```

(Continued)

```
wire.Input.Link(OOPic.Hz1)        'Link in the 1Hz pulse
wire.Output.Link(LED)             'Link it to the LED

smooth.Input.Link(button)         'Link in button
smooth.Output.Link(wire.Operate)  'Link to wire operation
smooth.InvertIn = cvTrue          'Invert the button logic
smooth.Period = 2                 '32ms button debounce
smooth.Operate = cvTrue           'Turn debounce on

End Sub
```

Listing 4-2

That's it. No procedural code is used in there at all. The objects are defined and linked, and the main() subroutine ends. The OOPic objects are all running in the background with no other supervision needed. Other VCs can be added to this program without having to change anything; they would work independently of each other, not even aware that the other VCs exist.

Project: Push a Button and Toggle an LED

You've seen the program you're going to use for this project in Listing 4-2. Now you're going to examine a methodology for designing VCs and building the hardware interfaces you need. The first lesson involves hardware.

HARDWARE LAYOUT

Most newcomers to the OOPic start to pick up the basics of programming pretty quickly because writing poor code doesn't hurt the OOPic microcontroller at all. You can simply start over and do it better next time. When it comes to a hardware interface, however, if you do something wrong *enough,* you will fry your OOPic and need to get another one. This is a potentially expensive and disheartening turn of events.

My first rule is this: Don't plug anything in backwards! Always check it three times. Sure, this seems obvious, but even I have plugged my battery or OOPic chip in backwards at one time or another when I'm in a rush. My second rule is: Don't be in a rush. It's funny how often a shortcut takes so much longer than the slower path.

One great way to fry an I/O line on an OOPic, or any other microcontroller, is to connect an output to ground and have that output then try to change to a logic 1. The magic smoke gets out and the chip stops working. This can be avoided in a simple and easy way. Always put a 330 ohm resistor in a series using the I/O line to the switch. By this simple precaution, you prevent hazardous currents from ruining your OOPic chip. This precaution is shown in Figure 4-1 on the button circuit. The LED already has a current-limiting resistor in it for the LED, so that's protected as well.

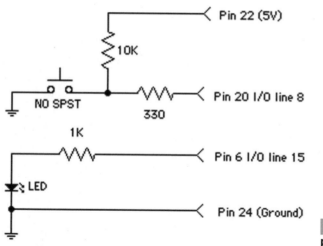

Figure 4-1 Chapter 4 project schematic

Figure 4-1 also shows the schematic of the circuits used to create this project. It's a simple schematic, but it illustrates I/O line protection safeguards, and it demonstrates that an LED will light up nicely using a 1K resistor. You don't need a current-sucking 330 ohm resistor to see an LED!

As mentioned in the introduction, solderless breadboards are a wonderful tool for prototyping and experimenting with hardware. A small RadioShack™ model is used for this project and is just big enough for your needs (see Figure 4-2). It is an inexpensive model and all you need to add is 22-gauge solid wire cut to various lengths to make connections.

Another item to remember when building external hardware interfaces to the OOPic microcontroller is that I/O line numbers *do not* equate to pin numbers on the OOPic I/O connectors. Be sure to check out Chapter 1, "OOPic Family Values," for the mapping between I/O lines and I/O connector pins before you apply power to your circuit. The pin numbers in Figure 4.1 are the I/O connector pins, not the OOPic pins.

DESIGN THE VC

I've been asked, "How do I decide what to use in a VC?" many times before. My answer is, "Get a pencil and start drawing pictures." Seriously! OO design techniques universally use a graphical format to help visualize objects and their relationships. This is a good way to design OOPic programs as well, especially because so many OOPic objects are already there and are well defined in their functionality.

Formal OO design processes use standardized OO templates of symbols, lines, arrows, diamonds, and . . . basically they get really complicated really fast. They are also unnecessary for your needs in designing an OOPic program. All you need is a way to represent objects and links between objects. There is no such thing as "program flow," because an OO program, like an event-based program, does not follow a logical flow or a data path flow. An event-based OO program just sends messages between the objects. Keep reminding yourself of that and eventually you'll get it.

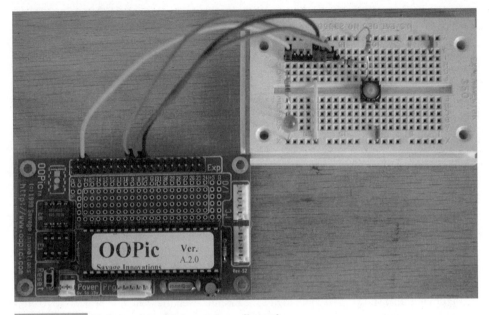

Figure 4-2 Chapter 4 project on a breadboard.

Regardless, use what you are comfortable with. I just draw boxes with link locations denoted on them, and lines with arrows are drawn between the boxes. It's simple and makes sense to me. I like to leave enough space in the boxes to write in various settings that need to be configured when the objects are set up. It also helps to write both the object type and the name you will be using for it in your program. Figure 4-3 shows my object mapping for this button/LED program.

Obviously, my original artwork isn't perfect, because my handwriting isn't so neat. All that is needed is a simple drawing where you can sketch in all the connections that need to be made and all the properties that need to be set to make your VC do what you want. It really is just that simple; you don't have to figure out any fancy timing, just make sure that your logic is correct. Drawing it out on paper and walking through the steps is a good way to see if you are capturing everything you need to get your VC working properly.

LINKING OBJECTS

The most confusing detail of OOPic programming is typically the linking process. What can and can't be linked? Chapter 3 discusses processing objects, which are the *only* objects that can link other objects together. All processing objects have a link pointer for their inputs. Almost all processing objects have both a link pointer for their output as well as a result value that can be read directly. Typically, the Operate flag is the default property for a processing object, but a few exceptions exist, and they are noted in the appendices where the objects are described. Linking is what creates a VC.

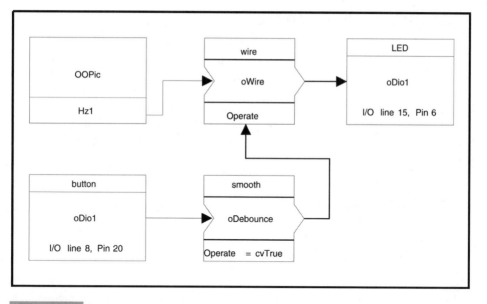

Figure 4-3 Object definition map

When a pointer link is used in a processing object, the proper way to link to another object is (as an Output property example):

```
Processing_object_name.Output.Link(other_object_name).
```

The other object may be a hardware object or a data object. If the flag you want to link to isn't the default flag of the other object, use this format:

```
Processing_object_name.Output.Link(other_object_name.Value).
```

Remember, flag pointers may only link flags. Number pointers and object pointers can link anything other than a flag, and those objects will *only* link the default property of the object being linked. The flag pointers can link any flag, whether or not it is a default property. Refer to Listing 4-2 for several examples of flag links in a VC.

Writing OOPic Programs

Most of this isn't anything specific to the OOPic; I would offer mostly the same advice to anyone that writes in any language. A few simple guidelines must be followed if you want to understand a program you wrote two months ago. Also, some requirements for proper OOPic operation need to be observed, and some potentially confusing details about OOPic programming need to be addressed. The following sections outline the most important ones.

VARIABLES AND DATA OBJECTS

Data can be represented in the OOPic microcontroller in two ways: One is as a data object such as oByte and oWord; the other is as a programming variable, which isn't an object. In the A.2.X firmware, both the objects and variables reside in the same memory space. In all subsequent OOPic firmware varieties, they reside in two different memory banks. Refer to Chapter 2, "The OOPic IDE and Compiler," for a thorough discussion of how much memory you have in each of these storage areas. If you need to link data to an object, you need to use the data objects. If you simply need variables for your program, use the regular variable types.

For instance, a data object would be dimensioned like this:

Dim dataOb as New oByte.

A variable would be dimensioned thusly:

Dim dataVar as Byte.

Variables take up less space than a data object usually, but most importantly they use a different memory storage area that allows more space for objects in your program. Single dimension variable arrays are supported. With the B.2.2+ firmware came the sByte data type. It stores data in the OOPic internal EEPROM and has some restrictions on its use, more information is in Appendix C.

The oWord and oByte data objects have a property called *Signed,* which, when set, enables the VCs and math to recognize that these variable objects are signed; that is, they have both plus and minus values. The Signed property does not confer any special capabilities to the data object, it is simply a flag telling the OOpic that it must consider the sign bit when moving data between objects and doing math operations. You can create arrays of data objects. Arrays of objects or data are 1 based, that is the first element in the array is element 1.

OOPIC MATH

The OOPic is strictly integer math, integer math means that fractional amounts are not supported, only whole numbers. Even functions like Sin() are integers, but in their own way. OOPic integers can be signed (+ or -), but support for signed integers is inconsistent among objects and the debugger, and signed integers are not supported by OOPic functions. Variables cannot be signed; they are unsigned integers only. In fact, signed integers are only used in some objects. Read the object descriptions carefully before you use a signed integer or you'll get some really odd results. The signed math works fine, and negative numbers are properly added and subtracted, but OOPic control constructs like do/loop will not recognize negative numbers, rather, they will look like really big positive numbers.

The OOPic compiler and firmware work with negative numbers, but you must keep track of the negative values yourself so that your program doesn't consider them to be really big positive numbers. You'll also need to understand two's complement math and the fact that any variable's MSB is a negative number. Table 4-4 outlines how you translate a negative number into binary format.

To translate a binary negative number to decimal, work in reverse, as shown in Table 4-5.

TABLE 4-4   BINARY NEGATIVE NUMBER REPRESENTATION

Description	Decimal	Binary (8 bit)
Start with	−25	
Make it a positive integer	25	00011001
Invert all bits		11100110
Now add 1 to LSB	−25	11100111

TABLE 4-5   TRANSLATING NEGATIVE BINARY TO DECIMAL

DESCRIPTION	DECIMAL	BINARY (8 BIT)
Start with	−25	11100111
Subtract 1		11100110
Invert all bits	25	00011001
Now add the − sign	−25	

You don't need to know the details of two's complement math; the OOPic will do that for you. You just need to know when you have a negative number and how to translate it to a decimal number you understand. You also need to remember that when you add 1 to 1, you get 0 with a 1 carry. Also, when you subtract 1 from 0, you get a 1 with a 1 borrow. It's easy when you get used to it. All binary math in the OOPic is done with no error checking and no overflow protection. It is up to the programmer to recognize when errors occur and to check and correct for them, because the OOPic will not.

All OOPic functions that perform divisions will return the integer result. The OOPic math function Mod returns the remainder after you divide, which is not a fraction, so that one's easy. If you want the fractional part, such as whatever is returned by the sin or cos functions, you have to think about it a little. The OOPic divides the circle into 255 parts instead of 360 degrees or 2p radians. You use this returned by Sin or Cos to describe the angle given to these functions, and it is commonly called the *binary radians*. The sin or cos functions return a number that, when used as the numerator over 255, is closest to the fraction of the circle being represented.

Let's say you use one of those functions and you get 90 as the result. The fraction this number represents is 90/255 binary radians. Figure 4-4 illustrates what this fraction means.

You may assign values to variables or objects by using decimal representation:

$$A = 125$$

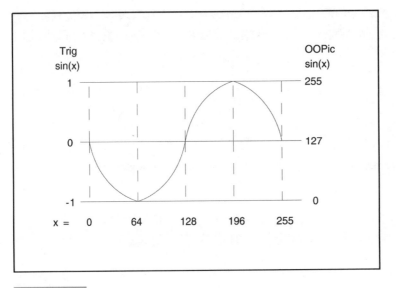

Figure 4-4 Sin return value meaning

Or hexadicimal representation:

$$A = \&HF5$$

Or binary representation:

$$A = \&B01000101$$

OOPIC INTERNAL FUNCTIONS

The OOPic doesn't have many functions. Most of the functionality is locked up in the objects for VC construction. OOPic internal math is 17 bit, so these functions will take 16-bit operands and place the result into whatever variable you want them stored in. However, if you stuff a 16-bit word result into an 8-bit variable, you'll get what you deserve. String support in the OOPic is not strong. A robot doesn't use strings; only we do. Therefore, the only string support is the minimum required to communicate with the operator.

Abs

This function returns the absolute value of a number. In other words, it strips the minus sign. It is not available on the A.X.X firmware.

Chr$

This function returns the string representation of the ASCII argument given to it.

Cos

This function requires its argument to be in binary radians (0–255 as discussed). It returns the trigonometric ratio of the length of the side adjacent to the angle divided by the length of the hypotenuse.

Sin

This function requires its argument to be in binary radians . It returns the trigonometric ratio of the length of the side opposite the angle divided by the length of the hypotenuse.

Sqr

This function takes a positive number between 0 and 65,535, calculates, and then returns the integer square root of that number. No fractions are used.

Str$

This function takes any numerical argument and returns the string representation of that number. All positions are represented and leading zeros are present. This means that five characters will be returned. If the number is 5, for instance, you'll get 00005 as the result.

LOOPING CONSTRUCT GOTCHAS

The OOPic microcontroller is an embedded controller, and like all embedded controllers, no operating system support exists for error handling. You, the programmer, must anticipate and catch the error before it occurs. The following code snippet demonstrates what this means by showing a for next loop that will never end:

```
Dim dummy as byte

For dummy = 1 255 step 45
    ...
Next dummy
```

In most of your programming environments, this is taken care of by the operating system. It checks the number and acts accordingly. If an overflow occurs, it throws up an error and the program stops. However, Table 4-6 illustrates what happens in the OOPic.

If you want to match against 255, you should use a Word variable. This will enable the counter to proceed past your maximum number so that your loop will terminate properly.

TABLE 4-6   LOOP CONSTRUCT OVERFLOW

LOOP	DUMMY =	DESCRIPTION
1	1	Normal loop through
2–6	46–226	Same
7	15	271–256 = 15. The 8-bit variable overflowed or wrapped around. After 255, it became 0, then 1, and so on. This loop will never end because every number will be less than 255.

FINAL OOPIC PROGRAMMING HINTS

Some less than obvious bugs can creep into your OOPic program if you aren't careful. This section offers some commonsense hints that can help you debug and upgrade your programs:

■ Remember to define an *IOLine* or *IOGroup* before you set the Direction bit. If you don't, your I/O lines will just be inputs (the default) because the compiler won't know which bits to set up.

■ Remember, if an object has an *Operate* flag, the object won't update until that flag is set to 1. An oEvent will not restart until Operate has returned to 0 and again been set to 1.

■ You have to give the OOPic a node number to debug it. OOPic.node at 1 is a good start.

■ Use descriptive variable names. The days of the "one alphabetic and one numeric character" variable names have been over for nearly 20 years. If the input is for the power-on button, then name the variable "power_on_button." You'd be surprised how easy this makes understanding and debugging a program.

■ Reuse temporary variables; it conserves memory.

■ Use comments liberally, but not stupidly. By this, I mean you should comment on what is going on within a line or block of code where the functionality or logic is not obvious. In other words, when you link to another object, comment about *why* you made the link, not about the fact that you made the link. Too many comments are almost as bad as too few when they are trivial comments.

■ Comment on a block of code to describe an algorithm you used to make a decision. Alternatively, you can describe the physical hardware that makes the VC connection necessary or successful.

■ Be consistent with your indenting. When you have a block of code that is running inside of a loop or decision statement, indent that block of code. I use a four-space indent; others just use the tab. It doesn't matter what you use as long as you are consistent throughout your program.

Using the Debugger

Because the OOPic is based around the *Inter-IC* (I2C) standards, it is very simple for a program to interrogate an OOPic VC while it is running to query any object to see what it is doing. The OOPic IDE does exactly that and then displays what it discovers in a nicely formatted window. It's a built-in debugging facility that is almost unrivalled in the embedded community.

At this book's writing, the OOPic IDE is going through a monumental rewrite that decouples the compiler from the user interface and adds support for *Serial Control Protocol* (SCP) handling, so not all the objects are in the debugger; only the hardware and data objects are there. This is why no more examples of debugging processing objects will be examined. Refer to Chapter 2 for more detail on the debugging process and for specifics on cable connections and setup.

If your VC is not working according to plan, it is often useful to look at the properties in the various objects to make sure they are what you expect. For this project, you'll see what the debug windows look like for the *button* object when the button is depressed and when it is not. Figure 4-5 shows the debug window for the button object when it's just sitting there. Figure 4-6 shows window when you press the *Refresh* button while holding the pushbutton down on your prototyping board.

Here are the steps required to begin debugging this project:

1. Compile and download your program to your OOPic. I'm using an OOPic I with the A.2.0 firmware. If you are using another version, make sure your *Target Device* is properly set. When you use the parallel port cable, make sure you are connected to the programming connector, not the networking connector.
2. If you are using the parallel port cable, move it from the programming connector to one of the networking connectors. If you are using a serial cable (OOPic R or C), you don't need to do anything. Make sure you set OOPic.Snode for serial port debugging.

Figure 4-5 The button object when no button is pressed

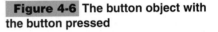

Figure 4-6 The button object with the button pressed

3. Make sure that *View Objects* is set so that you see the graphical representation of the OOPic objects in your right-side window.
4. Click the *Network Node* icon at the top of the object list, and fill in the blank with the node you assigned in your code.
5. Click the object that you want to investigate. The IDE resizes the source code window, and an object window pops up. You can click on any number of objects and have all their πwindows open at the same time.
6. To refresh the display, click any of the Refresh buttons on any of the debug windows, and all the debug object windows will be refreshed.

Have fun! I always enjoy looking at the real-time values of what is happening in my projects by using the debug facility in the OOPic IDE.

Now What? Where to Go from Here

This was the simplest program you'll find in this book. It's a good place to start if you are really unfamiliar with the OOPic paradigm or if you are new to programming at all. It is recommended that you play with this VC and add other things to it, such as one-shots, timers, or programmable dividers to change the LED blink rate—go crazy. Once you have the hardware parts set up, you can do whatever you want with the software. This will allow you a comfortable starting point to experiment with your own ideas. Although this project is a simple one, it is one that you can use in almost every program that you write, such as bumpers for robots, input switches for a user interface, start or stop buttons, and so on. It is a very versatile idea.

ANALOG-TO-DIGITAL AND HOBBY SERVOS (PROJECT: PUSH MY FINGER)

CONTENTS AT A GLANCE

This chapter shows you how to use *analog-to-digital* (A2D) ports and how to connect, power, and use *radio-controlled* (RC) hobby servos. You will also see some experiments that will help you understand servos and introduce you to the concept of the *Uniform Robotic Control Protocol's* (URCP's) robotic velocity instructions. The chapter's project is to create a simple remote actuator that controls a hobby servo by measuring the angle of a bend sensor.

Because you'll be linking data values in this chapter, not just flags, other (OOPic) processing objects will be introduced, such as oMath and oBus, which provide various means for linking data and modifying it at the same time. Finally, the OOPic A2D system is quite flexible and can be configured for any special requirements you might have.

OOPic A2D Conversion

The OOPic microcontroller has a very easy-to-use A2D system that consists of hardware objects that handle all the mundane functionality. All you have to do is enable them and read the values.

OOPIC A2D CHANNEL TYPES, CONFIGURATIONS, AND MAXIMUM VOLTAGE LIMITS

Two different types of A2D conversion ports exist, and an A2D port's type and number depend on the OOPic version. Table 5-1 illustrates each OOPic's type and number.

TABLE 5-1   A2D RESOLUTION AND CHANNELS

OOPIC VERSION	FIRMWARE REVISION	CHANNELS AND RESOLUTION
OOPic I	Firmware A.X.X	(4) 8-bit A2D channels
OOPic II	Firmware B.X.X	(7) 8-bit A2D channels
OOPic R and C	Firmware B.X.X+	(4) 10-bit A2D channels
OOPic II+	Firmware B.X.X+	(7) 10-bit A2D channels

The OOPic 8-bit A2D objects get their values from a single PICMicro hardware register that contains the entire reading. The OOPic 10-bit A2D objects have to assemble the value from two PICMicro hardware registers. One register contains the upper 8 bits of the reading, and this is the same PICMicro register that holds the 8-bit reading in the OOPic I and II oA2D objects. The lower 2 bits come from another PICMicro hardware register that is normally set to zeros. The oA2D10 object then assembles these bits together to give a full 10-bit reading.

If you define an oA2D10 object on a B.X.X firmware OOPic II, you get a reading that is essentially multiplied by 4. This is because the reading is left-shifted two bits and the two bits from the other PICMicro register (which are zeros) are placed in the 2 *least significant bits* (LSBs). If you wanted a quick way to multiply your A2D reading by 4, this is an option.

If you define an oA2D or oA2DX object on a B.X.X+ firmware OOPic II+, R, or C, you will only get the upper 8 bits of the reading, effectively dividing your reading by 4. This is because the lower 2 bits will not be retrieved from the other PICMicro hardware register and assembled into the completed value; you have to use oA2D10 or suffer the confusion of the lost bits.

The OOPic *input/output* (I/O) channels are all on the Group 0 ports and can be used either as A2D ports or digital I/O ports, depending on the objects defined. OOPic controllers with four channels have their ports on I/O lines 1 through 4. If you assign an oA2D object to any of those I/O lines, they will all be considered to be A2D input ports, and any digital reads on those lines will return 0.

OOPic controllers with seven channels have their ports on I/O lines 1 through 7. If you assign an oA2D object to any of the I/O lines 5 through 7, then all of those lines will be A2D input ports, and any digital reads on those lines will return 0.

SETTING EXTERNAL REFERENCE VOLTAGE ON I/O LINE 4

The OOPic object has a property called *ExtVRef* that affects the results of all A2D objects. If ExtVRef equals 0, the PICMicro uses 5V internally as the maximum voltage reference. If ExtVRef equals 1, the voltage on I/O line 4 of the OOPic is used for the maximum voltage reference. According to the PICMicro data sheets, the maximum voltage you can put on I/O line 4 is 5.3V, but 5.0V is realistically the maximum voltage you should use. The minimum voltage that will function properly depends on the PICMicro being used by that version of OOPic.

The following list details the minimum and maximum voltage reference that the OOPic devices can use as an external voltage reference.

- 16C74 (OOPic I) range = 3.0V minimum, 5.3V maximum
- 16F77 (OOPic II) range = 2.2V minimum, 5.5V maximum
- 16F877 (OOPic II+, R and C) range = 2.5V minimum, 5.3V maximum

Obviously, if you are attaching a reference voltage to I/O line 4, you can't use that I/O line as an A2D object.

OOPIC A2D HARDWARE OBJECTS

Three OOPic A2D hardware objects are available: oA2D, oA2DX, and oA2D10. The A2D10 objects are 10-bit A2D channels and are supported in the OOPic R, C, and II+. The oA2D and oA2DX objects are supported in all OOPic variants.

oA2D (All OOPic Firmware Versions)

The oA2D object is a basic 8-bit A2D channel. If you define an oA2D object on B.2.X+ firmware (OOPic R, C, or II+), you get an 8-bit result, not a 10-bit result. The upper 8 bits of the 10-bit result are returned, and you lose the two lowest bits, effectively dividing your result by 4. The oA2D memory size is 3 bytes and Table 5-2 lists the oA2D properties.

TABLE 5-2 OA2D PROPERTIES	
IOLine (nibble)	Assigns the I/O line to use. Essentially, 1 through 7 are I/O lines 1 through 7; however, on the A.X.X firmware, 1 through 3 are I/O lines 1 through 3, but 0 is I/O line 4.
MSB (bit, flag)	The *most significant bit* (MSB) of the *Value* property. If this bit equals 1, then Value is equal to or over half of the reference voltage. If it is 0, it is under half.
Operate (bit, flag)	Must be 1 to enable the object.
String (string)	The string representation of the Value property.
Value (byte)	The result of the A2D conversion (the default property).

oA2DX (B.X.X Firmware or Later)

The oA2DX object compares the *IOLine* defined as the A2D channel with the OOPic reference voltage that is applied to IOLine 4. It then sets this measurement as a percentage of the reference voltage and subtracts 128 from that. This gives a range of -128 to +127 as the result, and the *Center* property is added to this value. Because you could use an offset of -128 or +127, the actual range of the Value property of oA2DX is -256 to +255, so the Value property is a word so that values outside of the -128 to +127 may be represented.

The *Center* property is useful when you have a range of analog values that don't sit nicely centered around the zero center mark. An example using this property is shown in the "Experiments" section.

The oA2DX memory size is 5 bytes and Table 5-3 displays the oA2DX properties.

TABLE 5-3   OA2DX PROPERTIES

Center (byte)	Adjusts the center's compare point by -128 to 127.
IOLine (nibble)	Selects the A2D line to use, which is within 1 through 4 or 1 through 7 depending upon which OOPic is being used. 0 selects no A2D line.
Limit (bit)	When set to 1, this limits the Value property from -128 to +127, no matter where the center is set.
Negative (bit, flag)	Is set whenever the Value property is a negative number.
NonZero (bit, flag)	Is 0 whenever the voltage is 0.
Operate (bit, flag)	Must be 1 for the object to operate.
Unsigned (bit)	When set to 1, the percentage is from 0 to 255 and is not signed.
Value (word)	The adjusted A2D result.

oA2D10 (B.2.X+ Firmware or Later)

The oA2D10 object returns the full 10-bit A2D reading taken. The 10-bit A2D channels are only available on the B.2.X+ firmware revision. If you define an oA2D10 on the B.X.X firmware, it works, but bits 1 and 0 will be 0 and the two most significant bits of the reading will be lost. Thus, it's not very useful. Seven 10-bit A2D channels are on the B.X.X+ firmware, but on the OOPic R those last three ports are used for buttons and *light-emitting diodes* (LEDs). On the OOPic C, ports 5 through 7 are not brought out to pads like ports 1 through 4 are so they can't be used. Only the OOPic II+ uses all seven A2D10 channels. The oA2D10 memory size is 3 bytes and Table 5-4 outlines its properties.

TABLE 5-4   OA2D10 PROPERTIES

IOLine (nibble)	Assigns the I/O line to use. Essentially, 1 through 7 are I/O lines 1 through 7; however, on the A.X.X firmware, 1 through 3 are I/O lines 1 through 3, but 0 is I/O line 4.
MSB (bit, flag)	The MSB of the Value property. If this bit equals 1, then Value is equal to or over half the reference voltage. If it is 0, then it is under half.
Operate (bit, flag)	Must be 1 to enable the object.
String (string)	The string representation of the Value property.
Value (byte)	The result of the A2D conversion (default property).

MEASURING ANALOG VOLTAGES

Sometimes you'll be measuring voltages directly from a device, such as a Sharp GP2D12 *infrared* (IR) ranging module, but often you'll have to build up some kind of circuit to help you get the results you want and make it easy to interpret the readings. Circuits can be configured in two basic ways to give you meaningful analog signals: using the voltage divider or the Wheatstone bridge. The voltage divider circuit is perfect for making measurements that require the output voltage to be positive only. The Wheatstone bridge outputs both positive and negative values, so you need to be knowledgeable about electronics to use it.

The Voltage Divider Circuit

The voltage divider circuit is often called a resistance divider, or simply, a divider. It works according to a couple of electronics theorems that I won't examine in great detail, and by Ohm's Law, which I will discuss. Here a simple problem illustrates what a voltage divider is and how to use it.

One of the most common questions I've been asked about A2D is, "How do I measure my battery voltage so I know when to recharge my robot?" The answer: By using a voltage divider. Your OOPic is powered by 5V, that is supplied by the voltage regulator on the board. The battery clearly has to be higher voltage than 5V, or the regulator won't work. If you are using a 7.2V, 6-cell *nickel cadmium* (NiCad) battery pack, then the nominal charged voltage is 7.2V, or 1.2V per cell. As a rule, your 7.2V pack will actually read as high as 1.4V per cell when fully charged. A NiCad cell is considered fully discharged when its voltage is 0.9V per cell. Now you know your limits to look for: The highest voltage could be 8.4V and the lowest voltage could be 5.4V. Your OOPic can measure a maximum of 5V when you use the default internal 5V reference voltage, so you need to limit the voltage to an A2D port to 5V.

I can hear you asking, "Why not just use a higher reference voltage?" You can't, because where would you get that reference voltage? From the battery? Hmm, the battery voltage is slowly dropping; so is your reference voltage dropping. Bad idea. So how can you measure a higher voltage than 5V? Figure 5-1 shows a resistor network configured as a voltage divider that keeps the voltage on the A2D port at or below 5V.

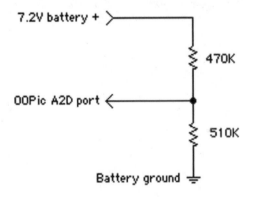

Figure 5-1 Battery voltage divider

Here is how I decided on those values. I'm a pessimist; I figure if I think the highest battery voltage I can expect will be 8.4V, it'll really be 9V. So just for a margin of safety, I used 9V as the maximum expected battery voltage. You need to have a maximum of 5V on the A2D inputs, and this is the ratio of desired to expected voltage:

$$\frac{5}{9} = 0.555 \ldots$$

This means you want to see 0.56 of the battery voltage at the A2D port. The resistor values 470K ohm and 510K ohm will provide a ratio of

$$9V \times \frac{510K}{470K + 510K} = 9V \times 0.52 = 4.7V$$

It pays to be careful. The resistors you'll use are 5 percent tolerance (gold fourth band), which means that in the worst case you could have resistors whose values are 446.5K ohms and 535.5K ohms, which would give the following number:

$$9V \times \frac{446.5K}{446.5K + 535.5K} = 9V \times 0.54 = 4.9V$$

Okay, now I feel comfortable enough to be safe. I've used really large value resistors because I don't want this circuit to draw very much current, which would drain the batteries even faster. Some PICMicro devices don't like using resistances in the 100K ranges for A2D measurements. If you find that your A2D values don't match your calculations, then use resistors in the 10K range; eg 47K and 51K.

To use this circuit, simply fully charge the battery and take a reading. This will be the maximum voltage that is detected. If, let's say, you read an oA2D value of 228 with the battery fully charged and you know that the battery voltage is 8.2V, you can now figure the reading that you'll get when the battery is at 5.4V (0.9V per cell) and fully discharged. That ratio would be

$$\frac{5.4V}{8.2V} = \frac{X}{228} \Rightarrow 228 \frac{5.4}{8.2} = X = 150.1$$

You can expect the battery to be near death when you have a reading of 150 from your oA2D object. The A2D readings I used in this example are fictitious; they were chosen for the sake of discussion. You will certainly get different values; don't use mine.

Another use for the voltage divider circuit is to find the value of an unknown resistance, or a variable resistance. This is the operating principle behind this chapter's project.

OOPic Hobby Servo Objects

The OOPic has many types of hobby servo hardware objects; some are only different in subtle ways from others. This section outlines them and their intended purposes.

SERVO CONTROL AND RESOLUTION

The OOPic has many different servo objects, each designed to control a servo for a different reason or to interface a servo with a different object. One of them will certainly accomplish what you need.

The oServo and oServoX objects are designed to control standard hobby servos that will move to a position and stay there. The oServoX object uses URCP notation and oServo does not. The oServo object is the more generic servo object, whereas oServoX would be ideal for controlling a robot arm joint, because it uses a (+) and (−) numbering scheme that revolves around a 0 point. I'll talk more about the URCP notation scheme later on in this chapter. The oServoSP1 object is designed to operate hacked hobby servos being used as drive systems for mobile robots. The oServoSE object is designed to streamline communications with a Scott Edwards™ *Serial Servo Controller* (SSC) board. You may see the oServoSP2 object listed and wonder what it is. Don't bother; it was an experiment and will no doubt be obsolete in future OOPic firmware revisions.

Even though many of these servo objects have data ranges from 0 to 255, or from -128 to +127, the OOPic servo objects are designed to have a resolution of 6 bits, which is 0 to 63. This means whichever number you use, when the 64 position span has been exhausted, you will start issuing pulses outside of some hobby servo's mechanical capability. Some objects have properties that allow excessive values to be set to a maximum limit at that extreme. Because each object has its own way of dealing with servo positioning, I'll discuss this in the description of the objects later.

Hobby servos are designed to operate by being sent a pulse that is approximately 1 millisecond to 2 milliseconds long every 20 milliseconds, as shown in Figure 5-2. In reality, most servos can handle pulse widths from about 0.8 milliseconds to about 2.2 milliseconds, and some even further. Use caution when going beyond these extremes. The OOPic servo objects cover that span with 64 settings. In reality, the OOPic breaks each step into increments of 1/36 of a second. This means if you have the servo centered (the Center property equals 22 on oServo, and Center equals 0 on oServoX), then the minimum pulse width equals 0.61 milliseconds and the maximum pulse width is 2.36 milliseconds, with 1.5 milliseconds being in the center. The oServoSE has no need for center position adjustment.. There is no reason *not* use the Center values given previously with the servos you use normally. However, not all hobby servos can move to the positions at the extremes of the OOPic servo objects. I suggest you avoid pulse widths lower than 0.85 milliseconds or larger than 2.2 milliseconds until you are certain your servos can reach those positions without stripping gears.[1]

---

1. You'll know if you've stripped a gear. Your servo will make a sickening "crack!" at the end of its movement.

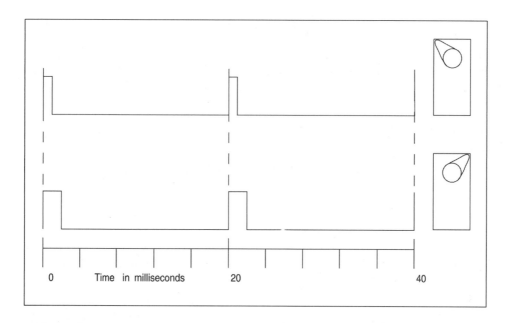

Figure 5-2 Hobby servo operation

The other variation from the normal servo control pulse stream is that the OOPic normally uses a 30-millisecond pulse refresh rate instead of a 20-millisecond one. You can use the Refresh property on oServo, oServoX, and oServoSP1 to decrease that time to 15 milliseconds, but it may not be possible to always provide the pulse every 15 milliseconds. In fact, it may not be possible to provide a pulse every 30 milliseconds, at which time the OOPic will stretch to its refresh rate to accommodate the extra time needed. The discussion that follows is the reason why the servo pulse refresh rate may change in your program.

THE OOPIC SERVO UPDATE QUEUE

Along with an object list loop described earlier, the OOPic has a servo update queue where it keeps track of the current and queued servos waiting for a refresh pulse. Each servo's pulse is handled by itself, in line, before the next servo's pulse is output. This means if you have four servos defined in your OOPic program labeled S1, S2, S3, and S4, the servo queue will first output a pulse for S1. When that pulse is done, it outputs a pulse for S2, and so on. Only one servo pulse is *ever* on at any given time. The time used to output those pulses is tracked. When all servos have been serviced, if any time is left before the next S1 servo pulse needs to be output (every 30 milliseconds, remember?), the queue is inactive until the proper time has elapsed for S1 to output its pulse again.

Can you see what this means to the pulse repetition rate for the servos? If each servo is set to output a 2.36-millisecond pulse (the maximum), 12 servos (30/2.36) may be running at the same time and still maintain the 30-millisecond repetition rate. If you need the faster refresh time of 15 milliseconds, only 6 servos can be maintained before the time is up to start a new loop of servo updates.

In reality, it is highly unlikely that every servo will be at its maximum value at the same time, which is what it would take to use up the entire queue time to service all servos. It is conceivable that you could run 12 servos at the higher refresh rate, if they were used for a walking robot, for instance, because it is not possible for every servo to be at the maximum pulse position at the same time. A robot arm, however, could conceivably have all servos at or near their maximum pulse length.

Your application will determine the servo refresh rate you can use. Most of the time the refresh will have plenty of time, and if it occasionally doesn't, you'll probably never notice it.

A servo being used with a low refresh rate is not as strong as a servo with a higher refresh rate. This is because when the servo stops getting a pulse, it stops controlling the motor, so if there is a long wait between pulses, the servo motor gets weaker and doesn't hold its position as well.

Under normal circumstances, the 30-millisecond pulse refresh rate is just fine for every hobby servo that I've used. It might not be acceptable for high-powered servos meant to hold heavy weights, such as a robot arm. In that case, you should use the Refresh property to increase the pulse refresh rate.

THE OOPIC SERVO OBJECTS

Four OOPic servo objects are available, and they are quite different from each other in their details and operation. Not all oServo objects are available in all OOPic firmware versions. I've left out the *String* property and the *Address* property, because these are in all OOPic objects.

oServo (All OOPic Firmware Versions)

 This is the original and most basic servo object. When the Center property is set to 22, the following is true:

Value =	Servo position and pulse width
0	Pulse = 0.61 ms; servo is full to one side.
32	Pulse = 1.5 ms; servo is at mechanical center.
63	Pulse = 2.36 ms; servo is full to opposite side.

Beware of the extremes; your servo may not be able to move that far. It is possible you may have a servo whose center is not at 1.5 milliseconds, and you could move the Center property to something other than 22, but you shouldn't need to for the majority of your applications.

The *InvertOut* property is very useful. When you are using hobby servos that have been hacked for continuous rotation, you will usually place one servo on one side of your robot and one on the other side. They will be turning in opposite directions, one facing left and one facing right. When you set this property to 1, that servo will reverse the direction of the pulse. In other words, if you sent a 0.8-millisecond pulse normally, and you sent this bit, that servo will really get a 1.2-millisecond pulse, effectively reversing the direction of rotation.

The standard refresh rate of about every 32 milliseconds works fine for most things, but some servos get weak as kittens at that rate. The oServo object has the Refresh property, which, when set to 1, doubles the refresh rate to about every 17 milliseconds. In many cases, I've seen the servos even run faster when this is done. The oServo memory size is 4 bytes, and its properties are detailed in Table 5-5.

TABLE 5-5   OSERVO PROPERTIES

Center (byte)	Used to move the servo's mechanical center position. 22 is the 1.5-millisecond servo pulse center when the Value property equals 32.
InvertOut (bit, flag)	1 = reverse the servo's direction. 0 = normal servo direction.
IOLine (byte)	Chooses the I/O line (1 through 31) for the object.
Operate (bit, flag)	Must be a 1 to enable the servo.
Refresh (bit, flag)	When set to 1, it doubles the refresh rate (to about every 17 milliseconds).
Value (byte)	The position to move the servo to, from 0 to 63 (default property).

oServoX (B.X.X Firmware and Later)

This servo object is the first servo object to use the URCP, which is just a fancy way of saying it uses both negative and positive Value settings that represent degrees of speed or directions to the left or right. URCP is a convenient way to denote direction for robotics projects. If every object says that positive numbers "go forward" and negative numbers "go backward," then things like compasses, line follower devices, and motor controllers would all take and use the same number ranges. This means no translation would need to be made, and that would be convenient, wouldn't it? Chapter 10, "OOPic Robotics and URCP (Project: A Robot that Toes the Line)," goes into more detail about URCP headings, speed, and distance. The OOPic supports URCP in many of its objects at Firmware versions B.X.X and later.

This object also has a Center property, but because it uses URCP, that property is from 128 to +127 instead of an unsigned integer like oServo. Obviously, this means that 0 is the center. I recommend you leave it there unless you need to tweak the position of a servo that is already built into a device.

5

ANALOG-TO-DIGITAL AND HOBBY SERVOS

The *InvertOut* property works as in oServo to reverse the direction of servo travel. This allows a mirrored pulse to be sent to the chosen servo.

The *Mode*, *Offset*, and *OutOfRange* properties are interesting modifiers, but I've never used them. Play with them to see what they do. Leave *Mode* and *Offset* set to 0 for a standard -128 to +127 servo setting range. The oServoX memory size is 6 bytes and its properties are listed in Table 5-6.

TABLE 5-6   OSERVOX PROPERTIES

Center (byte)	Can be used to adjust the mechanical center of your servo. If *Center* = 0, then 1.5ms is the center pulse width to the servo. The range is -32 to +31.
InvertOut (bit, flag)	1 reverses the servo's direction.
	0 is the normal servo direction.
IOLine (byte)	Chooses the I/O line (1 through 31) for the object.
Mode (nibble)	Specifies how the servo responds to values outside of its mechanical limit.
	0 = The servo splits the inaccessible area into 2 parts and will position itself to the limit that is closest to the Value property.
	1 = The servo will position itself on the low value side.
	2 = The servo will position itself on the high value side.
	3 = The servo shuts off.
Offset (byte)	This specifies the Value property that the object will consider to be mechanical zero of the servo. If Offset equals 64, then *64 will be center and the Value range will be 0 through 127.*
Operate (bit, flag)	Must be a 1 to enable the servo.
OutOfRange (bit, flag)	Will be a 1 to denote when the servo Value property is set outside the mechanical range of the servo.
Refresh (bit, flag)	When set to 1, it doubles the refresh rate (to about every 17 milliseconds).
Value (byte)	The position to move the servo to, which is from -128 to +127 (default property).

oServoSP1 (B.2.X Firmware and Later)

 This object is in most respects the same as oServoX, with a couple of exceptions. The first exception is that fewer properties to configure exist in this object. The second difference is subtle, but very useful. In oServoX if you set Value to 0, a 1.5-millisecond pulse is output to hold the servo at the center position.

In oServoSP1, when the Value property is set to 0, no pulses are sent out at all. This means that when using hacked *radio-controlled* (RC) servos, a Value property of 0 means the wheels stop. If you have used hacked servos, you know you can never quite get them to stop turning; they will turn ever so slowly, even when you send the center pulse value. This object solves that problem very simply by not issueing servo pulses when set to 0 (stop).

oServoSP1 includes the Center property, which is essential for tweaking the servo so that positive numbers make the servo spin one way and negative numbers make it turn the other way. The oServoSP1 memory size is 6 bytes and its properties are listed in Table 5-7.

TABLE 5-7   OSERVOSP1 PROPERTIES

Center (byte)	Can be used to adjust the mechanical center of your servo. If Center equals 0, 1.5 milliseconds is the center pulse width to the servo. The range is -32 to +31.
InvertOut (bit, flag)	1 reverses the servo's direction. 0 is the normal servo direction.
IOLine (byte)	Chooses the I/O line (1 through 31) for the object.
Operate (bit, flag)	Must be a 1 to enable the servo.
Value (byte)	The position to move the servo to, which is from -128 to +127 (the default property).

oServoSE (B.X.X Firmware and Later)

This object is radically different from the other servo objects in that it does not really control any servos at all. Instead it speaks to the Scott Edwards SSC board over a serial link.

To connect the SSC II board to the OOPic controller, locate the top 2 pins of the 10-pin header on the SSC II. The selected I/O line needs to be connected to the S pin and the ground line needs to be connected to the G pin on this connector. The oSerialSE communicates with the SSC II at 2400 baud, so be sure *not* to jumper the baud rate selector on the SSC II to any other baud rate value.

This is one of the very few objects that has a method as well as attribute properties. In this case, the *position* method is very important, as it is how you select the servo to move and the position in which to move it. Writing to the *Value* attribute will allow you to send commands directly to the SSC II (see Figure 5-3). The oServoSE memory size is 5 bytes and its properties are outlined in Table 5-8.

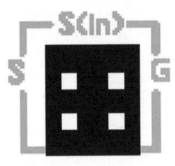

Figure 5-3 SSC II serial connection

TABLE 5-8 OSERVOSE PROPERTIES	
IOLINE (BYTE)	**CHOOSES THE I/O LINE (1 THROUGH 31) FOR THE OBJECT.**
Value (byte)	A write transmits the value, unmodified directly to the SSC II board.
METHODS	
Position (S, P)	Servo (byte) chooses the servo to move. Position (byte) chooses the position in which to move the chosen servo.

CONNECTING AND POWERING RC HOBBY SERVOS

You cannot power hobby servos from the OOPic 5V line. In fact, it is a good idea to power them from a separate power supply altogether from your OOPic. The OOPic R is the exception to this; it has a separate I/O bus on the board that you can use to power your servos, or you can choose to power them from another battery. For details on the OOPic R power supply options, please see Chapter 1, "OOPic Family Values," where this is discussed in great detail. It makes good sense to power hobby servos directly from a battery and not use a voltage regulator, servos take a lot of current and most voltage regulators won't power more that three or four servos.

Normally, RC servos are powered from 4.8V to 6V. I, however, am well known for living on the edge when it comes to servos. In most robots, servos don't run for more than about 5 minutes at a time, so I power my servos with anything from 7.2V to as much as 12V. Test your servos before you overpower them; any mistakes with servo timing are bad on unmodified servos at 6V. The same mistake is disaster on that servo at 12V.

Figure 5-4 shows the proper way to connect a servo to your OOPic controller. A separate battery for the servos is displayed, but you can run multiple power supplies from the same battery, or even run the servos directly off the battery without any regulator at all. Figure 5-5 shows how to power your OOPic board and your hobby servos using a single battery, which cuts down on weight for those 500-gram MiniSumo robots. Note in Figure 5-4 that the grounds *must* be connected together among the batteries, OOPic, and servos.

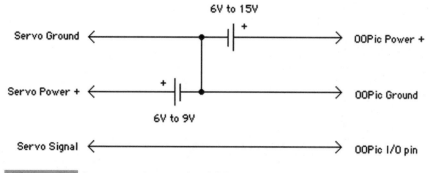

Figure 5-4 A separate battery for OOPic and servos

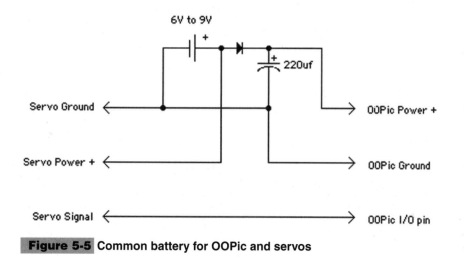

Figure 5-5 Common battery for OOPic and servos

Figure 5-5 shows a simple circuit that will protect the OOPic from voltage sags[2] caused by turning on servos. Basically, it works because the diode only enables current to flow in one direction. When the motor draws a great deal of current, a battery's voltage may suddenly drop momentarily while trying to supply the demanded current. When this happens, the voltage on the left side of the diode is lower than the voltage on the right side of the diode. Because a diode won't conduct in reverse, no current will flow in that direction. The large capacitor on the right side of the diode then supplies the energy needed to keep a proper voltage at the input to the OOPic voltage regulator. The end result is that the OOPic circuitry never notices that the battery dipped its voltage.

---

2. A voltage sag is caused when the voltage drops suddenly when the battery is trying to supply enough current for a sudden spike in demand.

Your life is considerably simplified if you use the OOPic R board. The I/O connector is already set up to handle Futaba™ and HiTech™ hobby servos; you can just plug them in. If you have chosen to use the 5V I/O voltage regulator as the power supply on the I/O pins for the bank of I/O that your servos reside on, you also have protection against voltage sags by virtue of that I/O voltage regulator. No further wiring is needed. See Chapter 1 for details on selecting your I/O power supply sources on the OOPic R board.

Data Bus Processing Objects

In this chapter's project, you'll link the A2D value to a servo to move it to different positions. To do this, you need another means for linking properties than you used for flags. Depending upon which OOPic firmware you have, different choices are available. The OOPic A.X.X firmware has the oMath object to link Value properties. The OOPic B.X.X firmware adds more, and this chapter introduces the oBus processing object, which uses the least memory space to link two Value properties. Note that the processing objects will *only* link the default Value between two objects.

The strings in the quotation marks in Table 5-9 denote OOPic constants that can be used instead of a number to aid in understanding object actions when you go back to look at your code. Remember that you can always use cvTrue for 1 and cvFalse for 0 for bit and flag property setting and checking.

OMATH PROCESSING OBJECT (A.X.X FIRMWARE AND LATER)

The oMath processing object has several functions, and this chapter looks at both its capability to link Value properties and how it handles math functions. The oMath object also has a clocked version that comes in handy. Its memory size is 4 bytes, and Table 5-9 illustrates its properties.

TABLE 5-9 OMATH PROPERTIES	
Input1 (pointer)	Links to an output Value property to use from another object.
Input2 (pointer)	Links to an output Value property to use from another object.
InvertC (bit, flag)	Used by clocked oMath object to invert the logic of the clock property.
Mode (nibble)	Selects one of eight math operations:
	0: "cvAdd" adds Input1 to Input2.
	1: "cvSubtract" subtracts Input2 from Input1.
	2: "cvLShift" bit shifts Input1 left a number of times given by Input2.
	(continued)

	3: "cvRShift" bit shifts Input1 right a number of times given by Input2.
	4: "cvAND" logically ANDs Input1 and Input2.
	5: "cvOR" logically ORs Input1 and Input2.
	6: "cvXOr" logically exclusive-ORs Input1 and Input2.
	7: "cvLatch" Input1 is copied to Output.
Negative (bit, flag)	Will be 1 if the result of Modes 0 through 3 is a negative number.
NonZero (bit, flag)	Will be 1 if the result of any operation is nonzero.
Operate (bit, flag)	Must be 1 for this object to function.
Output (pointer)	Links to an input Value of another object.
Value (byte)	Optional value when used with oMathI or oMathO variants in B.X.X or later firmware.

OBUS PROCESSING OBJECT (B.X.X FIRMWARE AND LATER)

This object's functionality is to link a Value property between two objects. The oBus object is introduced in the B.X.X firmware version. The clocked or special variants of this object are not discussed in this chapter, just the unadorned oBus. The oBus memory size is 3 bytes, and its properties are shown in Table 5-10.

Project: Read a Bend Sensor and Move a Servo

This is a fun little project. You are going to create a circuit to read a bend sensor and use that reading to position a hobby servo. Tape the bend sensor to your finger and the servo follows your movements. You can get a resistive bend sensor from a few places. A new one can be obtained from www.jameco.com; order the part number 150551 Flex Sensor. I got mine another way: by finding a surplus Nintendo Power Glove, which has four bend sensors in the glove. I'm going to do two variations on this project, one that uses OOPic I objects and one that experiments using OOPic II objects.

TABLE 5-10   OBUS PROPERTIES

ClockIn (flag pointer)	Used by clocked variants of this object for the signal that clocks data through.
Input (pointer)	Links to an output Value property to use from another object.
InvertC (bit, flag)	Used by clocked variants of this object to determine the clock direction:
	0: Clock changes from 0 to 1.
	1: Clock changes from 1 to 0.
InvertIn (bit, flag)	Input is bitwise negated if set to 1; it is unmodified otherwise.
Mode (bit)	Determines when the copy is performed:
	0: Data is copied continuously as long as Operate equals 1.
	1: Data is only copied when Operate changes from 0 to 1.
Operate (bit, flag)	Must be 1 for this object to function.
Output (pointer)	Links to the input Value property of another object.
Value (byte)	Signed 8-bit number that is optionally used by oBusI and oBusO variants.

HARDWARE LAYOUT

The bend sensor is interfaced to the OOPic by using a voltage divider (you do remember the voltage divider, right?). I've chosen my lower resistor value to get the maximum voltage swing I could with the bend sensor I have. Your sensor may be different, so choose your resistor similarly. Just about any value will work, but a well-chosen one will allow more detailed movements. See Figure 5-6 for the schematic and hookup diagram to use for this project. Try to find a resistor whose value is near the center of the sensor's resistance range.

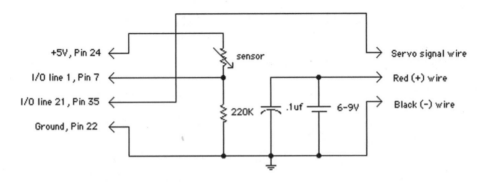

Figure 5-6 Project circuit and servo hookup

Remember this circuit; it will be modified later on in the project after you see how it acts in the real world.

DEFINE THE VIRTUAL CIRCUIT (VC)

Which objects do you need for this project? Obviously, you need the oServo oA2D hardware objects; these are your sensor input and mechanical output. You need to link the values from the former to the settings for the latter. In the OOPic I firmware revisions A.X.X, one object can link bytes of data: oMath.

I created a short *virtual circuit* (VC) to test my layout and to find the range of values I could expect from the A2D channel. The values at the extremes of my bend sensor were 88 to 190, but we're limited to the range of 0 to 63 for normal servo movements. If you just link the oA2D to the oServo, you'd be getting some really large pulse widths and you'd damage your servo. You need to restrict the maximum value to 63 or less.

The obvious choice is to divide the oA2D reading by 4, so you'd have a range of 22 to 47, which is a respectable range. The OOPic oMath object does not have a *cvDivide* mode, but it does have a *cvRShift* mode. In binary math, shifting a data value to the right effectively divides the number by 2 for each shift. So if you shift the input to the right 2 places, you are dividing by 4. That's it; you're done. This is a simple VC to design. Figure 5-7 is the proposed object map.

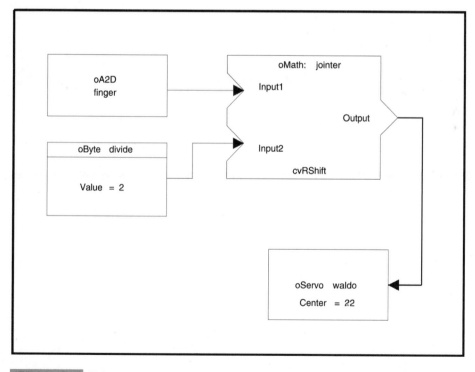

5

ANALOG-TO-DIGITAL AND HOBBY SERVOS

Figure 5-7 Object map

WRITE THE CODE

With the OOPic microcontroller, once you have your OOPic object map planned out, writing the code is almost trivial. This is especially so when you can do all your processing in a VC, as you are in this project. Listing 5-1 details the VC coding for your sensor/servo project.

The program looks almost as simple as the object map. I've set the servo Center property to the standard 22 because it wasn't important where the center might be in this application.

Experiments

Figure 5-8 is a picture of what this project looks like on a solderless breadboard. These boards are useful because they allow designs to be altered painlessly and quickly.

```
'FPushA.osc
'Chapter 5 Example "Push my Finger"
'Demonstrates oA2D, oMath and oServo objects and
'linking concepts.
'For OOPic I A.2.X firmware
'
'Copyright Dennis Clark 2003
'Permission is granted to use this code anyway you like
'as long as you say you got it from me.

Dim finger as New oA2D       'Bend Sensor
Dim joiner as New oMath      'Linking object
Dim waldo as New oServo      'Check out RAH's writing
Dim divide as New oByte      'sensor result divisor

Sub Main()

    OOPic.Node = 1                 'For debugging

    finger.IOLine = 1              'IOLine 1 for sensor
    finger.Operate = cvTrue

    waldo.IOLine = 21             'IOLine 21 for servo
    waldo.Center = 22             'The "standard" center value
    waldo.Operate = cvTrue

    divide = 2                     'Shift two bits

    joiner.Input1.Link(finger)    'Get sensor data
    joiner.Input2.Link(divide)    'will divide by 4
    joiner.Output.Link(waldo)     'Put it to servo
    joiner.Mode = cvRShift        'Effectively a divide
    joiner.Operate = cvTrue

End Sub
```

Listing 5-1

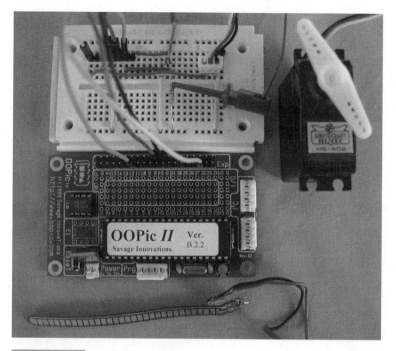

Figure 5-8 Breadboarded project

EXPERIMENT #1: JITTER REMOVAL

Did you notice that the servo tended to jitter and shake when you moved the bend sensor or even just left it in one place? That is called *noise*, and most analog systems have some kind of noise in them. How can you remove the noise and reduce the jitter in the servo? This problem can be approached in a couple of different ways. One method would be to hack off one or two LSBs from your readings so that minor variations in the A2D value won't be passed on to the servo. This approach works reasonably well, but you will end up losing resolution, and each change of servo position will be a larger percentage of the overall swing of the servo arm, making it seem jerky, not smooth.

The other method is electrical. You don't have to solve *all* your problems in software. The most common means used to remove noise in electronic systems is a capacitor. A capacitor basically stores energy and fills in the gaps, so to speak, where noise spikes raise their ugly heads. This application is especially well suited for a capacitor filter because it creates an RC delay circuit when used in conjunction with the resistors in our voltage divider. The RC time constant for a simple resistive circuit is the time it takes to charge approximately 63 percent of the difference between the current voltage and the maximum voltage, or the time it takes to discharge that amount. This number gives you a rough idea of how fast the voltage can change in your circuit. Figure 5-9 shows where to put the filter capacitor.

A capacitor resists changes in voltage. Electrical noise is typically fast-changing voltages caused by devices switching on and off. A proper noise filter capacitor must be chosen such that it will (in our case) allow our desired voltage change to occur when the bend

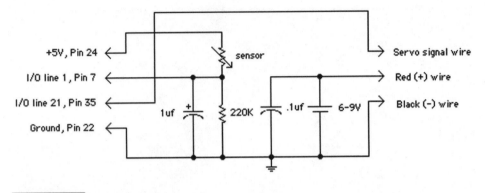

Figure 5-9 Filtered project circuit

sensor is moved, without allowing a much faster and undesired change in voltage. Noise (and jitter) will be essentially eliminated.

I've used a 1uf capacitor in my circuit and found it works quite well to remove nearly all servo jitter in this project. I chose this value somewhat empirically[3] and by using a little math. I found that a .1uf capacitor helped, but not much, so I deduced that the noise wasn't very high in frequency. A quick trip to the calculator showed that the RC time constant for a 220K resistor with a 10uf capacitor was 630 milliseconds, over half a second. That would sure stop noise, and everything else too! I then split the difference, by basically seeing that the only thing I had in the middle of these two extremes was a 1uf cap, for a 63-millisecond RC time constant. Since human reflex is usually no faster than 200 milliseconds in trained athletes, I felt that this was a good compromise, and it turned out to be so. If you are interested, substitute higher values of capacitors into the circuit and see what happens; it can be highly entertaining.

EXPERIMENT #2: GETTING A LARGER SERVO MOVEMENT

The servo arm moves quite nicely when you bend the sensor. If you did the first experiment, it also moves quite smoothly, but it doesn't seem to move very far. In this experiment, you'll try changing the code quite a bit to get a longer servo throw.

You've seen that your A2D range is about 88 to 190. Dividing by 4 brought your numbers into the 0 to 63 range with a simple VC. If you complicate the VC by doing more math, you can get a lot more motion. Here's how: Let's subtract 80 from the A2D reading and then divide that number by 2. This gives you a range of 4 to 55, a *much* larger change of values, and still within the common servo range. Figure 5-10 shows what I came up with to handle all this math in a VC.

An intermediate oByte object has been added to hold the measurement between the two math operations done in the OOPic I. It is called *tweenie*. Two other oByte objects hold the

---

3. That means I swapped a few values and checked to see what happened.

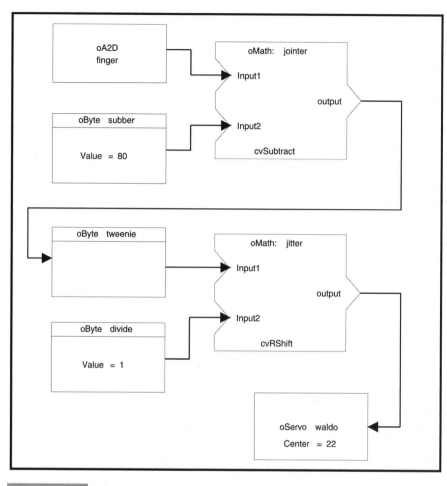

 **Figure 5-10** Getting more motion

data that determines the subtraction and division values. Again, all this is done in the VC with no other code required. Listing 5-2 is the program for this VC.

The program is more complex, but it is still easy to follow. Note that I like to use descriptive names for my objects. It is downright disingenuous to use names like A and X for object names. You've got 255 characters per name as a limit, so use 'em! This program will move the servo a *lot* more than Listing 5-1. The price you pay is complexity and using up more object memory space.

The 5.01 compiler has a bug that causes incorrect data to be read from oMath when oA2D is linked to it. The solution is to use and intermediate oByte object to hold the data. Link the output of oA2D to this oByte and then link the input of oMath to the oByte to get around the bug.

```
'FPushA_2.osc
'Chapter 5 Example number 2 of "Push my Finger"
'Demonstrates oA2D, oMath and oServo objects and
'linking concepts.
'For OOPic I A.2.X firmware
'
'This generates an expanded range of servo values over the
'FPushA.osc example.
'
'Copyright Dennis Clark 2003
'Permission is granted to use this code anyway you like
'as long as you say you got it from me.

Dim finger as New oA2D          'Bend Sensor
Dim joiner as New oMath         'Linking object
Dim waldo as New oServo         'Check out RAH's writing
Dim subber as New oByte         'sensor subtract argument

Dim tweenie as New oByte        'intermediate value holder
Dim divide as New oByte         'Noise masking argument
Dim jitter as New oMath         'Used to mask off lower bits

Sub Main()

    OOPic.Node = 1              'For debugging

    finger.IOLine = 1          'IOLine 1 for sensor
    finger.Operate = cvTrue

    waldo.IOLine = 21          'IOLine 21 for servo
    waldo.Center = 22          'The "standard" center value
    waldo.Operate = cvTruc

    subber = 80                'Subtract 80

    joiner.Input1.Link(finger) 'Get sensor data
    joiner.Input2.Link(subber) 'scale value down
    joiner.Output.Link(tweenie)'Put it to temp. storage
    joiner.Mode = cvSubtract   'Subtract 2 from 1
    joiner.Operate = cvTrue

    divide = 1                 'divisor

    jitter.Input1.Link(tweenie)'Get result of last math
    jitter.Input2.Link(divide) 'get divisor
    jitter.Output.Link(waldo)  'finally, to the servo
    jitter.Mode = cvRShift     'divide by two
    jitter.Operate = cvTrue

End Sub
```

Listing 5-2

EXPERIMENT #3: B.X.X FIRMWARE AND URCP

I said I was going to stick to OOPic I code in this chapter so everyone could do the examples. Well, maybe I won't. Some really nice objects in the B.X.X (OOPic II and later) versions make a person tap their chin and say, "What if?"

The OOPic is moving toward being a robotics controller platform. One of the interesting adoptions is the URCP. More about URCP is covered in Chapter 8, but here you're going to use it to simplify the connection between the A2D and servo objects. In fact, for this version of our servo moving code, you'll use only an A2D object, a servo object, and a new object called oBus. The oBus object only connects data on the input link to a receiver for that data on its output link. I've chosen oA2DX as the input because it can be set up to deliver URCP values that have 0 as its center position. I've also chosen oServoX because it too takes URCP values as valid ranges. Figure 5-11 shows the object map for this experiment.

Because I'm averse to risk when it comes to using my hobby servos, I wrote an abbreviated version of this program so I could look at the oA2DX output values with the OOPic debugger, and modify the output that will be sent to my servo. When running the code, I discovered that my output values with all the default settings on the oA2D object were in the -28 to $+61$ range. Not a very even distribution. To even up the numbers on either side of the 0, I set the Center property to -16. This meant that my range was now -44 ($-28 - 16$) to 45 ($61 - 16$). This had a pleasing symmetry as well as a *very* wide range of motion. This is more of what is good about URCP numbers: the OOPic's capability to set the ranges without using math objects.

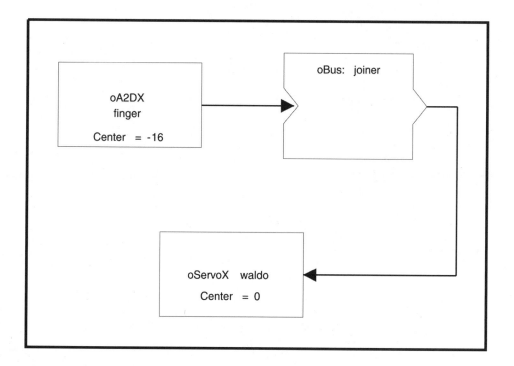

Figure 5-11 OOPic II object map

Notice that this is a range of about 90; the OOPic servo objects only guarantee safe settings for a range of 64. This means that this VC can and will generate ranges of values that might be outside the mechanical limitations of your servos. Be warned and be careful when using it. Listing 5-3 shows the code for this experiment.

This is the simplest VC of them all, and it gives the greatest range of motion, too. On my servos, it is over 180 degrees of movement. Tape your bend sensor to your index finger and flex it. Doesn't that give you ideas?

```
'FPushB.osc
'Chapter 5 Example 3 for "Push my Finger"
'Demonstrates oA2D, oMath and oServo objects and
'linking concepts.
'For OOPic II B.2.X firmware
'
'Care must be taken with this example because it can and
'will generate servo pulses in excess of 2.36ms, which is
'a bit more than 180 degrees on a normal servo.  An oMath
'addition can cut down this range if desired.
'
'Copyright Dennis Clark 2003
'Permission is granted to use this code anyway you like
'as long as you say you got it from me.

Dim finger as New oA2DX          'Bend Sensor
Dim joiner as New oBus           'Linking object
Dim waldo as New oServoX         'Check out RAH's writing

Sub Main()

    OOPic.Node = 1               'For debugging

    finger.IOLine = 1            'IOLine 1 for sensor
    finger.Center = -16          'observed offset needed
    finger.Operate = cvTrue

    waldo.IOLine = 21            'IOLine 21 for servo
    waldo.Center = 0             'The "standard" center value
    waldo.Operate = cvTrue

    joiner.Input.Link(finger)    'Get sensor data
    joiner.Output.Link(waldo)    'Put it to servo
    joiner.Operate = cvTrue

End Sub
```

Listing 5-3

Now What? Where to Go from Here

I hope these experiments have stimulated your imagination, along with your sense of caution. Try modifying properties on the oServo object to limit the range of motion or use a math object to set limits on the value given to the servo. Fiddle with the filter cap to see what kinds of delayed actions you can get from the system. There's lots to play with here.

Did you notice that I called the servo "waldo"? A waldo is a remote manipulator device that enables an operator to remotely control appendages connected electronically to the operator's station. Robert Heinlein proposed such devices in a book written 50 years ago called *Waldo and Magic Incorporated*. The engineering world simply adopted the name when we started creating them for real. You could create your own waldos by using several bend angle sensors controlling several servos, maybe even an entire arm. Wouldn't that be fun?

OOPIC TIMERS, CLOCKS, LCDS, AND SONAR (PROJECT: SONAR PING)

The OOPic microcontroller gives the user access to one 16-bit timer and several clock sources. This chapter details how to use those resources. SONAR has need of a fast timer so that accurate timing can be done with range-finding applications, and this chapter's project gives you a good grounding in how to use the oTimer object for timing events. It is pretty boring to do something and not be able to see the results, so this chapter also shows you the details of using the common *liquid crystal display* (LCD) as a display device. Finally, in many applications, a system of time keeping is needed that uses our human scale of hours, minutes, and seconds. The OOPic oRTC object enables us to keep track of time in those units, and that too will be investigated in this chapter.

This chapter's project is also a fun exercise. You will learn how to use a Devantech® SRF04 SONAR ranging unit and an LCD to measure distance using either an OOPic I or II variant. A great deal of basic information is presented in this chapter that you will use over and over again in your projects.

Timers in the OOPic

The OOPic has one 16-bit timer that is available to the programmer for use. Two objects enable generic access to that timer and some objects use the timer implicitly.

The two objects that enable access to the 16-bit timer are oTimer and oTimerX. The only difference between them is that oTimerX enables linkable access to the *most significant bit* (MSB) of the timer value as a flag; oTimer does not. Access to the sixteenth bit of the oTimer object is important if you want to extend the 16-bit timer into a wider 24- or 32-bit timer by using the oTimer's MSB as the clock source into oWord or oByte objects. Because oTimerX takes two more bytes of object memory than oTimer does, you should only use oTimerX if you need access to the MSB for clocking purposes. To create an oTimerX object instead of an oTimer object, simply add the X to the end of oTimer when you use "Dim" to create an create an object.

YOU ONLY GET ONE 16-BIT TIMER

Even though you may create any number of oTimer objects (up to your memory limits), only one 16-bit timer exists in the OOPic microcontroller. This means that every instance of oTimer has the same value as every other instance of oTimer. It also means that the last *PreScale* set is the prescale value used in all oTimer objects. When you clear an oTimer object, you clear *all* oTimer objects, and so on. For this reason, care is required when using oTimer objects that one *virtual circuit* (VC) does not adversely affect another by its use of the oTimer object.

PRESCALING AND GETTING THE TIMER PERIOD OR FREQUENCY YOU WANT

The oTimer and oTimerX hardware objects have the capability to be scaled to your needs. Without any prescaling, the oTimer increments at a 5 MHz rate, or a clock period of 200 nanoseconds. This rate is the PICMicro controller's 20 MHz crystal value divided by 4, which is the internal instruction rate of the PICMicro used by the OOPic microcontroller. You have the choice of four options to slow down that counting rate, and they are displayed in Table 6-1.

The clock period is the inverse of the frequency. The period of a waveform is the time it takes to make one entire cycle from 0 to 1 and back to 0 again. Frequency is the number of times that the waveform cycles between 0 and 1 per second.

Why prescale the clock? Sometimes you don't need the full clock resolution for your timing, and 16 bits may not be enough to capture the time you need. For instance, this chapter's project uses a SONAR to measure the distance to objects. The maximum range for the Devantech SRF04 SONAR is about 3 meters (about 10 feet). Sound will take approximately 148 microseconds (a microsecond is one-millionth of a second) to travel 1 inch out and back, or about 1.8 milliseconds per foot (a millisecond is one-thousandth of a second). At 10 feet, that would be about 18 milliseconds to measure that SONAR ping. If you used the default 5 MHz clock of the OOPic, with the 200-nanosecond time period (a nanosecond is one-billionth of a second) you could only time out to 13.1 milliseconds. You'll need to divide the timer down to be able to measure out to 18 ms using the standard oTimer object.

TABLE 6-1   PRESCALE RESULTS

PRESCALE VALUE	CLOCK FREQUENCY	CLOCK PERIOD
0	5 MHz	200 ns
1	2.5 MHz	400 ns
2	1.25 MHz	800 ns
3	625 KHz	1.6 us (microseconds)

OOPIC EXTERNAL CLOCK SOURCE

So far our discussion has been limited to the internal 5 MHz PICMicro clock with the oTimer object. You may, if you want, attach an external clock source for the oTimer object to get certain specific clock values. You can use two types of clock sources: another crystal connected to *input/output* (I/O) lines 16 and 17 of the OOPic (pins 15 and 16 on the PICMicro chip respectively) or a digital clock source connected to only I/O line 16. This is the way to get a specific frequency clock for your OOPic oTimer operations. Use the *ExtClock* and *ExtXtal* properties to select the optional oTimer clock source that you will be using. Figures 6-1 and 6-2 show how you would connect your external clock source to the OOPic PICMicro chip.

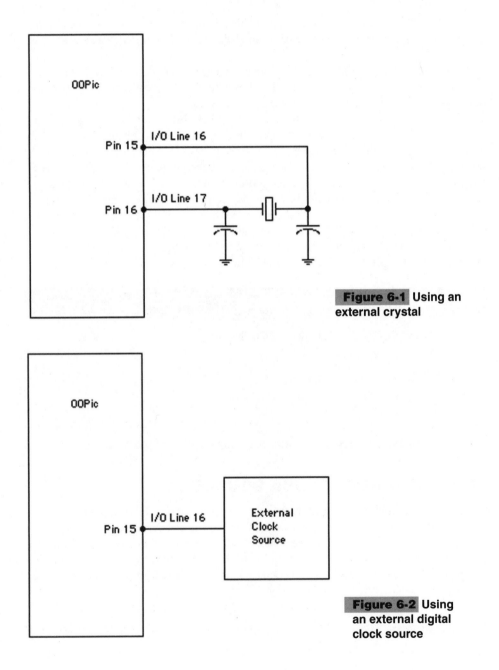

Figure 6-1 Using an external crystal

Figure 6-2 Using an external digital clock source

When using an external crystal, you will usually need bias capacitors to get the crystal circuit to oscillate properly. The values you use depend on the type of crystal, its frequency, and how fast you want it to start up. Larger capacitor values make the oscillator more stable, but slower to start. Consult the crystal manufacturer's specifications to choose the correct capacitor values.

THE OTIMER AND OTIMERX HARDWARE OBJECTS

The oTimer object is the basic timer object used in OOPic VCs. The oTimerX object enables the MSB flag to be linked to other objects to extend the maximum number that can be recorded.

oTimer(X) (All OOPic Versions)

 This object provides a 16-bit, resettable timer for OOPic objects. oTimerX is available only in B.X.X or later. The oTimer memory size is 1 byte, and for oTimerX, it is 3 bytes. The oTimer properties are outlined in Table 6-2.

TABLE 6-2   OTIMER(X) PROPERTIES

ExtClock (bit)	0 = Use the internal 5 MHz clock.
	1 = Use I/O line 16 or lines 16 and 17 for the clock.
ExtXtal (bit)	0 = Use external clock input on I/O line 16 only.
	1 = Use external crystal circuitry on I/O lines 16 and 17.
MSB (bit, flag)	MSB of oTimer's Value property. This bit cycles once for every count to 65,535. *This is only in OOPic B.X.X or later.*
Operate (bit, flag)	Must be a 1 to enable the oTimer object.
PreScale (nibble)	Specifies the divisor to the clock signal:
	0 = Divide by 1.
	1 = Divide by 2.
	2 = Divide by 4.
	3 = Divide by 8.
String	The Value property represented as a string.
Value (word)	The current count of the timer (the default property).

Clock Resources in the OOPic

If you need an internal source for clock pulses, a few are available that are strictly for use by VCs. You can use *pulse width modulation* (PWM) outputs, but they are really not for internal use with a VC and can be inconsistent due to timing or race[1] conditions. The clock resource objects for the OOPic are the OOPic.Hz1, OOPic.HZ60, and the oClock object. All these clock resources are low-frequency clocks. You can use the oTimerX.MSB property for a high-frequency clock resource.

A clock is used to sequence a VC through actions or to record time or counts when something is detected. Clocks are commonly used in stepper motors, *real-time clocks* (RTCs), heartbeat indicators, or timed events. Each one of these needs a different clocking rate. The following section discusses the clock resources used in OOPic VCs and event programming.

OOPIC HZ1 AND HZ60 CLOCKS

The OOPic object has many functions beyond reading and interpreting the *electrically erasable programmable read-only memory* (EEPROM) instructions that set up and run OOPic programs. One of those functions is to provide clocks for a variety of reasons. These two OOPic Hz1 and Hz60 clocks are commonly used for RTCs in the OOPic, and an example using Hz60 is provided later in this chapter.

The Hz1 and Hz60 clocks are flags and can be linked to any flag property of an OOPic VC. The Hz1 clock cycles once each second, and the Hz60 clock cycles 60 times a second. Both clocks are useful for simple time keeping. Using either clock is simple, as shown in this code snippet:

```
RTC.ClockIn1.Link(OOPic.Hz1)
```

This code links the oRTC clock input with the Hz1 clock so that the RTC increments every second.

THE OCLOCK PROCESSING OBJECT

The oClock processing object is somewhat unique because it is the only processing object that originates a clock rather than consuming one. This object starts with a base clock frequency of 283 Hz. Yes, it's a weird value, but a number of useful frequencies can be generated with it. The *Rate* property is used to divide the base clock frequency like so:

$$Frequency = \frac{283}{(256 - Rate)}$$

Another interesting feature of oClock is that by setting the *Mode* property to 1, the output is a single pulse at the specified frequency that lasts for one object list loop time period, instead of appearing as a 50 percent duty cycle square wave. This is useful for initiating an oEvent trigger or to trigger a VC on a regular basis.

1. A race condition is a condition where two processes are trying to use a single resource before the other does.

oClock (B.X.X Firmware or Later)

oClock provides a slow, configurable clock for VCs or other timing events. Its memory size is 5 bytes and its properties are outlined in Table 6-3.

TABLE 6-3   OCLOCK PROPERTIES

InvertOut (bit, flag)	0 = The Output and Result properties are the same.
	1 = The Result property is inverted before it appears on the Output property.
Mode (bit)	0 = Clock output is a 50-percent duty cycle square wave.
	1 = The clock only pulses once every time period for the duration of one object list loop.
Operate (bit, flag)	Must be a 1 to clock its output.
Output (flag pointer)	Points to a property that will be updated with the value of the Result property.
Rate (byte)	Defines the divisor for the 283 Hz clock.
Result (bit, flag)	The result of the clocking operation (the default property).

USING OTIMERX AS A CLOCK RESOURCE

Usually, oTimer objects are used to time a pulse, SONAR's return echo, and so on. The oTimerX object opens up new worlds for VCs in need of a clock source. Because the MSB property is a flag in oTimerX, it can be linked to a VC in such a way as to create a single pulse (through an oOneShot object) or to output a fixed frequency squarewave on an I/O line. For instance, you could use oTimerX in conjunction with oPWM to output a 38 KHz modulated signal in an *infrared* (IR) beacon circuit. Or you could simply use oTimerX as a source of high-frequency pulses for a fast-clocking VC. The oTimer MSB flag is linkable in the same fashion as any other OOPic flag.

The OOPic Real-Time Clock (RTC)

When humans time events, we tend to use units of hours, minutes, and seconds, instead of microseconds or milliseconds for tasks such as timing car races or how long you are jogging. Your program may have a functionality you want to occur every night at 7 P.M., such as an alarm clock. For these tasks, an RTC is needed. The OOPic provides the oRTC object for these purposes. It's not battery backed up, and it'll reset to zeros when the board is reset or powered up, but oRTC is still very useful for timing purposes.

6

OOPIC TIMERS, CLOCKS, LCDS & SONAR

The OOPic RTC requires the use of an oRTC processing object and an oBuffer data object to perform its operation. Multiple RTC objects can be made active, and they can each have their own times set independently of each other. The oRTC object provides fields for hours, minutes, seconds, and 1/60 of a second.

Three options are available for clocking the oRTC. It can be clocked with the OOPic Hz1 property, the OOPic Hz60 property, or an oTimerX clock. Because it is designed to be used with the internal Hz1 and Hz60 clocks, it makes sense to just use one of these. An experiment is included later in this chapter, showing how to configure the oRTC object.

OOPIC RTC OBJECTS

Two objects need to be used to create an OOPic RTC: the oRTC processing object and the oBuffer data object.

oRTC (All OOPic Firmware Versions)

 oRTC is the OOPic RTC object. To function, it requires linking to the oBuffer object, described in the following section. The oRTC memory size is 4 bytes and its properties are outlined in Table 6-4.

TABLE 6-4 ORTC PROPERTIES	
ClockIn1 (flag pointer)	Links to a flag in another VC supplying clock ticks.
Direction (bit, flag)	0 = Increment clock. 1 = Decrement clock.
Operate (bit, flag)	Must be 1 for this object to function.
Output (buffer pointer)	Links to an oBuffer object to hold the RTC values.
PM (bit, flag)	0 = AM; 1 = PM.
Tick (bit)	0 = Each tick increments the 1's position of the Seconds field. 1 = Each tick increments the 1's position of the 1/60 Second field.

oBuffer (All OOPic Firmware Versions)

 The oBuffer object is an array of bytes that enables the array to be referenced as either a string of characters or individually by address. The oBuffer object has a special property that returns the array formatted for oRTC use as a string. The number of bytes in the buffer is specified when the object is declared, as shown here:

```
Dim buffer as New oBuffer(8)    'Creates an 8 byte buffer.
```

TABLE 6-5   OBUFFER PROPERTIES

Direction (bit, flag)	0 = Increment the Location property after each read or write to the Value property.
	1 = Decrement Location after each read or write to Value.
Location (byte)	The location of the desired byte in the array, zero-based, from 0 to 31. This address is incremented or decremented each time Value is read from or written to.
NonZero (bit, flag)	Reads 0 when every byte in the array is 0; it reads 1 otherwise.
RTCString (string)	The contents of the buffer as a string in RTC format.
String	The contents of the entire buffer as a string.
Value (byte)	The value of the byte pointed to by the Location property (default property).
Width (byte)	The number of elements specified when the object is defined.

The oBuffer memory size is 4 plus the number of bytes in the array (or 12 bytes for an oBuffer(8) object). The oBuffer properties are shown in Table 6-5.

The oBuffer methods are as follows:

- **Clear**   Clears the Value property to 0.
- **Dec**   Decrements the Value property by 1.
- **Inc**   Increments the Value property by 1.
- **Invert**   Inverts the bits in the Value property.
- **Lshift**   Shifts the bits in the Value property left.
- **Rshift**   Shifts the bits in the Value property right.
- **Set**   Sets the Value property to 255.

CREATING AN RTC OBJECT

To create an OOPic RTC, you need an oRTC and an oBuffer object that is linked and configured as shown in the code snippet that follows:

```
'RTC objects
Dim RTC as New oRTC
Dim buffer as New oBuffer(8)

Sub Main()

    'Set up RTC VC
    RTC.ClockIn1.Link(OOPic.Hz60)
    RTC.Output.Link(buffer)
    RTC.Tick = 1                    'every 60th of a second update
    RTC.Operate = cvTrue

Sub End
```

The RTC object stores the time in the 8 bytes of the buffer object. The 1's and 10's positions of the hours, minutes, seconds, and 1/60 of a second positions are included. The buffer addresses start at 0 and end at 7 for the RTC data. The hours data is in locations 0 and 1, the minutes in locations 2 and 3, the seconds in locations 4 and 5, and the 1/60 in locations 6 and 7. They are stored as regular numbers, not as strings. To get the RTC data out of the buffer as a string, use the *RTCString* property of the buffer.

OOPic LCD Programming and Control

Four different objects enable an OOPic program to control an LCD device that uses the standard 44780 chip set . These objects are oDataStrobe, oLCD, oLCDSE, and oLCDWZ. The oDataStrobe object is in all OOPic firmware versions; oLCD, oLCDSE, and oLCDWZ are in firmware versions B.X.X and later. ODataStrobe and oLCD deal directly with LCD devices, whereas oLCDSE and oLCDWZ are objects that communicate with a Scott Edwards® serial LCD controller and Wirz® SLI-OEM LCD controller boards respectively.

A number of different LCD configurations are available both new and on the surplus market. They all use the same 44780 chipset, and they all use the same commands and connections. They may have slightly different ways to connect the signals, but the interface is standard and unwavering. If you get your LCD program working with one LCD device, it will work with all LCD devices. Figure 6-3 shows 2×16-, 4×20-, and 2×24-character LCD

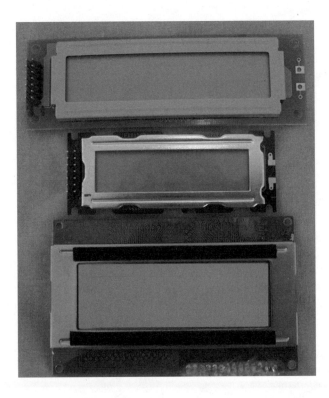

Figure 6-3 LCD device varieties

displays. They are all wired slightly differently, but in every case the pin numbers are the same for the standard signals and power connections.

LCD HARDWARE CONFIGURATION

The traditional LCD connection is via a 14-pin dual in-line connector that works nicely with a 14-pin ribbon cable connector, as shown in Figure 6-4.

Even though the cable pinout consists of eight data lines (DB0 through DB7), traditionally everyone uses the LCD in 4-bit mode to save on data lines and to control signal lines. Figure 6-5 shows the traditional hardware connections for every LCD, no matter what its number of lines and columns may be. Your LCD may have a pinout that is in a single line, rather than a 14 pin connector standard. To understand its pinout, you can usually just look for the pin that is grounded, which will be pin 1; the other pins align with the pin numbers shown in Figure 6-4, even though they aren't in the standard layout. If you have an LCD that doesn't have 14 pins in its connector, you've got a nonstandard one, and you'll have to find and read the specifications sheet for that LCD to discover the pinout. The OOPic enables an LCD to be connected to any I/O group where you have a nibble of consecutive bits available for the LCD data lines.

Note that the *Read/NOT Write* (R/W) line is tied to ground. Typically, the LCD is used as an output-only device. You're not interested in reading the location of the cursor or what is in the special character buffers. You keep track of that information in your program.

LCDs require a bias voltage to be applied to pin 3 of the connector to configure the visibility and contrast of the display. I've found that most LCDs work reasonably well if you connect a 1K resistor from there to ground. On some units, this makes them too dark; on others, it's just right. If you don't like how your LCD looks when you follow that suggestion, use a 100K trim pot instead, as shown in Figure 6-5, and adjust the display to your

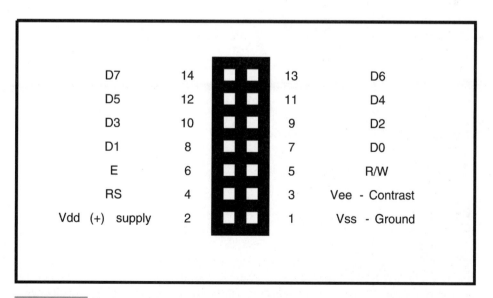

Figure 6-4 LCD cable pinout

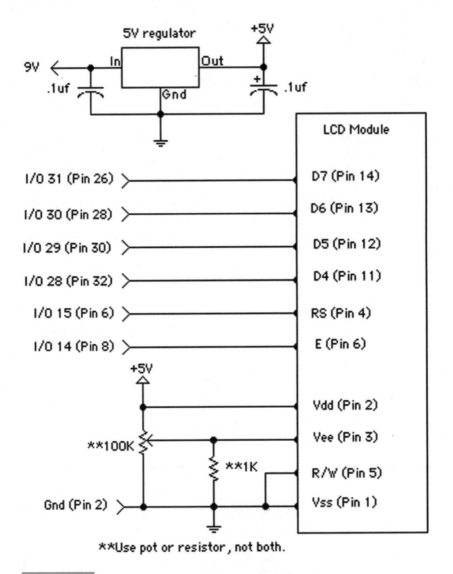

Figure 6-5 LCD hardware connections needed

liking. Using the 1K resistor shortcut is a good way to find out if your LCD is even working. I've seen values as low as 330 ohms required to get some LCD displays to be visible. Experiment with yours to see what works best.

The OOPic requires that four consecutive bits in a configurable nibble be used as the data lines to the OOPic. You can use any other available I/O lines for the *RS* and *E* control signal lines. Figure 6-5 shows just one suggested configuration, which is usable on all OOPic varieties; those I/O lines appear on them all.

LCD CONTROL AND CUSTOMIZATION

An LCD can do more than just display ASCII characters. It can scroll back and forth in the display, each character position can be individually addressed in the display buffer, and you can download and use up to eight custom characters of your own design. This section explains each of the commands and gives the detailed LCD initialization procedure customized for the OOPic microcontroller.

LCD Initialization à la OOPic

LCD devices are simple to use, but they require a strict initialization procedure. A code example provided here explains the process and what is going on. Because only the upper four bits of the data bus are connected, it can be difficult to visualize what is being sent. For this reason, the &H prefix on the data values is used to signify that they are hexadecimal numbers. Hex numbers are useful in this case because you can see the upper and lower nibbles of a byte individually in a way that a decimal number won't show. The following code is designed to work with the oDataStrobe object. The oLCD object handles this initialization sequence automatically when you use the .Init method of the object.

```
Sub LCDInit()
'This initializes the LCD.

    OOPic.delay = 3         'Power has been on more than 15 ms
    LCD.Mode = cv8Bit       'only one strobe for a byte
    RS = 0                  'instruction mode
    LCD = &H33
    OOPic.delay = 2         'Something that the
    LCD = &H33
    OOPic.delay = 2         'data sheet says to
    LCD = &H33
    OOPic.delay = 2         'do three times.
    LCD = &H22              'Places LCD in 4 bit mode
    OOPic.delay = 2
    LCD.Mode = cv4Bit       'DataStrobe in 4 bit mode
    LCD = &H28              '2 lines, font 0
    OOPic.delay = 2
    LCD = &H08              'screen off
    OOPic.delay = 2
    LCD = &H01              'screen on
    OOPic.delay = 2
    LCD = &H06              'set cursor mode to advance right
    OOPic.delay = 2
    LCD = 12                'Display = on, cursor on, no blinking
    OOPic.delay = 2
    RS = 1                  'data mode

End Sub
```

Note that I delay 20 milliseconds between each command sent out. In most cases, the LCD should process the commands in less than 10 milliseconds, but you should be cautious because you are not looking at the busy bit (you can't; you've set the hardware to be write-only), so you don't know when the previous command has completed. LCDs require that the first part of the initialization be in 8-bit mode. You've only connected the upper 4 bits of the data bus to the LCD, so the &H33 could just as easily have been &H30; it doesn't matter.

LCD Instruction Set

The LCD instruction set consists of the commands you can send to your LCD. Remember that the RS line needs to be set to 0 to send instructions to the LCD hardware. When the RS line is set to 1, you are sending data to display memory or the *character graphics* (CG) memory. An X in any position means it doesn't matter what you enter there.

Clear Display   This command clears the display and returns the cursor to the home position (Address 0) and sets I/D to 1 in order to increment the cursor. Its line settings are as follows:

RS	R/W	D7	D6	D5	D4	D3	D2	D1	D0
0	0	0	0	0	0	0	0	0	1

Home Cursor   This returns the cursor to the home position, returns a shifted display to the correct position, and sets the *display data* (DD) RAM address to 0. Its line settings are as follows:

RS	R/W	D7	D6	D5	D4	D3	D2	D1	D0
0	0	0	0	0	0	0	0	1	X

Entry Mode Set   This command sets the cursor move direction and specifies whether to shift the display or not. These operations are performed during the data write/read of the CG or DD RAM. Its line settings are displayed here:

RS	R/W	D7	D6	D5	D4	D3	D2	D1	D0
0	0	0	0	0	0	0	1	I/D	S

When I/D equals 0, the cursor position is decremented (moves right to left). I/D set to 1 means the cursor position is incremented (moves left to right). S set to 0 means normal operation, the display remains still, and the cursor moves (use this one). S at 1 means the display moves with the cursor.

Display On/Off Control   This command sets the on/off display as well as the cursor and blinking capabilities (0 equals off; 1 equals on). D controls whether the display is on or off, C controls whether the cursor is on or off, and B controls whether blinking is on or off. The line settings are as follows:

RS	R/W	D7	D6	D5	D4	D3	D2	D1	D0
0	0	0	0	0	0	1	D	C	B

Cursor or Display Shift This moves the cursor and shifts the display without changing the DD RAM contents. S/C set to 0 means move the cursor. S/C set to 1 means shift display. R/L at 0 means shift to the left, and R/L at 1 means shift to the right. The line settings are as follows:

RS	R/W	D7	D6	D5	D4	D3	D2	D1	D0
0	0	0	0	0	1	S/C	R/L	X	X

Function Set This sets the interface *data length* (DL), the *number* of display lines (N), and the character *font* (F). When DL is 0, it means 4 bits are being used (the standard), whereas when DL is 1, it means a full 8 bits are being utilized. When N equals 0, it means 1 line; when it is set to 1, it means two lines. N at 1 is used for any display with more than 1 line. F set to 0 means that 5×7 dot characters are used (which is how 99 percent of all LCDs are set up). F at 1 means 5×10 dot characters are used. The line setting are as follows:

RS	R/W	D7	D6	D5	D4	D3	D2	D1	D0
0	0	0	0	1	DL	N	F	X	X

Set CG RAM Address This command sets the *Custom Graphics* (CG) RAM address. Setting RS to 1 sends data to the CG RAM instead of the DD RAM. Eight CG characters are available, and they reside in the ASCII codes 0 through 7. The data is sent in 8-bit bytes from the top row to the bottom row and is left justified, meaning that only the bottom 5 bits matter (it's a 5×7 dot matrix.). The line settings are as follows:

RS	R/W	D7	D6	D5	D4	D3	D2	D1	D0
0	0	0	1	MSB		CG RAM Address		LSB	

Set DD RAM Address This sets the DD RAM address. Setting RS to 1 sends data to the display RAM, and the cursor advances in the direction where the I/D bit was set to. The line settings are as follows:

RS	R/W	D7	D6	D5	D4	D3	D2	D1	D0
0	0	1	MSB		DD RAM Address		LSB		

6

OOPIC TIMERS, CLOCKS, LCDS & SONAR

Read Busy Flag and Address   This reads the *busy flag* (BF). If BF equals 1, the LCD is busy and displays the location of the cursor. With the R/W line tied to ground, you can't use this command. The line settings are as follows:

RS	R/W	D7	D6	D5	D4	D3	D2	D1	D0
0	1	0	0	0	0	0	0	0	1

Write Data to CG or DD RAM   This command's line settings are as follows:

RS	R/W	D7	D6	D5	D4	D3	D2	D1	D0
1	0	MSB		ASCII code or CG bit pattern data					LSB

Read Data from CG or DD RAM   This command's line settings are as follows:

RS	R/W	D7	D6	D5	D4	D3	D2	D1	D0
1	1	MSB		ASCII code or CG bit pattern data					LSB

LCDs have 80 display locations in them. Depending on the display, you might not be able to see all 80 locations without scrolling through the display, which is why LCDs have a scrolling mode. The lines in the LCD are strangely addressed though, so Table 6-6 shows the first character of each line in a display. You can use these addresses to set the LCD to the beginning of the line using the *Set DD RAM address* instruction. All addresses will be 128 + something because bit 7 denotes the DD RAM address set instruction. As you can see, if you are writing to line 1 on a 4×16 display and go past the sixteenth character, you'll end up writing on line 3. The same is true for lines 2 and 4.

TABLE 6-6   DISPLAY LINE ADDRESSES

LCD FORMAT	LINE	START ADDRESS
Any	Line 1	128 + 00 = 128
Any	Line 2	128 + 64 = 192
4×16	Line 3	128 + 16 = 144
4×16	Line 4	128 + 64 + 16 = 208
4×20	Line 3	128 + 20 = 148
4×20	Line 4	128 + 64 + 20 = 212

OOPIC OBJECTS TO CONTROL LCD DISPLAYS

Four OOPic objects are used to control LCD displays: oDataStrobe, oLCD, oLCDSE, and oLCDWZ. The oDataStrobe object is a generic object that can be used to write to a parallel port printer, an LCD, or even an 8-bit data latch like the 74HCT574 octal latch. The oLCD object encapsulates all essential LCD operations within the object and is specific to 4-bit configured LCDs. The oLCDSE and oLCDWZ objects are customized to simplify communicating with the Scott Edwards serial LCD controller boards and the Wirz SLI-OEM LCD controller boards respectively.

oDataStrobe (All OOPic Firmware Versions)

 This object is a generic data strobe object that puts 8 bits of data out on the specified I/O lines and then sends a clock pulse out to latch it into the hardware device getting the data. This object can be used with parallel port printers, digital data latches, or other hardware that needs a strobe pulse to latch data. In this case, you're using it with an LCD display. Common configuration requires that you choose four or eight contiguous I/O lines for the data and another available I/O line for the data strobe. The name in parentheses indicates an OOPic text alias for the specified setting. You can also use cvTrue for 1 and cvFalse for 0. The memory size is 5 bytes and the oDataStrobe properties are displayed in Table 6-7.

TABLE 6-7   ODATASTROBE PROPERTIES

Mode (bit)	0 (cv8Bit) equals a single 8-bit transfer and a single data strobe.
	1 (cv4Bit) equals two 4-bit transfers with the high-order nibble being transferred first and using two data strobes.
Nibble (bit, flag)	Indicates which half of a two-nibble transfer is currently taking place. 0 is a lower nibble and 1 is an upper nibble.
OnChange (bit)	Specifies how the data strobe is generated.
	0 means a strobe is used every time the Value property is written.
	1 means a strobe is used every time the Value property changes.
Operate (bit, flag)	Must be 1 for this object to function.
Output (number pointer)	Links to the I/O lines (either oDio8 or oDio4) being used.
Result (bit, flag)	1 means the data strobe is active; 0 means otherwise.
String (string)	The Value property is represented as a string, and multiple bytes are allowed.
Strobe (flag pointer)	Links to the strobe I/O line being used.
Value (byte)	The data that is to be transferred (the default property).

For LCDs, the Strobe flag would be connected to the I/O line being used as the E line to the LCD.

oLCD (B.X.X Firmware and Later)

 This object encapsulates all the functionality required by the hardware and software to control an LCD display. Because I/O lines are precious commodities, the oLCD object assumes that you are using a 4-bit connection interface; that is all you need. Speed is *not* an issue with an LCD. The time it takes to transfer two 4-bit chunks of data to the LCD is dwarfed by the time it takes the LCD to execute the instruction. The oLCD is a complete object for LCD control; it includes two methods essential for LCD use: the clear display and cursor location settings. This object initializes the LCD when it is first used. The oLCD memory size is 6 bytes and the oLCD properties are displayed in Table 6-8.

TABLE 6-8   OLCD PROPERTIES

IOGroup (nibble)	1 means that I/O lines 8 through 11 or 12 through 15 are used for data I/O.
	2 means I/O lines 16 through 19 or 20 through 23 are used for data I/O.
	3 means I/O lines 24 through 27 or 28 through 31 are used for data I/O.
	The Nibble property chooses the set of four I/O lines.
IOLineE (byte)	The I/O line to use for the E line to the LCD.
IOLineRS (byte)	The I/O line to use for the RS line to the LCD.
Nibble (bit)	0 means the lower 4 bits of the *IOGroup* selection are used.
	1 means the upper 4 bits of the *IOGroup* selection are used.
Operate (bit, flag)	Must be 1 for this object to function.
RS (bit)	Enables you to manually set the RS line to 0 or 1 so you can issue either instructions or data to the LCD.
String (string)	The Value property as a string. Whole strings may be sent here.
Value (byte)	Individual byte value to be sent to the LCD (the default property).

The oLCD methods are as follows:

- **Clear**   Clears the display and homes the cursor to line 0, location 0.
- **Locate(R,C)**   Sets the LCD cursor to the row and column specified. It is zero based, which means that (0,0) places the cursor in the first column of the first line.

The *Locate* method only explicitly deals with two-line displays. This is logical because lines 3 and 4 begin at addresses that depend on the line length of your LCD. To locate the cursor to lines 3 and 4, consult Table 6-6 to see what the address is. For example, if you have a 4×20 display, you can locate the cursor to lines 3 and 4 like this:

```
LCD.Locate(0,20)          For line three
LCD.Locate(1,20)          For line four
```

oLCDSE (B.X.X Firmware and Later)

This object deals with a third-party[2] LCD controller manufactured by Scott Edwards Electronics (www.seetron.com). These LCD controllers encapsulate all the hard I/O interfacing and can provide a lot more than simple text displays in many cases. Best of all, you only need to use one I/O line to control it. The oLCDSE memory size is 5 bytes and its properties are displayed in Table 6-9.

TABLE 6-9 OLCDSE PROPERTIES	
Baud (nibble)	2 equals the 2400-baud serial rate (default).
	3 equals the 9600-baud serial rate
	1, 4, 5, 6, and 7 are reserved for future expansions.
IOLine (byte)	Any I/O line you choose to connect your LCD controller to.
Operate (bit, flag)	Must be 1 for this object to function.
String (string)	The Value property as a string. Whole strings may be sent here.
Value (byte)	The individual byte value to be sent to the LCD (the default property).

The oLCDSE methods are as follows:

- **Clear**   Clears the display and homes the cursor to line 0, location 0.
- **Locate(R,C)**   Sets the LCD cursor to the row and column specified. It is zero based, which means that (0,0) places the cursor in the first column of the first line.

oLCDWZ (B.X.X Firmware and Later)

This object deals with a third-party LCD controller manufactured by Wirz Electronics (www.wirz.com). The Wirz SLI-OEM LCD encapsulates all the hardware I/O and software controls, and hides them from you. Best of all, you only need to use one I/O line to control it. The oLCDWZ memory size is 5 bytes and Table 6-10 outlines the oLCDWZ properties.

2. A fancy name for another company thinking up something useful to use with your company's product.

TABLE 6-10   OLCDWZ PROPERTIES

Baud (nibble)	4 represents a 19,200-baud serial rate (default). 1, 2, 3, 5, 6, and 7 are reserved for future expansions.
IOLine (byte)	Any I/O line you choose to connect your LCD controller to.
Operate (bit, flag)	Must be 1 for this object to function.
String (string)	The Value property as a string. Whole strings may be sent here.
Value (byte)	The individual byte value sent to the LCD (the default property).

The oLCDWZ methods are as follows:

- **Clear**   Clears the display and homes the cursor to line 0, location 0.
- **Locate(R,C)**   Sets the LCD cursor to the row and column specified. It is zero based, which means that (0,0) places the cursor in the first column of the first line.

Experiments

Many fun experiments can be devised just from the objects in this chapter. The first one I've chosen is to combine the RTC object and the LCD object to create an OOPic-based clock. The next several experiments deal with customizing your LCD interface on your next OOPic project, which should fascinate the crowds at your next robotics get-together.

DISPLAY THE RTC ON YOUR LCD

Have I used used enough *three-letter acronyms* (TLAs) yet in this chapter? (TLA, in itself, is an acronym—how droll.) I hope not, because the electronics and programming worlds are filled with them. This little aside will show you how to set up an RTC and how to configure and initialize an LCD. You'll be using the connections given in Figure 6-2 to interface the LCD to the OOPic I controller. This code compiles and runs on all OOPic versions. The program demonstrates how to write and use *Clear* and *Locate* functions on an LCD without relying on the later (and easier to use) oLCD objects. Listing 6-1 displays the entire OOPic program.

When you run this program, you'll notice that the 1/60-second digits are somewhat random when they appear. This is because the code doesn't send a new line out to the LCD every 1/60 of a second, more like every few hundred milliseconds. You can link to the OOPic.Hz1 and specify that increment location in the oRTC object if this annoys you. I've left that as an exercise for the reader.

```
'LCDRTC.osc
'Chapter 6 RTC and LCD Experiment.
'Demonstrates the OOPic Real-Time Clock
'
'For OOPic I A.2.X firmware
'
'Copyright Dennis Clark 2003
'Permission is granted to use this code any way you like
'as long as you say you got it from me.

'LCD objects
Dim LCD as New oDataStrobe
Dim nibs as New oDio4
Dim RS as New oDio1
Dim E as New oDio1

'RTC objects
Dim RTC1 as New oRTC
Dim buffer1 as New oBuffer(8)

Sub Main()

    'Set up RTC1 VC
    RTC1.ClockIn1.Link(OOPic.Hz60)
    RTC1.Output.Link(buffer1)
    RTC1.Tick = 1                 'every 60th of a second update
    RTC1.Operate = cvTrue         'Set up LCD VCs and display
    setupLCD                      'Build LCD VCs
    LCDInit                       'Perform the icky LCD initialization
    LCDClear                      'Clear the screen

    LCDLocate(0,0)                'Put the cursor in line 1 position 1
    LCD.String = "OOPic RTC"

    Do                            'Loop forever
        LCDLocate(1,0)
        LCD.String = buffer1.RTCString
        OOPic.delay = 10
    Loop

End Sub

Sub setupLCD()
    'This handles the VC setup.

    nibs.IOGroup = 3              'I/O lines 28-31
    nibs.nibble = 1
    nibs.Direction = cvOutput

    RS.IOLine = 15               'LCD RS line
    RS.Direction = cvOutput
    RS = 1                        'set to data (no instructions!)

    E.IOLine = 14                'LCD E line
    E.Direction = cvOutput
    E = 0                         'enable off
```

Listing 6-1 *(Continued)*

6

OOPIC TIMERS, CLOCKS, LCDS & SONAR

```
    'The oDataStrobe object set up in Nibble mode will
    'output the lower nibble first and then the upper
    'nibble.  Keep this in mind when reading LCD init
    'routines.
    LCD.Output.Link(nibs)          'Data to LCD
    LCD.Strobe.Link(E)             'LCD enable line
    LCD.Mode = cv4Bit              'doing 4-bit LCD
    LCD.Nibble = cvLow             'when doing single transfers
    LCD.OnChange = cvFalse         'transfer when value written
    LCD.Operate = cvTrue

End Sub

Sub LCDInit()
    'This initializes the LCD.

    OOPic.delay = 3                'make SURE more than 15 ms
    LCD.Mode = cv8Bit              'only one strobe here
    RS = 0                         'instruction mode
    LCD = &H33
    OOPic.delay = 2                'Something that the
    LCD = &H33
    OOPic.delay = 2                'data sheet says to
    LCD = &H33
    OOPic.delay = 2                'do three times.
    LCD = &H22
    OOPic.delay = 2
    LCD.Mode = cv4Bit              'now in 4-bit mode
    LCD = &H28
    OOPic.delay = 2
    LCD = &H08                     '2 lines, font 0
    OOPic.delay = 2
    LCD = &H01                     'screen off
    OOPic.delay = 2
    LCD = &H06                     'screen on
    OOPic.delay = 2
    LCD = 12
    OOPic.delay = 2
    RS = 1                         'data mode

End Sub

Sub LCDClear()
    'Clear the LCD screen and home cursor to 0,0

    RS = 0                         'instruction mode
    LCD = 1                        'clear display
    OOPic.Delay = 2
    LCD = 2                        'home cursor
    OOPic.Delay = 4
    RS = 1                         'data mode

End Sub

Sub LCDLocate(row as byte, colm as byte)
'This locates the cursor in a specific location
'This works with 2-line displays or 4x20 line displays.
```

```
'If you have a 4x16 display simply change the "+ 20"
'on rows 2 and 3 to be "+ 16".

    RS = 0                          'instruction mode
    If row < 2 Then                 'first two rows only
        LCD = (128 + (64*row) + colm)
    End If
    If row = 2 Then                 'third row
        LCD = 128 + 20 + colm
    End If
    If row = 3 Then                 'fourth row
        LCD = 128 + 64 + 20 + colm
    End If
    OOPic.delay = 2
    RS = 1
```

Listing 6-1

LCD TRICKS, SCROLLING, AND CUSTOM CHARACTERS

A while back I wrote that LCDs can do more than just display alphanumeric characters. Now I'll show you how to exercise those other cool features.

Rather than talk about how to code scrolling routines and custom character downloads, I'll show you some demonstration code that does all of this. Feel free to hack it apart to add your own fancy ideas or graphics expressions if the spirit moves you. Listing 6-2 shows all that this section discusses.

```
'LCD1.osc
'Chapter 6  LCD customization experiments.
'Demonstrates LCD initialization and some interesting
'customization examples.
'
'For OOPic I A.2.X firmware
'
'Copyright Dennis Clark 2003
'Permission is granted to use this code anyway you like
'as long as you say you got it from me.

'LCD objects
Dim LCD as New oDataStrobe
Dim nibs as New oDio4
Dim RS as New oDio1
Dim E as New oDio1

Sub Main()
```

(Continued)

```
Dim n as byte

nibs.IOGroup = 3          'I/O lines 28-31
nibs.nibble = 1
nibs.Direction = cvOutput

RS.IOLine = 15       'LCD RS line
RS.Direction = cvOutput
RS = 1               'set to data (no instructions!)

E.IOLine = 14            'LCD E line
E.Direction = cvOutput
E = 0                'enable off

'The oDataStrobe object set up in cv4Bit mode will
'output the upper nibble first and then the lower
'nibble.  Keep this in mind when reading LCD init
'routines.
LCD.Output.Link(nibs)    'Data to LCD
LCD.Strobe.Link(E)       'LCD enable line
LCD.Mode = cv4Bit        'doing 4-bit LCD
LCD.Nibble = cvLow       'when doing single transfers
LCD.OnChange = cvFalse   'transfer when value written
LCD.Operate = cvTrue

LCDInit             'Perform the icky initialization
LCDClear            'Clear the screen
LCDLocate(0,0)
LCD.String = "Oopic I LCD"

'On line 2 I'm writing a 26-character-long string;
'this means that I'll have to scroll the display
'10 characters to the left to be able to see all
'of the string that was written.
LCDLocate(1,0)
LCD.String = "abcdefghijklmnopqrstuvwxyz"
OOPic.delay = 200

'Shift the display to the left and back to the right
RS = 0              'instruction mode
for n = 1 to 10
    LCD = &H18       'Shift left one character cell.
    OOPic.delay = 40
next n
OOPic.delay = 200

for n = 1 to 10
    LCD = &H1C       'Shift right one character cell.
    OOPic.delay = 40
Next n
OOPic.delay = 100

'Build and display user-defined characters
'Because ASCII codes 0 and 1 are being displayed, you
'can see them being drawn as the data is written into
'the CG RAM addresses.
LCDClear                'Clear the screen
LCDLocate(0,0)
LCD.String = Chr$(0)     'Display user character 0
```

```
    LCD.String = Chr$(1)     'Display user character 1
    LCD.String = "C"         'Just to look good

    RS = 0                   'instruction mode
    LCD = &H40               'write first custom character address
    RS = 1                   'Now send data
    LCD = &H00               'Sending smiley face
    LCD = &H1B
    LCD = &H1B
    LCD = &H00
    LCD = &H00
    LCD = &H11
    LCD = &H0E
    LCD = &H00

    LCD = &H08               'Sending degree symbol
    LCD = &H14
    LCD = &H14
    LCD = &H08
    LCD = &H00
    LCD = &H00
    LCD = &H00
    LCD = &H00

    RS = 0
    LCD = &H83               'Back to display memory
    RS = 1

End Sub

    Sub LCDInit()
'This initializes the LCD.

    OOPic.delay = 3          'make SURE more than 15 ms
    LCD.Mode = cv8Bit           'only one strobe here
    RS = 0                   'instruction mode
    LCD = &H33
    OOPic.delay = 2          'Something that the
    LCD = &H33
    OOPic.delay = 2          'data sheet says to
    LCD = &H33
    OOPic.delay = 2          'do three times.
    LCD = &H22
    OOPic.delay = 2
    LCD.Mode = cv4Bit        'now in 4-bit mode
    LCD = &H28
    OOPic.delay = 2
    LCD = &H08               '2 lines, font 0
    OOPic.delay = 2
    LCD = &H01               'screen off
    OOPic.delay = 2
    LCD = &H06               'screen on
    OOPic.delay = 2
    LCD = 12
    OOPic.delay = 2
    RS = 1                   'data mode

End Sub
```

(Continued)

6

OOPIC TIMERS, CLOCKS, LCDS & SONAR

```
Sub LCDClear()

    RS = 0                  'instruction mode
    LCD = 1                 'clear display
    OOPic.Delay = 2
    LCD = 2                 'home cursor
    OOPic.Delay = 4
    RS = 1                  'data mode

End Sub

Sub LCDLocate(row as byte ,colm as byte)
'This locates the cursor in a specific location
'This works with 2-line displays and 4x20 displays.
'If you have a 4x16, then change the "+ 20" to be
'"+ 16" below.

    RS = 0                  'instruction mode
    if row < 2 then         'lines 1 and 2
        LCD = (128 + (64*row) + colm)
    End If
    If row = 2 then         'line 3
        LCD = 128 + 20 + colm
    End If
    If row = 3 then         'line 4
        LCD = 128 + 64 + 20 + colm
    End If
    OOPic.delay = 2
    RS = 1

End Sub
```

Listing 6-2

This experiment shows how easy it is to scroll the display back and forth, and how easy it is to download a custom character. But how do you create that character, you say? I'll show you . . .

The LCDs you use have a 5×8 dot matrix that makes up the space the character is defined in. The specifications, however, say that you have a 5×7 dot character, and that eighth line is to allow for an underline of a character. Thus, you really have a 5×8 matrix to work with. To create and download a new character, you must follow these steps:

1. Create a 5×8 matrix on paper, or in your head if you have really good visualization abilities. Lucky you.

2. Fill in the dots for the character you want to display.

3. Break the matrix up into 8 bytes of data, starting at the top of the character and working down. From right to left, fill in only bits 0 through 4, a 1 for a dot and a 0 for a void.

4. Set RS to 0 and put the LCD into CG RAM mode at address 0 for the first custom character.

5. Set RS to 1 and send the 8 bytes of data out to the LCD.

6. Set RS to 0 and put the LCD back into DD RAM mode.

7. Set RS to 1 and now when you send ASCII code 0, you get your custom character.

In Listing 6-2, two custom characters were created. They were placed in ASCII code locations 0 and 1 by first setting the CG RAM mode with hex instruction &H40 and then setting RS to 1 again, as shown in the following code snippet:

```
RS = 0                 'instruction mode
LCD = &H40             'first custom character address
RS = 1                 'Now send data
```

Because the custom characters are right next to each other in CG RAM, when the second custom graphic is sent out it just gets written into the next available RAM space, which is ASCII code 1. Notice that hex notation is used for the pattern data. This is done because hex digits are nicely broken into upper and lower nibbles, which are easy to distinguish from each other. Decimal numbers jumble the bits up. Look carefully at the numbers and the bit positions. Can you see exactly where the 1 and 0 bits are located in both the hex and binary representations? Table 6-11 shows the dot matrix I used for my smiley face and the hex values I sent. Table 6-12 is the dot matrix I used for my degrees symbol and the appropriate hex values.

That's all there is to it. You have eight custom graphics you can build, and they are in the ASCII code locations 0 through 7. The LCD can place them there safely because those ASCII codes are nondisplayable characters that have no affect on an LCD display.

Try some of your own experimentation. Can you see how to create a set of custom graphics that, when quickly viewed , can create an animation on your LCD? Try it.

TABLE 6-11   SMILEY FACE PATTERN

4	3	2	1	0	Hex
					&H00
•	•		•	•	&H1B
•	•		•	•	&H1B
					&H00
					&H00
•				•	&H11
	•	•	•		&H0E
					&H00

TABLE 6-12 DEGREES SYMBOL PATTERN

4	3	2	1	0	Hex
	●				&H08
●		●			&H14
●		●			&H14
	●				&H08
					&H00
					&H00
					&H00
					&H00

Project: SONAR Distance Display

You want robotics projects? I've got 'em. This project creates a SONAR VC and an LCD object on which to display the results. First, you'll create an OOPic I version, because that illustrates better how to create a VC that needs to use a timer in a precise way. Then I'll show you how to do the same thing much easier with an OOPic II VC where the SONAR objects have already been created at a higher level.

I live in the United States, so this program uses inches and feet as units of measurement. It takes sound approximately 148 microseconds to travel 1 inch out and back, or about 1.8 milliseconds to travel 1 foot at sea level. I live at 5,000 feet and either all my errors in calculation even out, or this speed doesn't change much at that altitude. This program uses a timer period of 0.8 microseconds per timer tic, so to calculate the *time of flight* (TOF) distance I use the following formula:

$$feet = tics \times \frac{0.8}{(148 \times 12)} = \frac{tics}{2220}$$

Because each tic isn't 1 microsecond, you need to scale the value read from oTimer to reflect that, which is what the X 0.8 is doing. You want feet, not inches, so multiply 148 by 12. Remembering your algebra, and the "invert and multiply" rule when dealing with fractions, divide (148 × 12) by 0.8 to get 2,220 as the approximate value to divide your otimer value by, which I call tics, and you get feet as the translated distance.

HARDWARE LAYOUT

First, a nice, compact SONAR device must be introduced: the Devantech SRF04 SONAR board. This device is available in a number of locations. Search Google (www.google.com) for it and you'll find one for sale. Figure 6-6 shows a picture of the SRF04 SONAR board.

Figure 6-6 SRF04 SONAR board

The SRF04 has a range of approximately 3 meters (about 10 feet for the metric challenged), but I've gotten mine to work out to at least 12.5 feet (the length of my lab.) It's a tiny board that doesn't use much power to operate, and it has a simple interface. You toggle the *INIT* line, which brings the *ECHO* line high. Start your timer, and when the ECHO line drops back to 0, the SONAR ping has been received. Now calculate the distance based on the time of the sound ping's flight measurement. The SRF04 has a minimum range of 3 centimeters, so you can't really measure anything closer than about an inch away; 3 centimeters to 3 meters is a pretty good range of values.

You'll now connect the SRF04 and an LCD display to your OOPic. Figure 6-7 shows the schematic you'll be using for the hardware on this project. Figure 6-8 shows my solderless breadboard layout that allows the LCD and OOPic to be connected to each other.

I have created a 14-pin male connector for LCD cables using a 2×14 header strip I soldered 22 AWG wires to. I plugged the male connector into a 14-pin DIP wire-wrap socket, which I then plugged into the breadboard. This enables me to move the connector around easily and the connector is elevated far enough off the board to allow easy wire jumper connections between points on the breadboard.

DESIGN THE VCS

This is the first project where you can't simply connect objects together into a single VC. In this case, you'll create a couple of VCs and glue them together with "regular" code. This

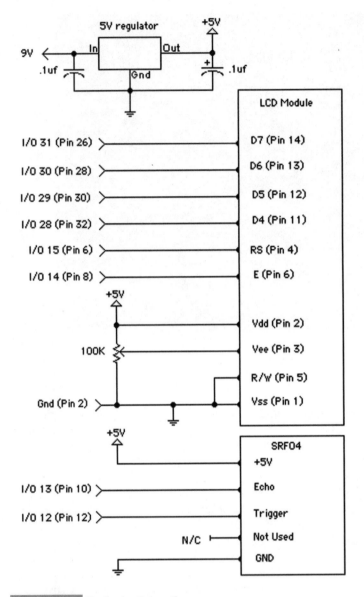

Figure 6-7 Project schematic

will be the norm for the rest of the projects in this book. The VCs needed for this project are shown in Figure 6-9. They include a VC for the LCD display and a VC for the SONAR, both of which are designed for OOPic I objects.

The LCD VC is straightforward. It provides the 4-bit output used by the hardware and the E strobe line used to clock data into the LCD display. The SONAR VC uses the Operate property to control when the timer is running. The SRF04 SONAR seems custom made for OOPic VC integration. When the *Init* line (which I call the *trigger*) is pulsed high, the

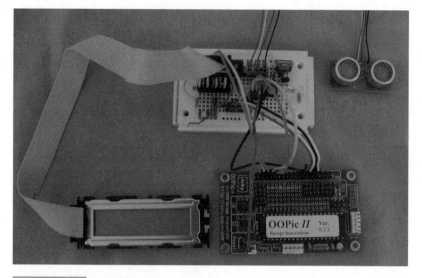

Figure 6-8 Breadboarded circuit

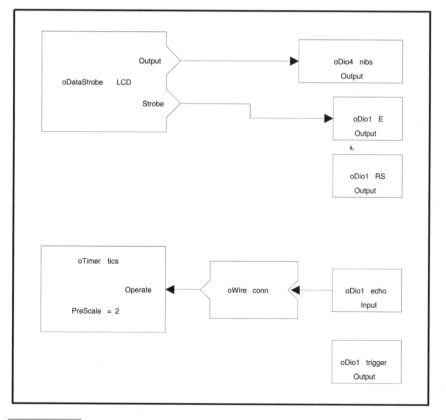

Figure 6-9 OOPic I SONAR and LCD VCs

SRF04 initiates a ping. When it pings, the board pulls the echo line high, which starts the oTimer object running (you've set it to 0 already), and oTimer starts timing the echo TOF. When an echo is detected, the SRF04 pulls its echo line back low, which stops the oTimer object. The oTimer object now has the TOF time of the SONAR pulse's trip out and back stored in its Value property for you to decode.

WRITE THE CODE

Listing 6-3 is the entire code that uses the SRF04 SONAR board and writes the SONAR ranging values to the LCD display. This code is written for OOPic I objects and will compile and run on any OOPic variant.

```
'SONAR1.osc
'Chapter 6 SONAR and LCD Example
'Demonstrates oTimer, LCD initialization and some interesting
'math examples used to calculate SONAR timing.
'
'For OOPic I A.2.X firmware
'
'Copyright Dennis Clark 2003
'Permission is granted to use this code any way you like
'as long as you say you got it from me.

'LCD objects
Dim LCD as New oDataStrobe
Dim nibs as New oDio4
Dim RS as New oDio1
Dim E as New oDio1

'SONAR objects
Dim TICS as New oTimer
Dim trigger as New oDio1
Dim echo as New oDio1
Dim conn as Neg oWire

Sub Main()

    'Set up the I/O pins for the LCD
    nibs.IOGroup = 3            'I/O lines 28-31
    nibs.nibble = 1
    nibs.Direction = cvOutput

    RS.IOLine = 15             'LCD RS line
    RS.Direction = cvOutput
    RS = 1                      'set to data (no instructions!)

    E.IOLine = 14              'LCD E line
    E.Direction = cvOutput
    E = 0                       'enable off

    'The oDataStrobe object set up in Nibble mode will
    'output the lower nibble first and then the upper
    'nibble.  Keep this in mind when reading LCD init
    'routines.
    LCD.Output.Link(nibs)       'Data to LCD
```

```
    LCD.Strobe.Link(E)           'LCD enable line
    LCD.Mode = cv4Bit            'doing 4-bit LCD
    LCD.Nibble = cvLow           'when doing single transfers
    LCD.OnChange = cvFalse       'transfer when value is written
    LCD.Operate = cvTrue

    'Set up SONAR I/O and timer
    tics.Prescale = 2            'divide clock by 4
    tics.Operate = cvFalse       'turned off
    tics = 0                     'clear counter

    trigger.IOLine = 12          'SRF04 trigger line
    trigger.Direction = cvOutput
    trigger = 0

    echo.IOLine = 13             'SRF04 echo line
    echo.Direction = cvInput

    conn.Input.Link(echo)        'connect echo to timer
    conn.Output.Link(tics.Operate)
    conn.Operate = cvTrue

    LCDInit                      'Perform the icky initialization
    LCDClear                     'Clear the screen
    LCDLocate(0,0)
    LCD.String = "Oopic I SONAR"
    OOPic.delay = 200            'Wait a little

    'Loop forever
    do

        'To take a reading, we first clear the timer to 0.
        'We are set to have 800 ns per timer tic; sound
        'takes about 148 us to travel 1 inch (both ways).
        'So to get inches we have: tics  *0.8/148 = inches,
        'or tics/185 = inches.  To get feet, tics/2220.
        'We trigger the SRF04, which takes the echo line
        'high, which enables the timer.  When an echo is
        'received, the echo line drops, disabling the timer
        'and leaving our distance in tics.Value.

        tics = 0                 'clear timer
        trigger = 1              'trigger SONAR
        trigger = 0

        while echo = 1           'wait for ping
        wend

        LCDLocate(1,0)           'Put value on second line
        printData(tics/2220)     'Feet
        LCD.String = "."
        printData(tics mod 2220 *10 /2220) 'fraction of a foot
        LCD.String = " feet  "                'tell the unit's name

        OOPic.delay = 100
    loop

End Sub
```

(Continued)

```
Sub LCDInit()
'This initializes the LCD.

    OOPic.delay = 3              'make SURE more than 15 ms
    LCD.Mode = cv8Bit           'only one strobe here
    RS = 0                      'instruction mode
    LCD = &H33
    OOPic.delay = 2             'Something that the
    LCD = &H33
    OOPic.delay = 2             'data sheet says to
    LCD = &H33
    OOPic.delay = 2             'do three times.
    LCD = &H22
    OOPic.delay = 2
    LCD.Mode = cv4Bit           'now in 4-bit mode
    LCD = &H28
    OOPic.delay = 2
    LCD = &H08                  '2 lines, font 0
    OOPic.delay = 2
    LCD = &H01                  'screen off
    OOPic.delay = 2
    LCD = &H07                  'screen on
    OOPic.delay = 2
    LCD = 12
    OOPic.delay = 2
    LCD = 6
    OOPic.delay = 2
    RS = 1                      'data mode

End Sub

Sub LCDClear()

    RS = 0                      'instruction mode
    LCD = 1                     'clear display
    OOPic.Delay = 2
    LCD = 2                     'home cursor
    OOPic.Delay = 2
    RS = 1                      'data mode

End Sub

Sub LCDLocate(row as byte, colm as byte)
'This locates the cursor in a specific location
'This only works with 2-line displays.

    RS = 0                      'instruction mode
    LCD = (128 + (64*row) + colm)
    OOPic.delay = 2
    RS = 1

End Sub

Sub printData(wt As byte)
    'Print out a word, byte, nibble, or bit value to LCD
    'This simply outputs one character at a time by finding the
    'digit in the 10,000's place (65,535 is highest number) and
    'works down to the 1's place.  No leading zeros are output.

    Dim wTemp as byte
    Dim bTemp as byte
```

```
    Dim mTemp as byte

    bTemp = 0
    mTemp = 0

    wTemp = 100
    While wTemp > 0
        mTemp = wt/wTemp
        if ((mTemp > 0) OR (bTemp = 1) OR (wTemp = 1)) then
            wt = wt - mTemp * wTemp
            bTemp = 1                    'Now print trailing zeros
            LCD.Value = mTemp + 48       'convert to ASCII character
        End If
        if wTemp = 1 then
            wTemp   = 0
        else
            wTemp = wTemp /10
        End If
    wend

End Sub
```

Listing 6-3

OOPIC II VERSION

The OOPic II defines objects for the LCD and the SRF04 SONAR board, which means you
don't have to define any VCs—just dimension the two necessary OOPic II objects that you
need for your program! As you can tell from Listing 6-4, this simplifies the code considerably and saves you some object memory in the process. In this program, as well as in all the
programs I've shown in this chapter, I've used my own subroutine called *printData* to display the string value of the oTimer data. You can use the OOPic built-in function *Str$* if you
like and save a little code space, but Str$ includes leading zeros that I've eliminated with
my subroutine.

```
'SONAR2.osc
'Chapter 6 SONAR and LCD Example
'Demonstrates oLCD and oSonarDV objects
'
'For OOPic II B.2.X firmware
'
'Copyright Dennis Clark 2003
'Permission is granted to use this code anyway you like
'as long as you say you got it from me.

Dim LCD as New oLCD
Dim SONAR as New oSonarDV

Sub Main()

    'Traditionally, an LCD is used in 4-bit mode to save on
    'I/O lines.  These will be the upper 4 bits (4-7) on
```

(Continued)

6

OOPIC TIMERS, CLOCKS, LCDS & SONAR

```
'the LCD module.  Typically, we tie the R/W line low so
'that all operations are writes.  The RS line chooses
'whether we are doing a command or data write, and the
'E line is the command latch to tell the LCD to use
'the information present on the data lines.

LCD.IOLineRS = 15          'RS line to the LCD module
LCD.IOLineE = 14           'E line to the LCD module
LCD.IOGroup = 3            'I/O lines 28-31 (pins 32-26)
LCD.Nibble = 1            'Upper nibble
LCD.Operate = cvTrue

OOPic.delay = 3           'wait for LCD to come up
LCD.Init                  'Perform the icky initialization
LCD.Clear                 'Clear the screen
LCD.Locate(0,0)
LCD.String = "OOPic II SONAR"
OOPic.delay = 100         'Wait a little

SONAR.IOLineP = 12        'Trigger to SRF04 board
SONAR.IOLineE = 13        'Echo back from SRF04
SONAR.Operate = cvFalse

do
    SONAR.Operate = 0
    SONAR.Operate = 1       'Trigger one reading

    'We'll check for the timeout bit, but at over 9M max. range
    'we need a pretty big room to get it!
    '
    'The LCD string that we are calculating will return the
    'distance value as feet and fractions of a foot in decimal
    'notation.  Mod 64 gives us the remainder after the foot
    'division.  I multiply this by 100 so that I can see the
    'fraction as an integer when I then divide that number by
    ' 64.  This is the easiest way to display the fractional
    'part of a division with integer math.  Both of these
    'numbers will have leading zeros, but you'll get the
    'picture.  Obviously, a robot doesn't need this conversion;
    'you'd just use the raw number, which is 64 tics per foot.
    'The range will be close, but not perfect.

    While SONAR.Received = cvFalse  'wait for echo
    Wend

    LCD.Locate(1,0)            'locate the line for next reading.
    if NOT SONAR.TimeOut then
        printData(SONAR/64)
        LCD.String = "."
        printData(SONAR Mod 64 *100 /64)
        LCD.String = " feet "
    else
        LCD.String = "Out of Range"
    End If
    OOPic.Delay = 100
loop

End Sub
```

```
Sub printData(wt As byte)
    'Print out a word, byte, nibble, or bit value to LCD.
    'This simply outputs one character at a time by finding the
    'digit in the 100 place (255 is highest number) and
    'works down to the 1's place.  No leading zeros are output.

    Dim wTemp as byte
    Dim bTemp as byte
    Dim mTemp as byte

    bTemp = 0
    mTemp = 0

    wTemp = 100
    While wTemp > 0
        mTemp = wt/wTemp
        if ((mTemp > 0) OR (bTemp = 1) OR (wTemp = 1)) then
            wt = wt - mTemp * wTemp
            bTemp = 1                      'Now print trailing zeros
            LCD.Value = mTemp + 48         'convert to ASCII character
        End If
        if wTemp = 1 then
            wTemp   = 0
        else
            wTemp = wTemp /10
        End If
    wend

End Sub
```

Listing 6.4

Now What? Where to Go from Here

This chapter's programs are fun, and they illustrate a lot of OOPic functionality and versatility. The LCD functionality is useful for remote control devices and games. Especially useful is the capability to scroll the display and create custom graphics characters that display information or icons that are unique to your creations. The SONAR objects are extremely useful in robotics applications if your creation needs to know how far it is from something. And I, for one, can't imagine any robot that doesn't need to know that information.

You can try certain experiments with OOPic II and later variants using the SONAR values in conjunction with oCompare objects and (to be seen later) motor objects to create VCs that keep a robot following a wall at a prescribed distance. Multiple SONAR units can be used to center a robot in a hallway or to find doorways. The possibilities are nearly endless, and because this book has to stop being written sometime, most are left as exercises for you to try on your own. Really now! Did you think I was going to ruin the fun of exploration for you?

6

OOPIC TIMERS, CLOCKS, LCDS & SONAR

OOPIC EVENTS, KEYPADS, AND SERIAL I/O (PROJECT: A MINI-TERMINAL)

CONTENTS AT A GLANCE

This chapter discusses the OOPic event system and you'll be instructed in how to do some simple event programming. Along the way you'll learn about keypad interfacing and buffered serial *input/output* (I/O). In the chapter's project section, you'll create a small serial terminal. You'll also look at the various serial communications objects the OOPic offers and play with a few more processing objects that add convenient functionality to your projects. First, let's talk about events and event programming.

OOPic Events

The OOPic has supported event programming at the core of its OO model from the first firmware version. Events can be initiated from *virtual circuits* (VCs) by the simple expedient of enabling the event flag *Operate*. Before you can use events, you really need to understand what an event is, as well as how to fire one off.

WHAT IS AN EVENT?

An event is an asynchronous, periodic, or externally initiated thread of execution within the program flow of a system. In laymen's terms, an event is something that happens either because of a timed trigger (a clock tick) or a user action. Event code does not run unless the event that it is tied to the code occurs. In most procedural programs, the code is in a loop waiting for something to happen. In the OOPic, you don't have any loops for waiting periods. All the waiting occurs in the VCs that you design to handle the events you want to process. You can't count on an event happening at a certain time; your program has to be able to handle it whenever it occurs.

Classic event-use cases involve keystroke processing, a serial port processing incoming data, and, for you robotics programmers, your robot running into something. Other common internally initiated events (programs) would be clocks signaling alarms to set off other regular program "housekeeping" duties, such as recording weather data at the top of the hour. Events are typically low-frequency, high-priority tasks that need to be performed immediately upon request.

HOW TO CREATE AN EVENT

The OOPic supports two modes of event instantiation: the nonmaskable event and, in B.X.X firmware, the prioritized event. Triggering either type of event is done the same way: by making the object's *Operate* property true (set to 1). Because the Operate property is a *flag* property, any VC can trigger an event by simply linking the appropriate flag to the event's Operate property. What happens after that depends on the type of event you have chosen to use.

Events are regular code because it is assumed that some kind of logical processing or task handoff needs to occur when an event is triggered. Otherwise, you'd simply have to put all the processing you need to do in the VC affected by the event. An event is programmed in a separate subprogram that is denoted by the _CODE() suffix in the subroutine definition, as shown in the following code:

```
Dim encode as New oEvent

<Other stuff you've defined>

Sub main()

<All the VCs and such defined>

End Sub

Sub encode_CODE()

<event task you've defined>

End Sub
```

OOPIC EVENT OBJECTS

The OOPic defines two types of events, oEvent(C) and oEventX(C), which are covered in the following sections along with the clocked optional definitions.

oEvent(C) (All OOPic Versions, oEventC only in B.X.X and Later)

The oEvent object interrupts any currently running program code to do its job, and, upon exit, the interrupted code resumes execution. You can think of oEvent as a nonmaskable interrupt triggered by an OOPic VC. The oEvent object is triggered when the Operate property transitions from 0 to 1.

The oEventC object is triggered when the property that *ClockIn* points to transitions in the manner set by the *InvertC* property, and only when the Operate property is set to 1. If Operate is set to 0, the event cannot be triggered. oEventC is only used in B.X.X and later.

The oEvent memory size is 3 bytes (oEventC's memory size is 5 bytes) and its properties are shown in Table 7-1.

TABLE 7-1   OEVENT(C) PROPERTIES

ClockIn (flag pointer)	Points to a property that determines when this object will perform its operation (B.X.X only).
InvertC (bit, flag)	0 means a transition from 0 to 1 will trigger this object.
	1 means a transition from 1 to 0 will trigger this object (B.X.X only).
Operate (bit, flag)	When set (cvTrue), the event is triggered. If this bit is still set upon completion of the subprogram, the event is not triggered again until this bit transitions back to 0 and then to 1 again (default property). (For oEventC in B.X.X only, the event can only be clocked when this bit is set to 1; otherwise, the *clockIn trigger has no effect*.)

Because the oEvent interrupts the normal program flow, you need to be concerned about its side effects after the event has been processed. This is especially vexing if you have several events that are being interrupted by other newer events that have just been triggered.

One way to coordinate overlapping events is to use a global flag that is read, and changed if needed, at the beginning of each event handler. This way you can assign priorities to events so that a low-priority event, upon seeing that a high-priority event is currently running (which this event just interrupted), can exit immediately. An even fancier event-handling scheme would then put the low-priority event into a queue that can be checked and triggered by the high-priority event upon its completion. Thus, the lower-priority event won't be ignored; it will simply be delayed.

Here is some code that illustrates the idea of checking priorities before executing the event subprogram:

```
Sub irpdEvent_CODE()
    //Chase something that I saw!
    if cPriority <= pChase Then
        cPriority = pChase

        <event handling code goes here>

        cPriority = 0
    End If

End Sub
```

In this code, the event checks the currently active priority *cPriority* to see if it is higher than its own priority *pChase*. If it is, this event handler exits without doing anything. If its priority is greater than the currently operating event, it sets the current priority to its own and executes its event-handling code. At the end of its code, it removes its priority so that lower-priority events can take over if they need to.

oEventX(C) (B.X.X Firmware and Later)

 The oEventX object interrupts any currently running event code with a lower priority (nonevent code is the lowest priority) to do its job. Upon exiting, the interrupted code resumes execution. If an event of a higher priority is currently running, the event queues up and runs after the higher-priority event code has completed. The oEvent object is the highest-priority event and interrupts any oEventX event code. The oEventX object is triggered when the Operate property transitions from 0 to 1.

The oEventXC object is triggered when the property that is pointed to by ClockIn transitions in the manner set by property InvertC, and only when the Operate property is set to 1. If Operate is set to 0, the event cannot be triggered. The OEventX memory size is 5 bytes (oEventXC is 7 bytes), and its properties are outlined in Table 7-2.

The oEventX objects handle some of the event-prioritizing housekeeping chores described earlier in the oEvent discussion relating to allowing prioritized events to interrupt each other. However, this priority is only half of the story.

TABLE 7-2 OEVENTX(C) PROPERTIES

ClockIn (flag pointer)	Points to a property that determines when this object performs its operation.
InvertC (bit, flag)	0 means a transition from 0 to 1 will trigger this object.
	1 means a transition from 1 to 0 will trigger this object.
Operate (bit, flag)	When set (cvTrue), the event is triggered. If this bit is still set upon completion of the subprogram, the event is not triggered again until this bit transitions back to 0 and then to 1 again (default property). (In oEventXC, the event can only be clocked when this bit is set to 1; otherwise, the clockIn trigger has no effect.)
Priority (nibble)	The range is 0 to 15. Lower numbers are higher priorities.

PRIORITIZING EVENT ACCESS TO RESOURCES

Although the oEventX events prevent low-priority events from interrupting high-priority events, they do not prevent priority inversion[1] issues altogether. You cannot prevent an event from occurring or cancel any interrupted event without doing more housekeeping. Why would you want to do this? Consider the following sequence of events in a sumo robot:

1. The priority 15 event (P15) begins in order to change the motor speed of your robot, so it makes lazy circles looking for evil sumo robots.
2. Priority 5 event (P5) fires and interrupts P15.
3. P5 changes the motor speed to chase the evil sumo robot it detected.
4. P5 completes the changes and exits.
5. P15 resumes the execution where it left off.
6. P15 changes the motor speed to make lazy circles looking for evil sumo robots.

You just had a priority inversion of the worst sort. Even though P5 interrupted P15 and went after the evil sumo robot, P5 started back up where it left off and removed all the changes that P5 put in place. That's not supposed to happen! P5 has higher priority! Actually, the event prioritizing prevents low-priority code from interrupting high-priority code; it doesn't prevent the low-priority code from executing. You still need to provide housekeeping functionality to prevent this kind of priority inversion from occurring. How? By having *more* housekeeping overhead.

To prevent two events from conflicting over a shared resource (like the motors in the robot), you need to *serialize* the code path that accesses that shared resource so you can lock

1. A priority inversion is where, contrary to all careful planning and programming, a low-priority event either blocks or nullifies the actions of a high-priority event. This is tricky logic to find and fix.

out any user until the current user has completed what it needs to do. By serialize, I mean that only one path through the code exists and that execution cannot be interrupted in any way. Although the OOPic is not a multithreaded[2] operating system, you can borrow some concepts from such operating systems to prevent the priority inversion sequence of actions you saw earlier.

In the OOPic, a programming device called a *mailbox* can be used to leave messages for other events to read and react to. The code snippet in the oEvent object discussion uses a form of mailbox with the *cPriority* variable. If you use a mailbox in conjunction with the oEventX(0) object, you can create a form of serialized access to the shared resource you need. You will not use oEvent at all in this schema because it is unclear which events can interrupt oEvent.

To create your mailboxes and your serialized access to a shared resource, some access rules must be defined:

■ Only one process can access the resource at a time, and it can't be interruptible.
■ While a high priority has control of the resource, no low-priority process can change it.
■ A priority mailbox defines which priority has control.
■ Data mailboxes describe a request to change the resource which include:
  ■ The priority of the requesting process
  ■ Data describing how to change the shared resource

The only means by which you can serialize access to the shared resource is with the oEventX objects. Let's further define the serialized access:

■ Access to the resource is provided by the oEventX object *ChangeResource* which is priority 0.
■ The variable cPriority defines the priority of the current owner of the resource.
■ The variable *newValue* defines the requested change to the resource.
■ The variable *rPriority* defines the priority of the requesting process.
■ ChangeResource checks rPriority against cPriority and exits if rPriority is lower.
■ If rPriority is higher, then the resource is modified, cPriority is updated to be equal to cPriority, and a timer is set for the duration of ownership.

Once a process has taken possession of a resource, how does the resource ever change ownership? The answer is by timing out. You will use a VC that is triggered by one of the OOPic system clocks (my favorite is OOPic.Hz60) to count down a value. When that value reaches 0, the ownership of the resource is revoked by the next operation of the Change-Resource event and is set to the priority of the next requestor that accesses it. Sometimes a process doesn't need ownership for that long; sometimes it needs it for a long time. You can decide that ownership needs to be periodically updated by your events, or you can have your events pass in a mailbox variable with the required ownership time to set.

By using this process, any event that decides to change the resource will trigger the ChangeResource event by toggling its Operate property. Since ChangeResource is the highest-priority event that can be triggered, all other events must queue up behind it. This means that the priority of ownership can be set, and the event won't be interrupted while doing it. Subsequent events will then trigger ChangeResource, which will check the relative priori-

---

2. Multithreaded means that more than one piece of code appears to be running at the same time.

ties and won't change the resource if the new requester isn't a higher priority. This allows high-priority events to cancel changes made by low-priority events, and it prevents a low-priority event from undoing a change made by a high-priority one. The only drawback is the amount of object space taken up by the ChangeResource event and the resource countdown VC.

Listing 7-1 shows one way the event and the ownership counter can be implemented. You will need to customize this boilerplate code to your needs by actually updating some valued resources.

Listing 7.1 is intended to be a template you can use and customize to your needs. Every event created to update a shared resource, such as motor speeds in a robot, needs to trigger this listing's event after it has set the mailbox variables to reflect the event's priority and needs. Because the combination of this event and the priorities of the other events take care of the housekeeping needs, your event processing can concentrate on the logic of what needs to be done, instead of trying to figure out if it should even run or not. More magic brought to you by OO thinking.

```
Dim owned as New oCountDownO    'Ownership timer
Dim chRes as New oEventX        'Resource change event
Dim resource as New oByte       'Dummy resource to update

'These global mailboxes are in variable RAM, not object RAM.
Dim cPriority as Byte           'Current owner's priority
Dim rPriority as Byte           'Requesting owner's priority
Dim rDuration as Byte           'Ownership timeout request
Dim newValue as Byte            'New resource value requested

Sub Main()

    'Create ownership expiration VC.  This will run in the
    'background and can never be interrupted accidentally.
    owned.Value = 0             'Initial value
    owned.ClockIn.Link(OOPic.Hz60)
    owned.Operate = cvTrue

    chRes.Priority = 0          'Highest priority

    'Let events take care of themselves.

End Sub

Sub chRes_CODE()
    'This is the ChangeResource event.  If rPriority <=
    'cPriority then we just exit, changing nothing.
    'This event is triggered by another lower-priority
    'event that wants to change the resource in some way.

    If rPriority > cPriority Then
        cPriority = rPriority    'New ownership set
        resource = newValue      'New setting done
        owned.Value = rDuration 'set ownership timer
    End If

End Sub
```

Listing 7-1

YET ANOTHER GOOD PROCESSING OBJECT TO KNOW ABOUT

Now is a good time to introduce the oCountDown processing object that I've used above. This is a handy way to implement an expiration counter because it is completely self-contained. Just link to a clock source as shown in Listing 7-1.

oCountDown(O) (B.X.X Firmware and Later)

 This object enables the user to set a value that will be counted down to 0 and then stop. It can either be linked to another object's Value property or, in the case of the oCountDownO version, the Value property can be set within the object itself. It's perfect as a timer for your resource ownership expiration counter. Its memory size is 6 bytes and its properties are outlined in Table 7-3.

oCountDown is a higher-level object that is practically a VC all itself. It is a great way to start up a VC that will automatically shut off when the countdown reaches 0 by linking the NonZero flag to the Operate property of another object. If oCountDown is used, the Output property links to the Value property in another object as the number to be decremented; if oCountDownO is used, the Output property is instead replaced by Value, which is a byte.

TABLE 7-3 OCOUNTDOWN PROPERTIES	
ClockIn (flag pointer)	Links to a flag in another object supplying clock ticks.
NonZero (bit, flag)	Reads 0 when the Value property is 0, and reads 1 otherwise.
Operate (bit, flag)	Must be 1 for this object to function.
Output (number pointer)	
Value (byte)	Links to the object whose Value property is decremented or is the internal Value to be decremented (oCountDownO).
PreScale (byte)	Specifies what to divide the ClockIn rate by. ClockIn/(PreScale + 1) = final rate.

Keypad Entry for the OOPic

You read earlier how to use a button or switch as a means of user input for the OOPic. When you want a user to enter data into your embedded program, a series of individual buttons is costly in terms of I/O lines. A better input device is a keypad. A keypad is a symmetrical matrix of buttons that can be queried for a specific key being pressed. Figure 7-1 shows a

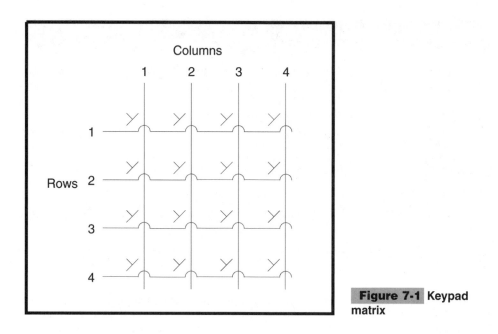

Figure 7-1 Keypad matrix

graphical representation of what is inside a keypad and suggests how you find out what key is pressed. For the smaller keypads of 16 keys or less, oKeypad is the object to use. If you have a keypad with more than 16 keys or a keypad with a different row/column matrix than 4×4, use oKeypadX, which can use up to 16 I/O lines to decode a keypad.

This chapter shows you how to interface a 4×4 matrix, 16-key keypad to the OOPic using the oKeypad object. You'll also see two different means by which you can connect the keypad to your VC. One is a simple press notification, and the second one allows an automatic key repeat similar to what a PC keyboard will do when you hold a key down.

CONFIGURING A KEYPAD WITH THE OOPIC

Keypad configuration on the OOPic is very straightforward. The oKeypad object requires the use of eight I/O lines: 8 through 15. This is one of the few objects that does not allow any flexibility as far as how lines are to be used; only one choice exists. These I/O lines are selected because the OOPic can be configured to use the internal pull-up resistors, which eliminates the need for you to have to wire them to your hardware circuit.

oKeypad (B.X.X Firmware and Later)

If you have the matrix map of the keys in your keypad, it is easy to calculate which key was pressed from the number returned from the object. Keypads are arranged in rows and columns, and the key that is pressed is represented by the following formula:

$$(((Row - 1) \times 4) + (Column - 1))$$

TABLE 7-4   OKEYPAD PROPERTIES

Mode (bit)	0 returns the encoded row and column.
	1 returns the phone-pad values
Operate (bit, flag)	Must be 1 for this object to function.
Received (bit, flag)	0 means no key is pressed.
	1 means a key is pressed, and a new value is stored in the Value property.
String (string)	The Value property is represented as a string.
Value (nibble)	The Value property is encoded as of the last key press (default property).

This formula assumes you have named the rows and columns 1 through 4. Sometimes you don't have the mapping of your keypad handy. In that case, you just press keys, write down the number that is returned, and create your internal key mapping by empirical observation.[3] If you should be so fortunate as to have a keypad whose layout resembles a telephone keypad, you can set the Mode property to 1, and the oKeypad object returns the number represented by the key instead of the encoded row and column number. The oKeypad memory size is 4 bytes and Table 7-4 displays its properties.

If a second key is pressed while the first detected key is being pressed, the oKeypad object ignores that second key. The Value property of oKeypad reflects the last valid pressed key. The Received flag remains set as long as any key is pressed.

oKeypadX (B.2.X Firmware and Later)

 Sometimes unusual layouts have keypads, such as a 6-row by 5-column keypad with 30 keys. The oKeypadX object enables up to an 8×8 keyboard to be scanned for key presses. This means that the OOPic can decode a QWERTY-style keyboard and have space left over for a numeric keypad and a few other special keys.

I/O lines 8 through 15 are used for the columns (because they have the internal pull-ups), and any I/O group of 8 lines is selectable by the programmer for the rows. Both the row and the column I/O groups have a mask that enables specific I/O lines within the groups to be selected for use. This option enables the keypad matrix to be any combination of rows and columns up to $8 = 8$. As an example, if only 5 I/O lines are selected for the row and 4 bits are selected for the column, the resulting scan pattern would scan a 5×4 keypad matrix that would contain 20 switches. Any I/O lines that are masked off in the oKeypadX matrix can be used for other OOPic I/O.

---

3. A fancy of way of saying you experimented and wrote down the results.

The oKeypadX memory size is 6 bytes and its properties are shown in Table 7-5.

To set the ColMask or RowMask, look at the bits as I/O lines. Bit 0 is the *least significant bit* (LSB) I/O line of the I/O group selected. In other words, bit 0 would be I/O line 8 of the columns, and bit 7 would be I/O line 15. Placing a 1 in any bit position means that oKeypadX will scan or read that line. Placing a 0 means it will be ignored.

TABLE 7-5   OKEYPADX PROPERTIES

ColMask (byte)	A bit mask determining which I/O lines are to be read.
IOGroup (2 bits)	1 means the oKeypad object uses IOGroup 1 for the row outputs.
	3 means the oKeypad Object uses IOGroup 3 for the row outputs.
	0 and 2 are not supported and cause odd things to happen.
Operate (bit, flag)	Must be 1 for this object to function.
Received (bit, flag)	0 means no key is pressed.
	1 means a key is pressed, and a new value is stored in the Value property.
RowMask (byte)	A bit mask determining which I/O lines are to be scanned.
String (string)	The Value property represented as a string.
Value (byte)	The Value property encoded as of the last key press (default property).

Serial I/O with the OOPic

Serial I/O is a means by which a single wire can carry information between remote locations. Unlike parallel port communications where all the data bits are transferred at the same time, serial transmission has each bit transferred one at a time, each following the last until the whole byte is done. Timing is important in serial communications because the receiver needs to know when to look at a bit to see if it is a 1 or a 0. Both the sender and the receiver need to have agreed upon a common transmission speed, or *baud rate,* so that the data can be reassembled at the receiver's end. Baud rates are specified in *bits per second* (BPS). This section details the OOPic objects that can send or receive serial communications.

THE OOPIC SERIAL COMMUNICATIONS OBJECTS

The OOPic has three serial I/O objects you can use. Two of them, oSerial and oSerialPort, use the PICMicro hardware *universal asynchronous receiver transmitter* (UART). The third, oSerialX, is a software UART and uses any available I/O line, but it has fewer supported baud rate values. If you are using the OOPic II+, R or C variants, you'll be doing most of your serial I/O using the oSerialX objects because the UART serial port will be taken over by *Serial Control Protocol* (SCP) communications.

Many differences exist between the hardware and software UART serial objects:

- The hardware UART has receive and transmit buffers so data isn't lost. The software UART has no buffers so one has to be more careful about pacing data transfers.
- The hardware UART has many available baud rates because the serial timing is done in the hardware. The software UART needs to create the baud rates in the software, and the amount of time available for this is limited.
- The hardware UART has both transmit and receive objects that work simultaneously. The software UART is defined in each case as either a transmit or receive object.
- The hardware UART can only use I/O lines 22 and 23, whereas the software UART can use any available I/O line for either transmit or receive objects.

oSerial (All OOPic Firmware Versions)

 This is the basic serial communications object for the OOPic. It utilizes the hardware UART for bidirectional serial communications without any kind of handshaking. The transmit line is I/O line 22, and the receive line is I/O line 23. If you are using an OOPic B.2.X+ firmware variant, this port is taken over by SCP communication processes and is unavailable until SCP is turned off. The baud values in parentheses are OOPic aliases that are easier to remember than a simple number. If the Mode property is 1, then the UART becomes a *universal synchronous receiver transmitter* (USRT), and transmission and reception are synchronous. Think of this as a clocked shift register in either direction at any one time, but not both. For details on using the synchronous serial data mode, see the PICMicro data sheet for the OOPic model you are using. The oSerial memory size is 4 bytes and its properties are shown in Table 7-6.

The output of the OOPic is serial data at *transistor-transistor logic* (TTL) voltage levels (the voltage levels that come out of your microcontrollers between 0V and 5V); it is *not* RS-232. RS-232 is a voltage level that is specified for long-distance serial transmission (more than a couple of inches), which will be explained later. The output is noninverted, which means that data bit 1 is logical 1 on either the transmit or receive I/O line. You must use an RS-232-level converter between the OOPic and a PC's RS-232 cable connection.

oSerialPort (B.X.X Firmware and Later)

The oSerialPort object encapsulates the OOPic's hardware UART similarly to oSerial. It has a 4-byte buffer and supports flow control. The transmit line is I/O line 22, and the receive line is I/O line 23. If you are using an OOPic B.2.X+ firmware variant, this port is taken over by SCP communications processes and is unavailable until SCP is turned off. The baud values in parentheses in table 7.7 are OOPic aliases that are easier to remember than a simple number.

The *Received* property can be used as a *Request to Send* (RTS) on the OOPic's side of communication in order to toggle the connected device's *Clear to Send* (CTS) line when it

TABLE 7-6 OSERIAL PROPERTIES

Baud (nibble)	0 (cvMidi) sets UART at 31,250 baud (MIDI).
	1 (cv1200) sets the UART at 1200 baud.
	2 (cv2400) sets the UART at 2400 baud.
	3 (cv9600) sets the UART at 9600 baud.
Mode (bit)	0 means the serial port operates asynchronously (default).
	1 means the serial port operates synchronously.
Operate (bit, flag)	Must be 1 for this object to function.
Received (bit, flag)	Is 1 when a byte of data has been received; it is cleared to 0 when that data byte has been read (the Value property has been read).
String (string)	The string representation of the data in the Value property.
Transmitting (bit, flag)	Is 1 when the UART is transmitting data; it is 0 otherwise.
Value (byte)	The data that is to be transmitted or the data that has been received (default property).

is connected through an RS-232-level converter (the CTS/RTS hardware handshake is discussed later). This serial port object has even more baud rates than oSerial, and because it has a receive buffer, it is more flexible and less prone to data overrun than oSerial,[4] even if flow control isn't used. The buffer in this object enables the OOPic to save data that is coming in for use later if the OOPic can't keep up with the speed of the data transmission. It's only 4 bytes though, so your program had better work fast or implement the Received property hardware flow control to prevent buffer overrun.

The *ReceivedOut* property is intended to be directly linked to an oDio1 object to be used in flow control. However, as of B.2.2, this flag pointer does not work properly.

If the Mode property is 1, then the UART becomes a USRT, and transmission and reception is synchronous. Think of this as a clocked shift register in either direction at any one time, but not both. Check the PICMicro datasheet for the PICMicro being used in your OOPic. The oSerialPort memory size is 10 bytes and Table 7-7 displays its properties.

This object is more complex than oSerial, and because of that, it is much larger. If you don't need flow control, use oSerial because it uses less object RAM space. You can only configure one oSerial or one oSerialPort object because only one hardware UART is used in the PICMicro chip.

---

4. Data overrun is caused when the UART starts receiving a new byte of data before the current byte has been read by the program. It is a bad thing.

oSerialX (B.X.X Firmware and Later)

With the advent of the B.2.X+ firmware (OOPic R, C, and II+) and the SCP taking control of the hardware UART in these variants of OOPic, another way to get serial data in and out of an OOPic was needed. Thus, the oSerialX object was created. This object can be linked to any available I/O line for serial input or output. Because the oSerialX object is a software UART, it is more limited in the baud rates that it can produce, and the possibility of data loss is present, so this object can also use flow control to reduce that likelihood.

The baud values in parentheses are OOPic aliases that are easier to remember than a simple number. The *IOLineF* property defines the I/O line that can be used as an RTS on the OOPic's side of communication to toggle the connected device's CTS line when it is connected through an RS-232-level converter. The oSerialX memory size is 5 bytes and its properties are shown in Table 7-8.

TABLE 7-7 OSERIALPORT PROPERTIES	
Baud (nibble)	0 (cvMidi) sets the UART at 31,250 baud (MIDI).
	1 (cv1200) sets UART at 1200 baud.
	2 (cv2400) sets UART at 2400 baud.
	3 (cv9600) sets UART at 9600 baud.
	4 (cv19200) sets UART at 19,200 baud.
	5 (cv4800) sets UART at 4800 baud.
	6 (cv50000) sets UART at 50,000 baud.
Mode (bit)	0 means the serial port operates asynchronously (default).
	1 means the serial port operates synchronously.
Operate (bit, flag)	Must be 1 for this object to function.
Received (bit, flag)	Is 1 when a byte of data has been received. It is cleared to 0 when that data byte has been read (the Value property has been read).
ReceivedOut (flag pointer)	Link to I/O line to use for flow control.
String (string)	The string representation of the data in the Value property.
Transmitting (bit, flag)	Is 1 when the UART is transmitting data; it is 0 otherwise.
Value (byte)	The data that is to be transmitted or the data that has been received (default property).

TABLE 7-8   OSERIALX PROPERTIES

Baud (nibble)	1 (cv1200) sets the UART at 1200 baud.
	2 (cv2400) sets the UART at 2400 baud.
	3 (cv9600) sets the UART at 9600 baud.
	5 (cv4800) sets the UART at 4800 baud.
InvertF (bit, flag)	1 means invert flow control logic.
InvertS (bit, flag)	1 means invert serial line logic.
IOLineF (byte)	Chooses the I/O line to be the flow control line.
IOLineS (byte)	Chooses the I/O line to be the serial data line.
Operate (bit, flag)	Must be 1 for this object to function.
Value (byte)	The data that is to be transmitted or the data that has been received (default property).

The flow control logic for oSerialX is a reverse of oSerialPort. When oSerialX is ready to accept data, it sets IOLineF to 1. IOLineF is dropped to 0 when a byte is received and remains at 0 until the Value property is read, whereupon it sets IOLineF to 1 again. This same flow control logic is used when oSerialX is used as a serial input and IOLineF is defined as a flow control line.

Because oSerialX uses the opposite states as oSerialPort for the flow control line, if you want the same logical flow control as oSerialPort, set InvertF to 1. If IOLineF is set to 0, no flow control will be used by oSerialX.

UNDERSTANDING SERIAL COMMUNICATIONS

Serial data communication is the transmission of data one bit at a time. Two kinds of serial communications exist: asynchronous and synchronous. The OOPic produces TTL voltage levels at the I/O pin of the PICMicro device; a serial cable must have RS-232 voltage levels to properly function. These two voltage levels don't play well together; you need a level conversion circuit to handle that. What do handshaking and flow control mean? Read on to learn about these terms.

Asynchronous Versus Synchronous Serial

Synchronous serial uses at least two I/O lines to transmit data: One is a clock and one is the data. It's known as synchronous because the clock line keeps both sides of the communication in sync. Many types of synchronous serial formats are used (*Inter-IC* [I2C] and *serial peripheral interface* [SPI] are the two most popular), but any serial communications that uses both a clock and a data line is synchronous. Figure 7-2 shows a form of a synchronous

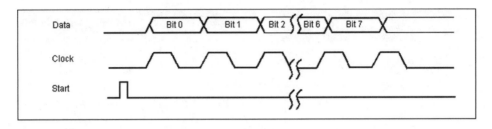

Figure 7-2 Synchronous serial data example

serial data transmission. This figure shows a clock signal line and a data signal line. The *Start* pulse simply shows when the transmission sequence begins. A receiver would start clocking in data when it started receiving the clock signal and would sample the data on the falling edge of the clock signal, because that is near the center of the data bit cell.

To allow two devices or computers that are communicating asynchronously to understand each other, they both have to agree on the speed of that communication and on the pattern of bits and the order in which they are transmitted. Because no clock line keeps the two sides of the communication in sync, this is known as *asynchronous serial*. Also, because no clock signal tells a receiver when it should start expecting data, a way for the transmitter to signal this intent must be established. With an asynchronous serial data transmission, this signal is called the *start bit*.

The quiescent or unused state of a serial line is logic level 1. To signal the start of a transmission, the data line is brought low, 0, for one bit cell time period. The receiver now knows that a transmission is about to occur, but how does it know when the transmission is over? That is called a *stop bit* and by standard acceptance, this is always a logic 1 at the end of a transmission of a byte. If parity is involved, the parity bit is placed between the last bit of the byte and the stop bit. The entire sequence of bits that make up the transmission of a single byte of data is called a *frame*.

Figure 7-3 shows the layout of an 8-bit data word with 1 start bit and 1 stop bit, for a total of 10 bits to transmit 1 byte. This type of asynchronous data transmission is called *not return to zero* (NRZ), because each bit is represented by a logic level, not a transition, so the data signal does not need to return to zero before each new bit. By tradition, a 1 is called a *mark*

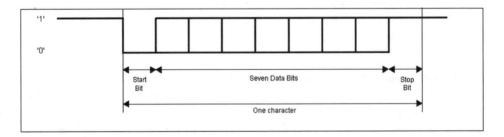

Figure 7-3 Asynchronous serial data example

and a 0 is called *space.* Data is transmitted with the LSB being sent first. The OOPic does not use parity so I'll not discuss that here.

What Does RS-232 Really Mean?

Many people use the term RS-232 to mean serial communications, but this is incorrect. Properly speaking, asynchronous serial communications is the correct term for this form of information exchange. EIA-RS-232 C is the *Electrical Industries Alliance* (EIA; www.eia.org) specification for the electrical and mechanical characteristics of the serial data, signals, and connectors used to transmit data between computers, printers, terminals, and modems.

RS-232 C signals have different electrical specifications than TTL or *complimentary metal oxide semiconductor* (CMOS, which consists of the voltage levels from many of your microcontrollers) signals. Table 7-9 compares the logic levels of these three specifications.

To use the OOPic's serial input or output ports on an RS-232 cable connected to another device, a level converter is required. The most common one used is the MAX232A chip. One chip and four capacitors are all you need to get two transmit and two receive buffers for your serial I/O projects. In most places, this combination is less than $2.50.

Now that you have your voltages correct, where do you put them? Figure 7-4 shows the common DB9 connector pinout from a workstation or PC. Figure 7-5 shows the DB25 cable

TABLE 7-9   SIGNALS' STANDARD VOLTAGE SPECIFICATIONS

STANDARD	LOGIC 0	LOGIC 1
TTL	0V to 0.4V	2.4V to 5V
CMOS	0V to 1.67V*	3.33V* to 5V
RS-232 C	3V to 25V	−3V to −25V

*Assumes that Vcc is 5V. The specification states that 0 is less than 1/3 Vcc, and 1 is greater than 2/3 Vcc. CMOS may operate with a Vcc of 3V to 15V. Vcc is the power line to the chips.

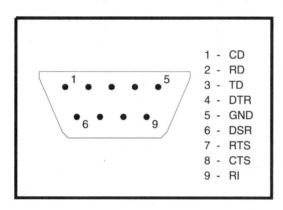

1 - CD
2 - RD
3 - TD
4 - DTR
5 - GND
6 - DSR
7 - RTS
8 - CTS
9 - RI

Figure 7-4 DB9 cable connector pinout from the computer

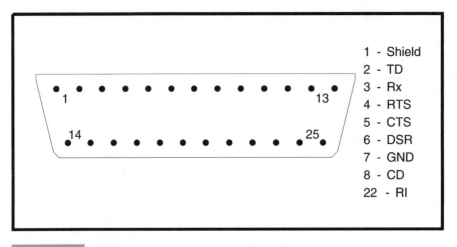

1 - Shield
2 - TD
3 - Rx
4 - RTS
5 - CTS
6 - DSR
7 - GND
8 - CD
22 - RI

Figure 7-5 DB25 cable connector pinout from the computer

pinout. These pins are shown from the perspective of looking into the male connector (called *Data Terminal Equipment* [DTE] from the computer) of the cable from your computer or, as you are looking at the back of the female connector (called *Data Communications Equipment* [DCE] from the device), which you would wire to your OOPic board.

The DB9 connector shown in Figure 7-4 is the most common cable you will likely find. It is a straight-through cable, which means the wires from the connector on the back of the PC run straight through the cable and come to pins in exactly the same place on the end of the cable. No wires are crossed. A DCE cable would have pins 2 and 3 crossed. Be careful to know which cable you use. Figure 7-5 shows a DB25 RS-232 C cable, which isn't as common, but it's still possible that you might want to use one.

What Is Flow Control?

The OOPic oSerialPort and oSerialX both mention flow control options that can be used with these objects, but what is flow control? Flow control, in this case, hardware flow control, uses RS-232-specified lines to handshake a byte of data across the serial link. The process works like this:

1. The OOPic raises the handshake line to 1, which says "OK, send me data."
2. The device or computer sees its request line go high and sends a byte of data.
3. The OOPic gets the byte of data and immediately drops its handshake line to 0, which says, "Wait, I want no more data."
4. The device or computer sees its request line go low and stops sending data.

Hardware handshaking, or hardware flow control, is most commonly used by devices that don't have a lot of memory to store incoming data or they are too slow to process the data as it comes in without stopping the data flow every so often. The most common set of lines to use for hardware flow control is the RTS/CTS pairs. Like the *Received Data and Transmit Data* (RD/TD) pairs, these lines are swapped when referring to the device connected to the computer. By this, I mean that the OOPic's RTS line would be connected to the computer's CTS line and vice versa. This chapter's project uses the oSerialPort object and the CTS line to the computer to implement a hardware flow control protocol.

Experiments

Here are some experiments to try with the objects I've introduced in this chapter. These experiments use events, the keypad, and your old friend the *liquid crystal display* (LCD) to create a simple user interface for your projects.

SIMPLE KEYPAD EVENT HANDLING

This experiment allows you to figure out what your row/column mapping is with your keypad. It shows how you can use an event to do some preprocessing of your keypad input before you display it, or any results, on your LCD.

For the wiring hookup, use the schematic in Figure 7-6 that shows how the LCD and keypads are connected to the OOPic. It's the same connection and code that are used in this experiment, just without the serial port section.

The Received property flag is used to trigger an event that reads the oKeypad Value and converts it into a displayable character or word. In the case of the Clear and Enter keys on my keypad, screen clears and a line feed are also performed. The code shown in Listing 7-2 simply waits for a key press and displays the key value on the LCD:

```
'keypad1.osc
'Chapter 7 Keypad and events Example
'Demonstrates oKeypad and oEvent usage
'
'For OOPic II B.2.X firmware
'
'Copyright Dennis Clark 2003
'Permission is granted to use this code anyway you like
'as long as you say you got it from me.

Dim LCD as New oLCD
Dim KBD as New oKeyPad
Dim encode as New oEvent
Dim wire as New oWire

Sub Main()

    'LCD initialization
    LCD.IOLineRS = 27            'RS line to the LCD module
    LCD.IOLineE = 26             'E line to the LCD module
    LCD.IOGroup = 3              'I/O lines 28-31 (pins 32-26)
```

(Continued)

```
    LCD.Nibble = 1                  'Upper nibble
    LCD.Operate = cvTrue

    OOPic.delay = 3
    LCD.Init
    LCD.Clear
    LCD.Locate(0,0)
    LCD.String = "OOPic Events"
    LCD.Locate(1,0)
    OOPic.delay = 300

    'oKeypad uses I/O lines 8-15 to create a rows/columns
    'matrix keypad encoder.  I/O lines 8-11 are rows and
    'I/O lines 12-15 are columns.
    KBD.Operate = cvTrue            'Not much to configure

    wire.Input.Link(KBD.Received)      'fires the event
    wire.Output.Link(encode.Operate) 'This fires the event
    wire.Operate = cvTrue

End Sub

Sub encode_CODE()
    'This is the event handler to use when a button is pushed
    'on the keypad.  The key matrix is pretty random, so
    'I've just used a big case statement to handle the values.
    'Mine is kind of a weird keypad, your keys will probably
    'be labelled differently!  Lots of processing of special
    'key meanings can be done here.  In this case, there is
    'an "enter" and a "clear" function defined.

    Select Case KBD

    case 0
        LCD.String = "1"
    case 1
        LCD.String = "2"
    case 2
        LCD.String = "3"
    case 3
        LCD.Locate(1,0)
        LCD.String = "        "
        LCD.Locate(1,0)
        LCD.String = "Up"
    case 4
        LCD.String = "4"
    case 5
        LCD.String = "5"
    case 6
        LCD.String = "6"
    case 7
        LCD.Locate(1,0)
        LCD.String = "        "
        LCD.Locate(1,0)
        LCD.String = "Down"
    case 8
        LCD.String = "7"
    case 9
        LCD.String = "8"
    case 10
```

```
            LCD.String = "9"
      case 11
            LCD.Locate(1,0)
            LCD.String = "        "
            LCD.Locate(1,0)
            LCD.String = "2nd"
      case 12                                    'Clear the screen
            LCD.Clear
            LCD.Locate(0,0)
            LCD.String = "OOPic Events"
            LCD.Locate(1,0)
      case 13
            LCD.String = "0"
      case 14
            LCD.Locate(1,0)
            LCD.String = "        "
            LCD.Locate(1,0)
            LCD.String = "Help"
      case 15                                    'Clear the entry line
            LCD.Locate(1,0)
            LCD.String = "             "
            LCD.Locate(1,0)

      End Select
End Sub
```

Listing 7-2

When you press a button, its meaning appears on the LCD. Nice, but what if you would like to have a key repeat, such as what happens on your computer's keyboard when you hold down a key? The OOPic has a processing object called oRepeat that is used for just this purpose. The next experiment shows how to use oRepeat to get this functionality.

KEYPAD INPUT WITH REPEATING KEYS

This experiment is exactly like the last one, except instead of using an oWire object to link the keypad object with the event, an oRepeat processing object enables automatic key repeat. Because oKeypad leaves the last value encoded in its Value property, when the keypad event handler is called multiple times, that key value is still there to use. Before looking at that code listing, you'll examine the oRepeat object that is the star of the show.

oRepeat (B.X.X Firmware and Later)

 This object allows a static event, like a button held down, to cause multiple events to occur; in this case, an event handler is called that continues to process the same keypad press. The oRepeat object takes the place of an oWire or oGate processing object to link between the object generating the event and the object processing the event. The oRepeat memory size is 7 bytes and the properties are shown in Table 7-10.

TABLE 7-10   OREPEAT PROPERTIES

Input (flag pointer)	Links to the flag that will cause the event.
InvertIn (bit, flag)	1 inverts the flag property, and a 0 causes the repeat, not a 1.
InvertOut (bit, flag) property.	1 inverts the flag property before sending it to the Output
Operate (bit, flag)	Must be 1 for this object to function (default property).
Output (flag pointer)	Links to the object that will be triggered by the event Input flag.
Period (byte)	The number of 1/60-second tics to wait before repeating the event. The limit is between 0 and 127.
Rate (byte)	The number of 1/60-second tics to wait between repeats of the event. The limit is between 0 and 63.
Result (bit, flag)	The result of this object's work as it occurs.

The difference between the InvertIn and InvertOut properties may not be immediately obvious. If you invert the input, a 0 flag triggers the repeat functionality and a 1 is passed to the object that will be repeated. If you invert the output, a 1 is still used to trigger the oRepeat object, but a 0 is passed to the object that will be sent this flag repeatedly. The former enables a 0 flag to trigger the event, and the latter still has a 1 triggering oRepeat, but oRepeat sends a 0 to the object it is linked to.

Now that you know what oRepeat does, let's look at how it works in Listing 7-3. Experiment with the Period and Rate properties to find a comfortable key repeat timing:

```
'keypad2.osc
'Chapter 7 Keypad and events Example with auto key repeat.
'Demonstrates oKeypad and oEvent usage
'
'For OOPic II B.2.X firmware
'
'Copyright Dennis Clark 2003
'Permission is granted to use this code anyway you like
'as long as you say you got it from me.

Dim LCD as New oLCD
Dim KBD as New oKeyPad
Dim encode as New oEvent
Dim rpt as New oRepeat

Sub Main()

    'LCD initialization
    LCD.IOLineRS = 27              'RS line to the LCD module
    LCD.IOLineE = 26               'E line to the LCD module
```

```
        LCD.IOGroup = 3            'I/O lines 28-31 (pins 32-26)
        LCD.Nibble = 1            'Upper nibble
        LCD.Operate = cvTrue

        OOPic.delay = 3
        LCD.Init
        LCD.Clear
        LCD.Locate(0,0)
        LCD.String = "OOPic Events"
        LCD.Locate(1,0)
        OOPic.delay = 300

        'oKeypad uses I/O lines 8-15 to create a rows/columns
        'matrix keypad encoder.  I/O lines 8-11 are rows and
        'I/O lines 12-15 are columns.
        KBD.Operate = cvTrue          'Not much to configure

        'Instead of a simple oWire processing object, let's use an
        'oRepeat object so that we can get a key repeat functionality
        'for our keypad.
        rpt.Period = 10            'How long to wait before repeats
        rpt.Rate = 3              'How fast to repeat
        rpt.Input.Link(KBD.Received)
        rpt.Output.Link(encode.Operate)   'This link fires the event
        rpt.Operate = cvTrue

End Sub

Sub encode_CODE()
        'This is the event handler to use when a button is pushed
        'on the keypad.  The key matrix is pretty random, so
        'I've just used a big case statement to handle the values.
        'Mine is kind of a weird keypad; your keys will probably
        'be labelled differently!  Lots of processing of special
        'key meanings can be done here.  In this case, there are
        'an "enter" function and a "clear" function defined.

        Select Case KBD

        case 0
            LCD.String = "1"
        case 1
            LCD.String = "2"
        case 2
            LCD.String = "3"
        case 3
            LCD.Locate(1,0)
            LCD.String = "        "
            LCD.Locate(1,0)
            LCD.String = "Up"
        case 4
            LCD.String = "4"
        case 5
            LCD.String = "5"
        case 6
            LCD.String = "6"
        case 7
            LCD.Locate(1,0)
            LCD.String = "        "
```

(Continued)

```
          LCD.Locate(1,0)
          LCD.String = "Down"
      case 8
          LCD.String = "7"
      case 9
          LCD.String = "8"
      case 10
          LCD.String = "9"
      case 11
          LCD.Locate(1,0)
          LCD.String = "       "
          LCD.Locate(1,0)
          LCD.String = "2nd"
      case 12                             'Clear the screen
          LCD.Clear
          LCD.Locate(0,0)
          LCD.String = "OOPic Events"
          LCD.Locate(1,0)
      case 13
          LCD.String = "0"
      case 14
          LCD.Locate(1,0)
          LCD.String = "       "
          LCD.Locate(1,0)
          LCD.String = "Help"
      case 15                             'Clear the entry line
          LCD.Locate(1,0)
          LCD.String = "              "
          LCD.Locate(1,0)

      End Select
End Sub
```

Listing 7-3

Now let's move on to the project for this chapter. I'll be combining just about everything I've talked about in this chapter with this project.

Project: Keypad and LCD Terminal

This project implements a simple terminal based on RS-232 C with a very small display and keyboard. In this case, the display is a 2×16 LCD, and the keyboard is a 4×4 matrix keypad. I've used the oSerialPort object to get its buffered input port and a hardware hand-shake flow control via the CTS line to my computer.

By using CTS, this program tells the computer when to stop sending data so that nothing gets lost while it processes the data that it has already received. I tested my program by sending the source code to the OOPic and watching it echo back to my terminal program. A word to the wise: Don't use *Hyperterm*. Although it says it implements hardware flow control, it really doesn't. Use *ProCom* or some other terminal emulator that enables you to upload text files and use a hardware flow control where you can select the RTS/CTS control lines for the handshaking.

HARDWARE LAYOUT

Refer to Figure 7-4 for the DB9 cable pinout. If you are wiring your own connector to plug into your prototyping board, the pinout in Figure 7-4 is also what you'll see at the backside of the female connector you'll be using. Figure 7-6 shows the circuit layout when using a conversion from MAX232 TTL to RS-232 C level, as well as the LCD hookup and other important electronics details for this project. The CTS connection also goes through this chip, which is why the oSerialPort Received property is pulled high (to a 1) when a byte of data is received; it will be a 0 at the computer after going through the level converter chip. Note that this LCD hookup is *not* the same as that used in Chapter 6, "OOPic Timers, Clocks, LCDs, and SONAR (Project: SONAR Ping)." The RS and E lines had to be moved so that the oKeypad object could use I/O lines 8 through 15.

DESIGN THE VCS

Two VCs need to be designed for this project: one for the keypad and its associated event handler, and one for the serial port and its associated event handler. Figure 7-7 shows what I used for VCs on this project. The oFanout object was needed to handle a bug in the B.2.2 firmware with the oSerialPort ReceivedOut property, so here is the information on oFanout too.

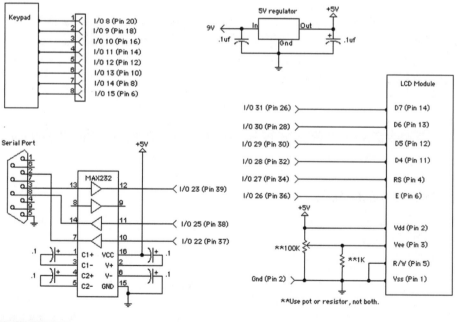

Figure 7-6 Project circuit layout

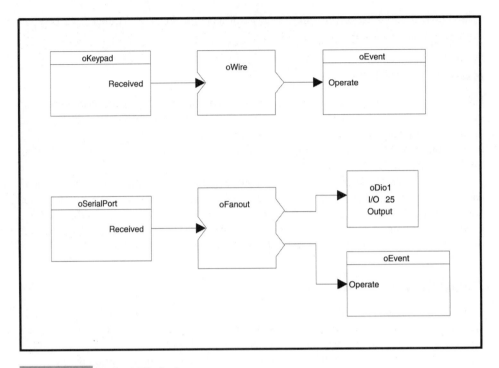

Figure 7-7 Project VC design

oFanout(C) (All OOPic Firmware Versions, oFanoutC B.X.X, and Later)

 This object has been created to allow a single flag to be linked to multiple objects. You could do this with several oWire objects, but oFanout has a lower object memory overhead to do the same thing. A single flag may be linked to up to four different flag inputs. The oFanoutC object enables the flag value to be sent to the output links only upon receiving a clock transition. The oFanout(C) memory size is two plus the number of outputs (oFanoutC is four plus the number of outputs). The oFanout(C) properties are shown in Table 7-11.

WRITE THE CODE

Listing 7-4 is the program written from the VC designs. Because I wanted to send a key press to both the LCD and the serial port, I stored the key value in a temporary variable. I did this because it takes a longer time to process straight code, and the VC is running in the background. Sometimes the value in the keypad buffer would change between writing to the LCD and writing to the serial port. That is why I couldn't just grab the value twice from the keypad object. The OOPic has no string storage variable, so the keypad value is stored in a byte instead. I prefer using hex values over decimal because the bit patterns are more readily discernable. Note the ASCII pattern for the 0 through 9 characters. You can't see that in a decimal number, but it really stands out in hex.

TABLE 7-11   OFANOUT(C) PROPERTIES

ClockIn (bit, flag)	(oFanoutC) When this property transitions from 0 to 1, the Input link is propagated to the Ouput[1-4] links.
Input (flag pointer)	Links to the flag property that is to be linked to other objects.
InvertC (bit, flag)	(oFanoutC) This inverts the sense of the ClockIn property when set to 1.
InvertOut[1-4] (bit, flag)	Enables the Input flag to be individually inverted before being passed to the flag specified by Output[1-4].
Operate (bit, flag)	Must be 1 for this object to function (default property).
Output[1-4] (flag pointer)	Links to flags that will receive the Input flag property's value.
Width (nibble)	Holds the number of output links defined when this object is created.

```
'serialport.osc
'Chapter 7 Keypad, events, and serial port project
'Demonstrates serial port flow control usage as well as
'Event programming.
'
'For OOPic II B.2.X firmware
'
'Copyright Dennis Clark 2003
'Permission is granted to use this code anyway you like
'as long as you say you got it from me.

'Objects used
Dim LCD as New oLCD
Dim KBD as New oKeyPad
Dim encode as New oEvent          'keyboard handler
Dim wire as New oWire
Dim receive as New oEvent         'Serial port handler
Dim port as New oSerialPort
Dim CTS as New oDio1              'Flow control pin
Dim rec as New oFanout(2)         'Link to serial port handler

'Global variables used
Dim kbdcursor as byte            'cursor location for kbd data
Dim sercursor as byte            'cursor location for serial data

Sub Main()

    OOPic.node = 1               'Ever present debug aid

    'Initialize the cursor values (will be incremented first)
```

(Continued)

```
        kbdcursor = 15                  'first location, first line
        sercursor = 15                  'first location, second line

        SetupLCD                        'Initialize LCD display

        'This links the keypad event handler
        KBD.Operate = cvTrue            'Not much to configure
        wire.Input.Link(KBD.Received)   'fires the event
        wire.Output.Link(encode.Operate) 'This fires the event
        wire.Operate = cvTrue

        'Using CTS to avoid receive buffer overruns at this baud.
        'In reality, the is the CTS line on the device that the
        ''OOPic connects to.  On the OOPic, this pin would be
        'more correctly labeled RTS.  When CTS/RTS is raised
        'high, it means the OOPic is requesting that data be
        'sent.  When pulled low, it means don't send data.  Since
        'the OOPic oSerialPort pulls the Received line high when
        'data is received, this means that the sense of this
        'signal must be reversed.
        CTS.IOLine = 25                 'Pin 38 on I/O connector
        CTS.Direction = cvOutput

        'Configure the serial port for use
        port.Baud = cv9600              '9600 baud is a good speed
        '*** Workaround code ***
        'port.ReceivedOut.Link(CTS) "Link to the CTS line

        'Link serial port data input to event handler for display
        '*** Workaround code ***
        rec.Input.Link(port.Received)   'Fire event handler
        rec.Output1.Link(CTS)           'And CTS for flow control
        rec.Output2.Link(receive.Operate)
        rec.Operate = cvTrue

        port.Operate = cvTrue           'Now turn serial port on

        'Now we just wait for things to happen, main() is done.

End Sub

Sub SetupLCD()
        'Handles the LCD layout
        'LCD initialization
        LCD.IOLineRS = 27               'RS line to the LCD module
        LCD.IOLineE = 26                'E line to the LCD module
        LCD.IOGroup = 3                 'I/O lines 28-31 (pins 32-26)
        LCD.Nibble = 1                  'Upper nibble
        LCD.Operate = cvTrue

        OOPic.delay = 3
        LCD.Init
        LCD.Clear
        LCD.Locate(0,0)
        LCD.String = "OOPic Serial"

End Sub

Sub encode_CODE()
        'This is the event handler to use when a button is pushed
```

```
'on the keypad.    Lots of processing of special
'key meanings can be done here.  In this case, there is
'an "enter" and a "clear" function defined.  All text
'from the keypad will be displayed on line 1 of the LCD
'as well as sent out on the serial port TD line.

Dim tmp as byte                 'save KBD data

'Set cursor location, only 0-15 is allowed on my 2x16 LCD
kbdcursor = (kbdcursor +1) mod 16
LCD.Locate(0,kbdcursor)         'we wrap the cursor after 16

'Note that all characters are saved as hex numbers and not
'as strings.  The OOPic has no way to store string variables,
'but it's just as easy to store them as their ASCII values.
'Hex is easier to use than decimal in programming; you can
'see patterns in Hex that you can't see in decimal.
Select Case KBD

case 0                          ' 1
    tmp = &H31
case 1                          ' 2
    tmp = &H32
case 2                          ' 3
    tmp = &H33
case 3                          ' ^
    tmp = &H5E
case 4                          ' 4
    tmp = &H34
case 5                          ' 5
    tmp = &H35
case 6                          ' 6
    tmp = &H36
case 7                          ' #
   tmp = &H23
case 8                          ' 7
    tmp = &H37
case 9                          ' 8
    tmp = &H38
case 10                         ' 9
    tmp = &H39
case 11                         '%
   tmp = &H25
case 12                         'Clear the screen
    LCD.Clear
    kbdcursor = 15
    sercursor = 15
    tmp = 13
case 13                         ' 0
    tmp = &H30
case 14                         ' ?
    tmp = &H3F
case 15                         'Clear the entry line
    LCD.Locate(0,0)
    LCD.String = "              "
    LCD.Locate(0,0)
    kbdcursor = 15
    tmp = 13
```

(Continued)

```
        End Select

        if tmp <> 13 Then     'LCD doesn't like control codes
            LCD = tmp
        End If

        'Just in case the UART is transmitting another character,
        'wait until it is finished so we don't lose this character.
        while port.Transmitting = 1
        wend
        port = tmp
        if tmp = 13 Then                '13 = CR, 10 = LF
            port = 10
        End If

End Sub

Sub receive_CODE()
        'This is the serial port event handler.  It is fired when
        'a byte comes in via the serial port and needs to be
        'displayed on the LCD.  This handler will place all text
        'on line 2 of the LCD that comes in via the serial port.
        'It's a nice test of the system if the character is echoed
        'back out to the terminal.

        Dim tmp2 as byte

        'There may be more than a single character in the UART receive
        'buffer.  If there is, this loop will get all that are there
        'before it exits.  There may be up to four characters in the
        'buffer.
        While port.Received = 1
            'map the character to the cursor location
            sercursor = (sercursor +1) mod 16
            LCD.Locate(1,sercursor) 'locate the cursor
            tmp2 = port                         'save the incoming data
            if tmp2 >= 32 Then                  'LCD doesn't like control chars
                LCD = tmp2                      'display the data
            End If
            port = tmp2                         'and echo it back to the terminal
            if tmp2 = 13 Then                   'CR will clear the line
                LCD.Locate(1,0)
                LCD.String = "              "
                LCD.Locate(1,0)
                sercursor = 15
                port = 10                       'let's issue a LF as well
            End If
        wend

End Sub
```

Listing 7-4

A couple of lines deserve comment:

```
'*** Workaround code ***
'port.ReceivedOut.Link(CTS)              'Link to the CTS line

'Link serial port data input to event handler for display
'*** Workaround code ***
rec.Input.Link(port.Received)           'Fire event handler
rec.Output1.Link(CTS)                   'And CTS for flow control
rec.Output2.Link(receive.Operate)
```

The oSerialPort object has ReceivedOut property which is supposed to be a flag pointer that can link directly to an oDio1 I/O line. In Firmware B.2.2, this link does not work. My workaround is to use an oFanout processing object to link the Received property to both the CTS handshake line and the receive_CODE() event handler.

Notice also that even though my temp variables (tmp and tmp2) are defined in separate subprograms, they occupy global memory space and need to have different names in OOPic programs. All variables, no matter where they are defined, are global and need unique names.

I've not used the oRepeat object here to get automatic key repeats on the keypad, as I did in Listing 7-3. I've left that as an experiment for the reader. It's more fun to hack the code yourself, isn't it?

Now What? Where to Go from Here

This chapter introduces some complex issues when dealing with priority encoded events and serial port programming. These are both difficult project areas, so use the code presented here as a template to solve your problems. The priority encoding of events is especially useful to robot builders. This makes behavioral programming via subsumption architectures much simpler to handle, and you'll see some of that in Chapter 8, "OOPic Interfacing and Electronics (and Steppers and Seven-Segment LEDs)."

Experiment with the serial ports and the keypad encoder objects. You'll find that no matter how carefully you plan your project, Uncle Murphy will inject a bit of humor into your life anyway. Knowing how these objects work will help you find the gremlins deposited in your code. Knowing your options will help you shoo those gremlins away!

OOPIC INTERFACING AND ELECTRONICS (AND STEPPERS AND SEVEN-SEGMENT LEDS)

CONTENTS AT A GLANCE

▌hadn't originally intended to write this chapter. I wanted to concentrate on OOPic issues specifically and not embedded electronics in general, but as you all know, I hang out a lot on the OOPic newsgroups and other robotics lists. I see the same questions come up again and again for robotics and embedded computing of all kinds. Cogitating on these queries for help, I decided that a chapter dedicated to basic electronics, and leaning heavily towards the OOPic, of course, would do a world of good for the OOPic hobbyist community. With that in mind, I have chosen the first topic in this chapter to be as follows:

How to Avoid Zapping Your OOPic on the First Day

Those of you who have participated in the OOPic discussion list, or even just lurked there, know when this topic came up in early 2003. I found the heading so humorous I couldn't resist using it to head a section of this chapter.

There are a few basic precautions that you can take that will prevent you from ever blasting a static-sensitive device again and they aren't onerous. The following section is a list of precautions and suggestions to ponder before you touch an electronic device.

PRECAUTIONS TO USE IN YOUR WORK AREA

The following precautions can help you minimize the exposure of your sensitive electronics to static in your work environment:

■ *Remove your shoes.*   Rubber soles insulate you from the ground and allow you to build up a static charge deadly enough to zap an *integrated circuit* (IC). Staying in your bare feet or socks allows static to bleed off onto the floor and other objects instead of building up. I'm not kidding!

■ *Remove your wool or polyester sweater.*   These things can generate thousands of volts of static potential just by waving your arms.

■ *Similarly, don't work in an area with a carpeted surface.*   Wood or even tile is less susceptible to static generation.

■ *Don't work on a Formica™-like or plastic surface.*   Again, these surfaces don't conduct electricity and will allow a great static buildup just looking for a part to zap. A smooth (but unpolished), wooden, metal, or antistatic workstation top is the best surface. If you have to work on a nonconducting surface, cover it with antistatic plastic (those blue or pink bags) or even aluminum foil first. Then ground that surface by doing something as simple as attaching a chain to it and letting the end of the chain (or wire) lay on the floor.

■ *Use metal storage bins for your parts.* If you only have plastic ones (quite common), store your parts in antistatic foam. Either the black or pink stuff works great.

PRECAUTIONS TO USE WHILE WORKING ON YOUR PROJECTS

The following are precautions to take while you are working on your projects. In addition to static, excess heat and shock can damage electronic components. Caution is always in order:

- *Always touch the surface a device is in contact with before you touch the device.* This allows any static to dissipate harmlessly into the PC board or parts bin instead of going through your parts first. This includes touching things like screwdrivers and pliers.
- *Use a grounded soldering iron when repairing or adding to a board with components loaded onto it.* If your soldering iron isn't a three-pronged cord, it isn't grounded. I know this means you have to pay more, but how many $40 OOPics does a decent soldering iron cost? Answer: about one.
- *Use a 60/40 Resin core solder.* This is a standard electronics solder. You don't need a silver solder, and you don't want an acid core. The former is for jewelry and the latter is for plumbing. I have heard that 63/37 is an exceptional solder too.
- *Keep the soldering iron on an IC or component lead for only as long as it takes to get the solder to flow.* Anything more than two to three seconds is asking for damage. In the case of surface mount parts, that time is reduced to about a one-second contact with the soldering iron tip.
- *Keep fresh laser copier paper and any plastic wrap away from components.* Both of these are heavily charged materials that can zap a part faster than you can say, "Ouch." After a number of hours, the paper becomes safe, but right after it is printed it is heavily charged; that is how the ink is transferred to the paper. Any kind of adhesive tape is also problematic; cellophane tape is the worst.

PROTECT YOURSELF WHEN WORKING ON A PROJECT

You should take a few safety precautions when working on electronics projects for your own benefit. Not many are needed, but none of them should be ignored:

- *Keep a fan blowing across your workbench while you are soldering.* Lead and tin fumes are not pleasant, and they're not safe for long-term exposure either. You don't need much airflow, just a little, and you can breathe easier while you work.
- *Make sure all your instruments are grounded, especially your bench power supply.* Electrocution is not a pleasant experience, and because you've set yourself up to bleed off static electricity, you've become a pretty good conductor for your power mains as well. Don't *ever* ground yourself with a wire to the ground of an electrical outlet! Use a true industrial grounding strap and follow its instructions for installation.
- *Wear safety glasses.* Solder resin pops and sputters, and hot solder splashes and flies about. Those leads that you are clipping off have to go somewhere as well, and it is better if these things did not end up in your eyes. Those of us with glasses have built-in shields, and because we always have a minimum of eye protection, we don't usually worry about this precaution. However, anyone with lightweight plastic lenses will only

have to have that solder blob melt his or her $150 lenses once[1] before the wisdom of this precaution sinks in.

■ *Let parts sit and cool before you handle them after soldering or desoldering.* Parts get really hot, over 400 degrees F, so let them sit and cool before you handle them and don't be in a hurry.

All these precautions are simple and in a short period of time will become second nature to you. After that, you'll never zap another part.

Some Electronics Basics

Resistors, capacitors, transistors, and diodes are all used to interface with embedded processors like the OOPic. You'll use them to turn on relays, run motors, power *light-emitting diodes* (LEDs), and initiate a host of other applications. I hope this is a gentle introduction to the art and science that is electronics for those of you who are new to the craft. For those of you who are more experienced, maybe there will be something here for you too. I'm going to stay away from anything more math intensive than Ohm's law and stick to the basics of what will be useful to the new embedded engineer. For more in-depth explanations, try Horowitz and Hill's *The Art of Electronics*, the bible for the craft.

UNITS OF MEASURE

The first thing you need to know are the basic units of measure. They are as follows:

$$\mathbf{V} \text{ or } \mathbf{E} = \text{Voltage, measured in Volts}$$
$$\mathbf{I} = \text{Current, measured in Amps}$$
$$\mathbf{R} = \text{Resistance, measured in Ohms}$$
$$\mathbf{F} = \text{Capacitance, measured in Farads}$$
$$\mathbf{P} = \text{Power, measured in Watts}$$

Rarely will any of these values be in their full units but volts. Most of the time they will be measured in some large or small quantity. The abbreviations for those very large and very small quantities are as follows:

m as in milli for *millivolts* (mv), *milliamps* (ma), or *milliwatts* (mw), which is $\times 10^{-3}$

u or **μ** as in micro for *microfarads* (μf), which is $\times 10^{-6}$

n as in nano for *nanofarads* (nf), which is $\times 10^{-9}$

p as in pico for *picofarads* (pf), which is $\times 10^{-12}$

K as in kilo for *kilo-ohms* (KΩ), which is $\times 10^{3}$

M as in mega for *mega-ohms* (MΩ), which is $\times 10^{6}$

---

1. But I'm not bitter.

Capacitors are usually shown looking like this, 100 uf, whereas resistors leave off any reference to the ohms and be shown like 10K.

RESISTORS

As the basic device for limiting current flow, the resistor is as ubiquitous to electronics as the axe is to a lumberjack. Here you'll be provided with the basic rules for resistor use with no explanations beyond what is required to do the math.

Ohm's Law and the Power Law

The resistor puzzle has two essential pieces: the value and how much power the resistor needs to dissipate. To decide the former, you use Ohm's law; to decide the latter, you use the power law. Here is Ohm's law:

$$V = IR \Leftrightarrow I = \frac{V}{R} \Leftrightarrow R = \frac{V}{I}$$

The above formulae are just three ways to say the same thing, I just did the algebra for you. You can use this formula to decide many design issues. If, for instance, you need to limit the current flow of a circuit (such as to your OOPic *input/ouput* [I/O] pins) to 20 milliamps (or 20×10^{-3} amps) and you know that the V maximum on the I/O pin is 5V, this is how you find that resistor value:

$$R = \frac{5V}{0.020} = 250\Omega$$

So some resistor larger than 250 ohms would be in order. Keep your units in terms of Volts, Ohms, and Amps, and the formula will find the voltage drops across a certain resistor at a certain current or the current flows through a resistor with a certain voltage across it.

The power law simply describes the power dissipated by a device with current flowing through it (meaning that a voltage drop exists across it, as mandated by Ohm's law). Here is the power law:

$$P = VI \Leftrightarrow I = \frac{P}{V} \Leftrightarrow V = \frac{P}{I}$$

Substituting in R, you also get

$$P = I^2R \Leftrightarrow P = \frac{V^2}{R}$$

Again, I did the algebra to show all the common forms for this law. This is most often used to find the power dissipated in a device like a *metal-oxide semiconductor field-effect transistor* (MOSFET), another transistor, or a resistor. For example, let's say you have a resistor that limits the current through a high-powered LED at 100 ma. You know (as you'll see later) that the LED uses 2V (called *drop 2V*) of the 5V available, which leaves 3V that will be dropped across the resistor (the sum of the voltages *must* equal the total voltage in a series circuit). Therefore, the resistor you choose must be able to dissipate the following power:

$$P = VI = P = 3.0 \times 0.1 = 0.3 \; Watts$$

You will need a 1/2-watt resistor for this application (resistors come in 1/10, 1/8, 1/4, 1/2, and 1 watt and then larger ratings).

The Voltage Divider

Sometimes you'll want to measure a voltage that is higher than the one allowed on the OOPic *analog-to-digital* (A2D) I/O line (5V). This is easy to do with a voltage divider. In a series circuit (one in which each component is connected to the next in line, end to end), the sum of the individual voltages dropped equals the total voltage. Thus, you put two resistors in a series and measure the voltage off of the lower leg of the voltage divider. Because the voltages are proportional, you can determine the higher voltage by what you read on the lower leg of the voltage divider. For example, Figure 8-1 shows a simple voltage divider off of a source (Vin) battery input that is connected to an OOPic A2D I/O line (Vout). It illustrates how you would measure the voltages that will be read at the various voltage levels of the battery as it discharges.

The voltage read is determined by the ratio of the upper and lower resistor:

$$Vout = Vin\left(\frac{R2}{R1 + R2}\right)$$

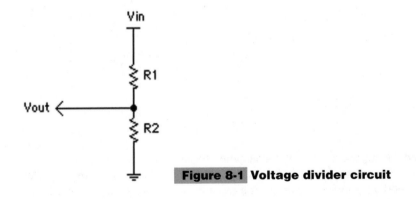

Figure 8-1 Voltage divider circuit

Therefore, to find what Vin is when Vout is read, you would use

$$Vin = \frac{Vout}{\left(\dfrac{R2}{R1 + R2}\right)}$$

Remember, this only works when $R1$ and $R2$ are fixed values.

Use large values of resistors, between 10K and 100K, for your voltage divider, so you don't waste the current from that battery you are keeping track of. You may want to keep resistor values under 100K if you can since some PIC A2D converters may not perform well with higher than 100K values.

Resistors in Series and Parallel

It is easy to find the value of a group of resistors in a series (lined end to end) and in parallel (lined side by side) using the following formulae referring to Figure 8-2:

Series resistance: $R_{Total} = R_1 + R_2$ and $R_{Total} = R_1 + R_2 + R_3$ and so on

Parallel resistance: $R_{Total} = \dfrac{1}{\dfrac{1}{R_1} + \dfrac{1}{R_2}}$ and $R_{Total} = \dfrac{1}{\dfrac{1}{R_1} + \dfrac{1}{R_2} + \dfrac{1}{R_3}}$ and so on

Just keep extending the series to R4, R5, and so on, and the relationship holds true.

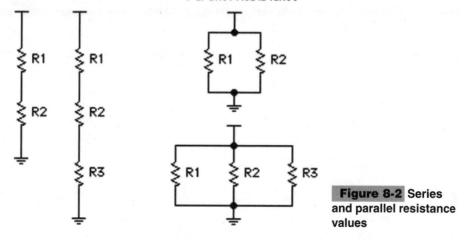

Figure 8-2 Series and parallel resistance values

What Value Is a Resistor?: The Resistor Color Code

As you've noticed, the manufacturer doesn't print number values on resistors. Two types of color bands are used, four- and five-color bands. Four-color bands are the standard resistors, and five-color bands are the precision resistors. In a four-band resistor, the first two bands are numbers, the third band is the multiplier, and the fourth band is the tolerance. In a five-band resistor, the first three bands are number bands, the fourth band is the multiplier, and the fifth band is the tolerance.

Table 8-1 displays how you can read those color bands to get the value of the resistor and the tolerance, or error factor, in the value. The first band to read is the one that is closest to the end of the resistor.

TABLE 8-1   RESISTOR COLOR CODE

COLOR	NUMBER BANDS VALUE	MULTIPLIER BAND VALUE	TOLERANCE BAND VALUE
Black	0	1	
Brown	1	10	1%
Red	2	100	
Orange	3	1000	
Yellow	4	10,000	
Green	5	100,000	
Blue	6	1,000,000	
Violet	7	10,000,000	
Gray	8	100,000,000	
White	9	1,000,000,000	
Silver		0.01	10%
Gold		0.1	5%

Let's use an example. If you have a resistor that is yellow-purple-red-gold, this is how it is decoded:

- **First band**   Yellow = 4, the first value digit
- **Second band**   Purple = 7, the second value digit
- **Third band**   Red = multiplier of 100

So, $47 \times 100 = 4700$, or 4.7 kilo-ohms. The fourth band is gold, which is 5 percent tolerance, meaning the value may be 4700 plus or minus 5 percent.

A variety of colorful pneumonics can be used to help remember these color-number relationships, but none of them are etymologically politically correct for a variety of reasons.[2] After a while, they become so ingrained you won't need the memory aids anyway.

CAPACITORS

A capacitor (*cap* for short) is a device that resists the change of voltage in a circuit. It doesn't limit anything; it actively resists it. Capacitors are used to dampen or remove the spikes in a circuit caused by high-speed devices switching state (such as your OOPic) or noisy mechanical and electrical devices switching state (such as motors and relays). Large capacitors filter slow signal changes like the 60-cycle U.S. power line AC ripple (or 50-cycle everywhere else). Smaller capacitors are used to filter digital and DC motor noise. When used in combination with resistors, they can form a circuit called an integrator or *Resistor-Capacitor* (RC) circuit, which will be discussed next.

Capacitors in a Series and Parallel

Placing capacitors in a series gives you values opposite of what you would get with resistors in a series, which is the same with capacitors in parallel. The formulae for finding these values are as follows, in reference to Figure 8-3:

Parallel capacitance: $C_{Total} = C_1 + C_2$ and $C_{Total} = C_1 + C_2 + C_3$ and so on

Series capacitance: $C_{Total} = \dfrac{1}{\dfrac{1}{C_1} + \dfrac{1}{C_2}}$ and $C_{Total} = \dfrac{1}{\dfrac{1}{C_1} + \dfrac{1}{C_2} + \dfrac{1}{C_3}}$ and so on

Again, just keep extending the series, C4, C5, and so on, and the relationship holds true.

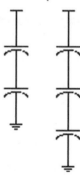
Series Capacitance

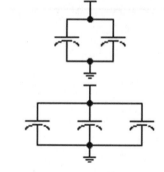
Parallel Capacitance

Figure 8-3 Capacitors in a series and parallel

---

2. I know of three myself. If you want to know one or more of them, send me an email and I might be convinced to tell you if you send me a six pack of the beer of my choice.

What Value Is a Capacitor? Well, That Depends

Unlike the resistor color code, the industry never got its act together to come up with a consistent coding for caps. The resulting mishmash means that if one of two things weren't done, you won't know what went wrong without buying the cap part specifically or measuring it.

One way, conveniently, is if the manufacturer printed the value on the cap. This is common with electrolytic (read bigger) capacitors and some tantalum caps. If you have a small ceramic capacitor, the value may be encoded on it as shown in Figure 8-4.

Look weird? Well, there is a reasoning behind it. Capacitors don't come with tolerances better than 5 percent, or more commonly 10 to 20 percent, and some are no better than 80 percent. For that reason, the tolerances aren't printed on them. You can read these values rather simply, if obtusely, as follows: Take the first two digits, and add the number of zeros after them indicated by the third digit. This is the value of the cap in *picofarads* (pf), which is $\times\ 10^{-12}$. To get the value in microfarads (which you are most comfortable with), simply move the decimal point six places to the left. Ick, but that is what we have. Figure 8.4 shows examples of two capacitors and their "decoded" capacitance values.

The RC Time Constant, or Tau

When you use a resistor and a capacitor so that the capacitor charges or discharges through the resistor, you are creating an RC circuit, which has some very unique characteristics. This type of circuit is often called an integrator (when the voltage is read across the capacitor) because it charges up to and holds a certain voltage dependent upon the frequency of the signal driving it and the RC time constant of the two parts. The circuit is called a *differentiator* if the voltage is read across the resistor. T or tau (another Greek letter) refers to the RC value of the circuit, which is R $\times$ C, and *that* has some special meaning. T is the time it takes the circuit to charge up to, or discharge down to, approximately 63 percent of the difference between the voltage that it started at and the maximum (or minimum) voltage it can reach. As you can see, it mathematically never reaches those extremes, but for our purposes, it does. Figure 8-5 shows the charge curve of an RC circuit; it is a nonlinear and

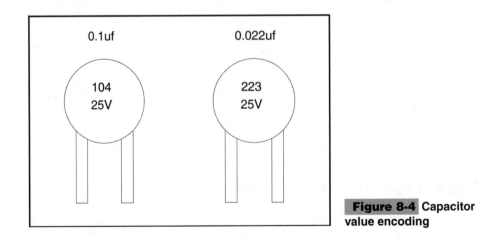

Figure 8-4 Capacitor value encoding

asymptotic (meaning it'll never get there in fancy math-speak) curve. A full charge or discharge is 5T, or 5 RC time periods.

Notations for 1T and 2T denote one RC time period, two RC time periods, and so on. The
time periods in Figure 8-5 represent 63, 87, 95, 98, and 99+ percent of the charge. The
actual time depends on the value of the R and C used. This book won't detail the use of this
type of circuit, , but it will give you a place to start when researching how to use it.

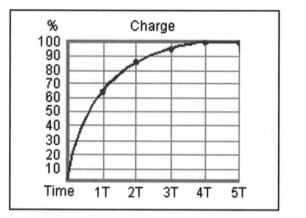

Figure 8-5 RC time
constant charge curve

SEMICONDUCTORS: DIODES, TRANSISTORS, MOSFETS, AND LEDS

The rest of this book could be filled with material on these devices, but that's not what you
want to know right away, so I'll give you the "what fors" on these instead. What follows is
the basic principles of their use. Figure 8-6 shows the schematic symbols for each of them.

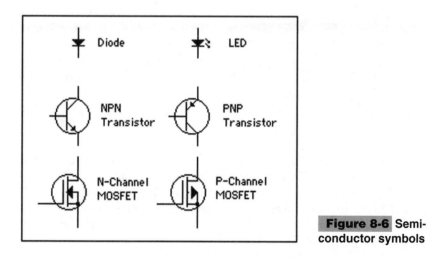

Figure 8-6 Semi-
conductor symbols

Diode

The diode enables current to flow in only one direction. I subscribe to the *electron current flow* position, which means that current flows from negative to positive. In other words, current flows against the arrow shown in the diode. The arrow points to the *cathode,* and the other side of the diode is the *anode.* Diodes are rated by forward current (I_F) and reverse-bias voltage (V_R), which is the maximum voltage they will tolerate blocking. Most diodes have a voltage drop of about 0.6V. Schottky diodes are much less, generally around 0.25V drop, which makes them better when you need as much voltage as possible. A standard diode would be a 1N4002; a Schottky diode might be a 1N5817.

LED

This is the light giver. LEDs come in multiple shapes, sizes, colors, and current limitations, so pick your favorite. An LED will drop between 1.7V and 2.3V across the resistor. I usually just pick the average of 2.0V to determine my correct resistor value for the current I want. Most of the time a 1K resistor does just fine as a current limiter. As already mentioned, the cathode is the side the arrow points to, and it's the flat side of the LED lens and the short lead. The anode is the other side of the LED. Current flows against the arrow like the diode.

Transistors

Transistors come in a variety of shapes, sizes, current limits, gains, and styles. They are primarily used as on/off switches. A transistor enables an OOPic with a low current-supplying capability to turn on and off a high-current device. You must place a resistor on the base of a transistor to the OOPic I/O pin to limit its current. You may be placing one on the collector or emitter of the transistor to limit the current of the device being switched.

An NPN transistor is normally used when you want a logical 1 to turn on a device, whereas a PNP transistor is normally used when you want a logical 0 to turn on a device. A PNP transistor has its emittter towards to the V+, or Vcc side of the power and the NPN transistor has its emitter towards the ground side of the power connections. Current flows against the arrow as in a diode. The *emitter* is the arrow lead of the transistor, the *collector* is the other side, and the *base* is the lead coming out of the side of the transistor. Where those pins are located on the transistor depends on the case style you are using. Check your data sheets for that pinout and for the current limitations of your transistor. In general, a TO92 transistor like the 2N3904, 2N3906, or 2N2222 is limited to 100 milliamps to 150 milliamps. A TO220 like a TIP120 may be 1 amp or more.

MOSFETs

The MOSFET is a special device. Unlike a transistor, voltage turns it on, not current. You don't really need to use a current-limiting resistor from the gate to the OOPic I/O line to limit the MOSFET current, but you should use one to limit the *inrush* current to the capacitor that is implicit in the gate. Any value over 330 ohms is sufficient as a rule. MOSFETs have very low resistance (R_{DS}on) between their source and drain, and as such they are the first choice to turn motors on and off.

The most important issue with a MOSFET is that in order to turn on, it needs 10V to 12V between the gate and source unless you have a logic level MOSFET, which isn't easy to find or inexpensive. This means you usually can't use an OOPic I/O line directly to turn on a MOSFET. The *source* of a MOSFET is the pin that the drawing connects the arrow to, the *drain* is at the other side, and the *gate* is the pin that comes out the side of Figure 8-6. These devices need to dissipate very little power, which is why they are used for motor control. A standard N-channel MOSFET is the IRFZ48, a logic-level MOSFET is an IRL3502, and a standard P-channel MOSFET is an IRF4905. The logic-level MOSFETs cost two or three times what a standard MOSFET would cost. A P-channel MOSFET is also more expensive.

OOPic I/O Line Protection

The OOPic is based on a PICMicro microcontroller, and which kind you use depends on the model that you have. These microcontrollers have a maximum current of 20 milliamps either sourced (pulled up to *Logic V+ power* [Vcc]) or sunk (pulled down to ground) before damage occurs to the chip. Because we all make mistakes, it is prudent to protect those inputs with a buffer resistor that limits the current, should you accidentally short an output to Vcc or ground. As calculated before, any value of 250 ohms or greater is sufficient to protect an OOPic output line from destruction. Figure 8-7 shows how to install one.

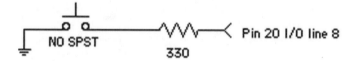

NO SPST   330   Pin 20 I/O line 8

Figure 8-7 OOPic output line protection resistor

Motor and Relay CEMF (Backlash) Protection

An OOPic can't drive a motor or relay directly. Driving these high-current devices requires the use of an external transistor. That transistor needs protection from the *counter-electromotive force* (CEMF, the official name for *flyback* or *kickback* voltage as it is sometimes known) generated by the inductors in relays, solenoids, and DC motors. Figure 8-8 shows how to protect a transistor from the kickback of a relay coil, and Figure 8-9 shows the positioning of flyback diodes in a DC motor H-bridge circuit. In both cases, these diodes either block or shunt away voltage that would otherwise damage or destroy the driver transistors. They are not optional; these diodes are needed to avoid damage.

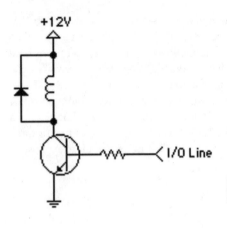

+12V

I/O Line

Figure 8-8 Protecting a transistor from CEMF

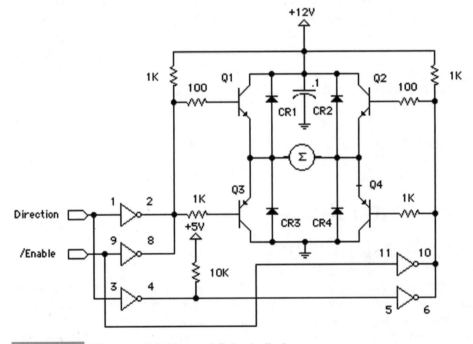

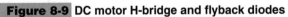

Figure 8-9 DC motor H-bridge and flyback diodes

Note carefully the direction of the diode in Figure 8-8. It's there to block the negative voltage that is generated by the coil of the relay. This same circuit is used with unipolar stepper motors. You'll never see an H-bridge without these diodes. This circuit is also used with bipolar stepper motors.

Supplying Power

This is a topic that comes up a lot, so I felt it deserved some coverage. First, let's look at some basics about power supplies and electronics:

- A battery supplies power, which is current and voltage. It does *not* force a circuit to accept current. A circuit requires current, which will be requested from a battery. If the battery can't supply that current, its voltage drops until it can supply the required current. Voltage and current are tied at the wrists and ankles of a battery; they are related, but not the same.
- Polarize your power connectors so that you can't plug them in backwards. It sounds simple, but very few people do this, and when they plug a battery in backwards, the voltage regulator and perhaps a capacitor or two go snap, crackle, and pop.
- When building connectors and cables, especially the three-pin variety, put the V+ line in the middle. This minimizes the chance that plugging a connector in backwards will zap a board by putting V+ on a device's output line.
- Some batteries can supply more current than others. Here is the order, from lowest current to highest current, that common batteries can supply:

  1. **Alkaline cells**   Low current (100 to 300 milliamps), average duration
  2. **Lithium ion**   A little better current (50 to 350 milliamps), very long duration
  3. **Nickel Metal Hydryde (NiMh)**   Good current (2 to 10 amps), long duration
  4. **Nickel cadmium (NiCd)**   Great current (5 to 20+ amps), fairly long duration
  5. **Lead acid**   Excellent current (30 to 100 amps), very long duration

The duration, or power density, of battery cells depends on the cell size and style. I've given some averages, and on any given day, I'll be wrong because something new will have come out. For instance, a higher-current *Nickel-Metal Hydrid* (NiMh) cell is released every day, and it's the same with lithium ion cells.

SEPARATE SUPPLIES AND COMMON GROUNDS

For every OOPic variant but the OOPic R, you will probably use two power supplies for your projects: one for the OOPic's logic board and one for the rest of your electronics. It's often a good idea to use two supplies when you are using DC motors, hobby servos, or Sharp *infrared* (IR) ranging modules, as all of these demand lots of current and can inject noise onto your power line, which computers like the OOPic don't like.

Figure 8-10 Upgraded voltage regulator orientation

No matter what your reasoning, remember this: Always connect all the grounds (negative supply lead) of all your battery packs together. Electronics require current flow from negative to positive and that creates the various logic levels. If the grounds are not connected together, the common reference point is lost and two isolated circuits won't understand each other's logic levels.

REPLACING THE OOPIC S BOARD'S VOLTAGE REGULATOR

The OOPic S boards all come with a small voltage regulator of the 7805L variety that can only supply 100 milliamps of current at 5V. This won't be enough if you are attaching Sharp IR units, SONAR boards, and other higher-current devices to your OOPic's +5V line. A simple solution is to upgrade the voltage regulator on the board to one that can supply more current. You have basically two choices: You can use a 7805 (or equivalent) regulator that requires at least 7V input voltage and delivers 1 amp of current, or you can use a *low dropout* (LDO) regulator that will work on as little as 5.6V input voltage such as the LM2940. LDO regulators are two or three times as expensive as a 7805, but you can run your OOPic on four AA cells if you use one. The choice is yours.

You will need to be proficient with your soldering iron and a solder sucker to make this upgrade to your OOPic board. I feel that solder braid is a poor joke foisted upon the electronics hobbyist community, so I recommend you go out and spend $4 and get a spring-loaded vacuum solder sucker if you're going to play with electronics. I also recommend a soldering iron of the 35- to 50-watt range with a grounded end and a small spade soldering tip.

To upgrade, carefully remove the small, black, three-lead regulator located directly above your power connector. This is called a TO-92 package (in case you're interested). Replace it with the higher-capacity regulator, which is called a TO220 package (for yet more trivia). The orientation of the new regulator is shown in Figure 8-10. Don't put it in backwards or it will heat up and not work very well. That's it; you're ready to go.

Reading Schematics

Creating schematics that are legible is an art. Some of us have it; some of us don't. Hopefully, mine are easy to read. I've certainly agonized over their organization to make them so.

The key to remember is that in both European schematics (where resistors look like boxes) and in American ones (where resistors look like squiggle lines), when lines cross at a dot, a connection exists; where lines cross without a dot, no connection exists. That's the way you're supposed to do it. Sometimes you'll see many lines connect to a larger, heavier

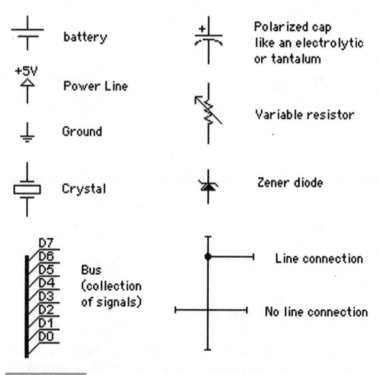

Figure 8-11 Schematic figures and their meanings

line; that indicated a bus. Individual lines are not differentiated, but the *name* of each line is the same where it enters the bus and exits the bus. This places the burden of discovery on the reader, but it takes up less space and doesn't look so confusing and cluttered. Figure 8-11 shows various things you'll see on a schematic and what they are (in addition to those symbols you've seen already in Figures 8-2, 8-3, and 8-6).

It can be confusing sometimes to tell which signal line is V+ and which is ground on some ICs. A V+ pin is typically labeled as either Vcc or Vdd. A ground pin is labeled as either G, Gnd, or Vss. Sometimes the V+ and Gnd pins aren't labeled at all, as is often the case when logic gates are being used. In those cases, it is assumed that the ground is pin 7 on a 14-pin chip, pin 8 on a 16-pin chip, pin 10 on a 20-pin chip, and so on. V+ is also considered the highest pin number of that chip. All *dual in-line package* (DIP) chip pins are numbered from 1 at the lower-left corner of the chip to the highest pin number moving counterclockwise around the perimeter of the chip. Surface mount chips have a variety of numbering schemes, but it's assumed you won't be installing any of those. Figure 8-12 illustrates the pin numbering pattern for DIP devices.

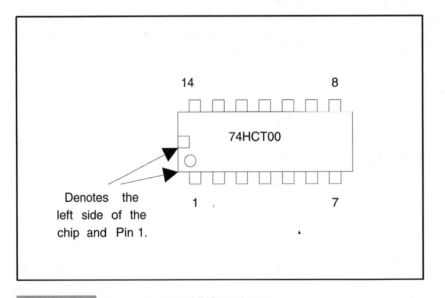

Figure 8-12 Pin numbering of DIP packages

Seven-Segment Displays

Sometimes you just want to display numbers. An LCD is overkill and reading binary from a bank of LEDs is too confusing. For this job the seven-segment display is the hands down winner.

OOPIC DIRECT CONTROL

The OOPic has an object called oConverter, that will translate a number into the proper outputs to create numbers on a seven-segment display. It is very convenient.

oConverter(C) (All OOPic Firmware Versions, oConverterC B.X.X, and Later)

The oConverter object has a variety of uses, and seven-segment display decoding is one of them. The oConverterC variant is the same with the addition of a clocked capability. The oConverter memory size is 3 bytes (oConverterC is 5 bytes) and its properties are displayed in Table 8-2.

The input or output Value property that is linked may be a Nibble or a Word property; the OOPic truncates the result or extends the result to fit the target numbers. Mode 3 is especially useful linked to an oDio4 as an address decode bit. Naturally, the Mode 1 setting links to an oDio4 to handle stepper motor drivers. In the case of the

TABLE 8-2 OCONVERTER(C) PROPERTIES

Blank (bit, flag)	If this bit is set to 1, the output is cleared to 0s instead of being updated. 0 has normal operation.
ClockIn (flag pointer)	Link to the flag that will clock this object (oConverterC only).
Input (number pointer)	Link to the Value property that will be used as the input to the conversion chosen.
InvertC (bit, flag)	If this bit is set to 1, the clock sense is inverted, which means that a 0 will clock the object, not a 1 (oConverterC only).
InvertOut (bit, flag)	If this bit is set to 1, the bits of the converted value are inverted before written to the Output object.
Mode (nibble)	0 (cv7Seg) is binary to or from the seven-segment display.
	1 (cvPhase) is binary to the eight-phase stepper motor.
	2 (cvSin) is binary to Sin. This can also be done in program code with the sin function.
	3 (cvDecimal) is binary to or from Decimal. Same as 74150 IC.
	I.E. Binary:011 is Decimal:00000100; Binary:111 is Decimal:10000000.
Operate (bit, flag)	Must be set to 1 for this object to operate (default property).
Output (number pointer)	Link to the Value property of the target object that will receive the result of the conversion.

seven-segment display, you'll need an oDIO8 to handle the seven bits of data that turn on the LEDs in the display.

Hardware Layout

A variety of seven-segment LED displays exist; some are common cathode (where the ground side of all segments is common) and some are common anode (where all the V+ sides of the segments are common). In order to keep current below the maximum that all I/O lines of the PICMicro can handle combined, it is recommended that you use the largest value resistor you can that will enable you to see the display comfortably. You can also use another driver chip to power the display, and the OOPic will use that device as a driver. One such driver chip is the Allegro UCN2003, which will drive common anode displays. Figure 8-13 shows my hardware setup schematic to drive a seven-segment display; mine turned out to be a common cathode device.

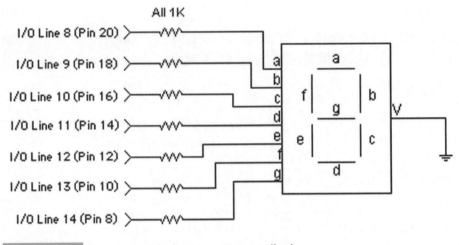

Figure 8-13 Direct control of seven-segment display

Writing the Code for a Seven-Segment Display

the following code demonstrates how to use an OOPic to drive a seven-segment display.

USING TRANSLATOR CHIPS

Directly driving a seven-segment display wastes I/O lines, but fortunately chips out there can handle the decoding. All you have to do is feed them the number you want them to display. The 7447 or 74LS47 drives common anode, seven-segment displays nicely using a single oDio4 port. This is a *binary coded decimal* (BCD) display driver, which defines only the numbers 0 through 9 as valid inputs. Other drivers are available, but they are more expensive.

```
'7segment.osc
'This shows using the oConverter object to drive a seven-segment
'display.
'
'Each display seems to have its own pinout; this one is a
'common cathode one, which means that the ground is common
'and each line must be pulled high to turn on a segment.
'
'For OOPic I A.2.X firmware
'
'Copyright Dennis Clark 2003
'Permission is granted to use this code anyway you like
'as long as you say you got it from me.

Dim seg7 as New oConverter
Dim holder as New oByte
Dim disp as New oDio8

Sub Main()

    disp.IOGroup = 1              'I/O Lines 8-15
    disp.Direction = cvOutput

    'The segments are decoded logically.  I/O Line 8 is
    'segment a,... I/O Line 14 is segment g.  If you are
    'using a common anode display, then set the property
    'InvertOut = 1.  This will enable the correct segment
    'by setting that segment to 0 instead of 1.

    seg7.Mode = cv7Seg            'Convert binary to seven-segment
    seg7.Output.Link(disp)        'Connect to I/O lines
    seg7.Input.Link(holder)       'Where we hold the value
    seg7.Operate = cvTrue

    Do
        For holder = 0 to 15      'oConverter knows hexadecimal!
            OOPic.delay = 100
        Next Holder
    Loop

End Sub
```

Listing 10-1

Stepper Motors

Everyone wants to use stepper motors, they are precise and predictable. They are also complex to control. The OOPic has an object that can handle all the common stepper stepping phase modes easily.

OOPIC DIRECT CONTROL

The OOPic can directly control stepper motors; all it needs are driver chips, because the PICMicro does not have the current capability to drive stepper windings, which take a lot of power and usually high voltages as well. Here you will see some objects that will do the grunt work of outputting the correct winding phases so that steppers will march.

oConverter (Again)

This object is quite versatile. Not only can it encode seven-segment LEDs, but it can also encode for stepper motors. Here you are setting Mode equal to 1, or cvPhase, which provides all your stepping patterns for stepper motors. If you use all eight phases (from 0 to 7), you get the Wave step pattern, which alternates two-phase and half-step modes. Phases 1, 3, 5, and 7 result in two-phase stepping; phases 0, 2, 4, and 6 result in half-step mode. Listing 10-2 shows you how to set up and use this nifty object.

oStepper (B.X.X Firmware Version and Later)

 The B.X.X firmware versions added high-level objects for many devices that included stepper motors (good for us). Using this high-level object saves on object memory because it is one object. All the stepper patterns are also worked out for you. Listing 10-3 shows you how simple this object is to use. The oStepper memory size is 8 bytes and its properties are shown in Table 8-3.

TABLE 8-3 OSTEPPER PROPERTIES	
Direction (bit, flag)	0 is forward; 1 is backward. Of course, these are kind of arbitrary.
Free (bit, flag)	0 is engaged. When the motor isn't turning, the windings are left energized so the motor holds its position.
	1 is off. The windings are not energized and the motor will turn freely.
InvertOut(bit, flag)	0 means logic 1 turns a coil on.
	1 means logic 0 turns a coil on.
IOGroup (nibble)	0 is I/O lines 1 through 7.
	1 is I/O lines 8 through 15.
	2 is I/O lines 16 through 23.
	3 is I/O lines 24 through 31.
Mode (bit)	0 means turn the number of steps specified.
	1 means turn continuously.
Nibble (bit)	0 means use the lower 4 bits of the selected I/O group.
	1 means use the upper 4 bits of the selected I/O group.
	(continued)

NonZero (bit, flag)	Is 0 when the Value property is 0 (finished stepping). Operate (bit, flag) Must be set to 1 for this object to function.
Phasing (nibble)	0 means one of the four coils is active while the other three are inactive (Wave).
	1 means two of the four coils are active while the other two are inactive (two phase).
	2 alternates between 1 and 2 active coils (half-step).
	3 means three-phase stepper motor phasing.
Rate (byte)	The step frequency is 1132.246Hz/(128 − rate).
Unsigned (bit)	0 means signed Value numbers are allowed.
	1 means only unsigned Value numbers are allowed.
Value (word)	Sets the number of steps to take, either forward (positive number) or backward (negative number). Value also reflects the number of steps left to take from the last setting (default property).

This object basically has two active modes; either the stepper is always turning and all you can do is adjust the speed, or the servo is always set to move a number of steps in a configured direction. Value holds the step number. If in Mode 0, it reflects the number of steps left to take from the last setting; if in Mode 1, it reflects the number of steps taken since last sent to 0. If the Unsigned property is 0, this number may be negative or positive.

oStepperSP (B.2.X Firmware Version and Later)

 The oStepperSP object adds a *Uniform Robotic Control Protocol* (URCP) capability to stepper motors, whereas essentially all other aspects of the oStepper and oStepperSP are the same. The mode settings of this object enable us to link to the Value property to modify either the number of steps to take (Mode is 0) or the speed and the direction of stepper motor (Mode is 1) by using a VC. This was not possible before this object. Cool, huh? The oStepperSP memory size is 10 bytes and Table 8-4 outlines its properties.

The oStepperSP object is nearly the same as the oStepper object, except that URCP values can be used to set the speed of the stepper motor similarly to a DCMotor object when the stepper is being used in continuous turn mode.

TABLE 8-4   OSTEPPERSP PROPERTIES

Direction (bit, flag)	0 is forward; 1 is backward. Of course, these are kind of arbitrary.
Free (bit, flag)	0 means engaged; when the motor isn't turning, the windings are left energized so the motor holds its position.
	1 means off; the windings are not energized and the motor turns freely.
Invertout (bit)	0 means logic 1 turns a coil on.
	1 means logic 0 turns a coil on.
IOGroup (nibble)	0 is I/O lines 1 through 7.
	1 is I/O lines 8 through 15.
	2 is I/O lines 16 through 23.
	3 is I/O lines 24 through 31.
Mode (bit)	0 means turn the number of steps specified.
	1 means turn continuously.
Nibble (bit)	0 means use the lower 4 bits of the selected I/O group.
	1 means use the upper 4 bits of the selected I/O group.
NonZero (bit, flag)	Is 0 when the Value property is 0 (finished stepping).
Operate (bit, flag)	Must be set to 1 for this object to function.
Phasing (nibble)	0 means one of the four coils is active while the other three are inactive (Wave).
	1 means two of the four coils are active while the other two are inactive (two phase).
	2 alternates between one and two active coils (half-step).
	3 means three-phase stepper motor phasing.
Rate (byte)	The step frequency is $1132.246\text{Hz}/(128 - \text{rate})$.
Steps (word)	Mode 0 sets the number of steps to take in the direction specified at the rate specified.
	Mode 1 keeps track of the number of steps taken.
Value (byte)	-128 to 127, URCP speed setting (default property).

Hardware Layout

Figure 8-14 shows a unipolar stepper driver setup and a bipolar stepper driver setup. The unipolar circuit drives up to 500 milliamps per winding, and the bipolar drives up to 1 amp per winding. Both of these chips are inexpensive and easy to come by, both new and in surplus shops.

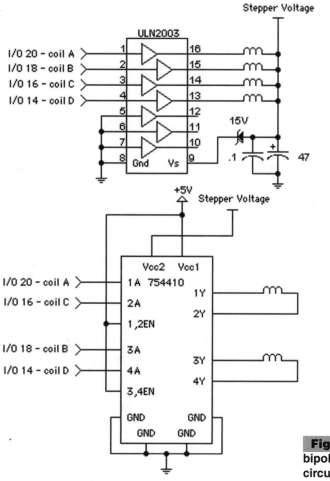

Figure 8-14 Unipolar and bipolar stepper driver circuits

Writing Stepper Motor Code for the OOPic I

The following code in Listing 10-2 illustrates how to use the OOPic I to control a stepper motor.

```
'twostep.osc
'Demonstrate controlling a stepper motor using the oConverter
'processing object.  Shows that all stepper patterns are
'available even though there is only one Mode for steppers.
'
'For OOPic I A.2.X firmware
'
'Copyright Dennis Clark 2003
'Permission is granted to use this code anyway you like
'as long as you say you got it from me.

Dim Stepper as New oDio4
Dim Pattern as New oConverter
Dim Wave as New oNibble
Dim s as Byte
Dim n as Byte

'All variables and objects in the OOPic are global

Sub Main()

        setup                   'Setup the oConverter object

        stepUp2P                'Step Up using 2-phase pattern
        stepDown2P              'Step down using 2-phase pattern

        stepUp                  'Step up using half-step pattern
        stepDown                'Step down using half-step pattern

        'I could show you how to do the wave pattern, but
        'it should be obvious how to do it from these two
        'examples.  HINT: The half-step pattern is alternating
        'wave and 2-phase.  Get it?

End Sub

Sub setup()
'Set up the VC
        Stepper.IOGroup = 1
        Stepper.Nibble = 0                      'Use I/O 8-11 (pin 14-20)
        Stepper.Direction = cvOutput
        Stepper.clear                           'clear bits

        Pattern.Mode = cvPhase                  'stepper pattern
        Pattern.Output.Link(Stepper)            'point to output bits
        Pattern.Input.Link(Wave)                'Var we're counting on
        Pattern.Operate = cvTrue                'turn it on
```

```
End Sub

Sub stepUp2P()
'This uses the two-phase pattern
'Steps the stepper forward
'The default pattern for oConverter is half-step.  We want to use
'the faster and stronger 2-phase pattern.  To do that, we skip
'every other number in the pattern.

    for n = 1 to 10
        for s = 1 to 7 step 2
            Wave.Value = s              'Increment step count
        next s
    next n
End Sub

Sub stepDown2P()
'This uses the two-phase pattern
'Steps the stepper backward
'The default pattern for oConverter is half-step.  We want to use
'the faster and stronger 2-phase pattern.  To do that, we skip
'every other number in the pattern.

    for n = 1 to 10
        for s = 1 to 7 step 2
            Wave.Value = 8-s            'decrement step count
        next s
    next n
End Sub

Sub stepUp()
'This implements the half-step stepper pattern.
'Steps the stepper forward
    for n = 1 to 10
        for s = 1 to 7
            Wave.Value = s              'Increment step count
        next s
    next n
End Sub

Sub stepDown()
'This implements the half-step stepper pattern
'Steps the stepper backward
    for n = 1 to 10
        for s = 7 to 1 step -1
            Wave.Value = s              'decrement step count
        next s
    next n
End Sub
```

Listing 10-2

Writing Stepper Motor Code for the OOPic II

The OOPic II and later OOPics have objects that greatly simplify writing stepper motor code. Listing 10-3 is a simple example of these objects.

```
'B2Stepper.osc (C language syntax)
'Demonstrates the use of the oStepper object to control a
'stepper motor.  The speed is adjustable via a pot on A2D 1.
'
' Stepper controller for either unipolar or bipolar stepper
' motor - It just depends how you connect the dots...
' From bits 0 to 3, the phases are A-B-C-D, simple.
' For bipolar steppers, coil1 is 1 & 3, coil2 is 0 & 2.
' Put a pot on A2D line 1 (pin 7 of connector) and make
' sure that you can't short to ground or +5 and you can
' use this pot to adjust the stepping speed.  I used a 100K
' pot with 330 ohm resistors to ground and +5V at each end
' and the wiper to A2D port.  Steps <value> steps in one
' direction; then steps <value> steps in the other direction.
'
'For OOPic II B.2.X firmware
'
'Copyright Dennis Clark 2003
'Permission is granted to use this code anyway you like
'as long as you say you got it from me.

Dim stepA As New oStepper
Dim rate As New oA2D

sub main()

    OOPic.Node = 1                      'For debugging

    'Stepper ouputs
    'Bit 0 = I/O 8 = Pin 20 = Phase A (bipolar winding A1)
    'Bit 1 = I/O 9 = Pin 18 = Phase B (bipolar winding B1)
    'Bit 2 = I/O 10= Pin 16 = Phase C (bipolar winding A2)
    'Bit 3 = I/O 11= Pin 14 = Phase D (bipolar winding B2)

    stepA.IOgroup = 1                   'I/O lines 8-15
    stepA.nibble = 0                    'I/O lines 8-11 chosen
    stepA.Mode = 0                      'Turn number of tics
    stepA.Phasing = 1                   'Two phase pattern
    stepA.Rate = 122                    'about 188 Hz
    stepA.Operate = 1                   'enable the object
    stepA.Unsigned = 0                  'Allow signed numbers

    'A2D input
    'A2D 1 = I/O 1 = Pin
    rate.IOLine = 1                     'A2D1
    rate.Operate = 1                    'Turn it on

    do
        stepA.Value = 100              'turn this way a bit...
        while(stepA.NonZero = 1)       'Wait until done
            stepA.Rate = rate.Value/2  'set new speed
        Wend
```

```
        stepA.Value = -100              'Then go the other way

        While (stepA.NonZero = 1)
            stepA.Rate = rate.Value/2  'set new speed
        Wend

    Loop
End Sub
```

Listing 10-3

USING STEPPER CONTROLLER CHIPS

Many driver chips are available out there for unipolar and bipolar steppers, but there aren't a lot of "smart" stepper chips that handle the stepping pattern, direction and drive the motor all in one chip.

The UCN5804 is the most popular of the all-in-one stepper decoder and driver chips for unipolar stepper motors. It can handle up to 35V on the stepper windings and up to 1 amp of current draw per winding. All you need to provide as signals are the direction, stepper mode, and step pulses.

The MC3479 is a bipolar stepper chip that handles all the chores of stepping and driving a bipolar stepper motor. This chip can handle between 7.2 and 16.5V per winding and up to 350 milliamps of current per winding. It's not as powerful as the UCN5804, and if you need more power, you will have to step up to the TI754410 or L298 drivers and use one of the many stepper motor controller chips out there like the Ferrettronics FT609 or EDE labs EDE1204 controller chip. These companies are listed in the appendices. Of course, you could just use your OOPic, because it only takes 4 bits to control a stepper directly.

UNIPOLAR, BIPOLAR, OR UNIVERSAL STEPPER MOTORS

Three basic kinds of popular stepper motors are available: *unipolar*, *bipolar*, and *universal*. Here is a brief description of all three.

Unipolar Stepper Motor

This motor is called unipolar because each winding carries current in one direction and thus has a single polarity when it is energized. The unipolar stepper has four windings that are energized in turn to get the motor to spin in one direction or another. This stepper can have either five or six wires. If it has five, then one side of each winding is attached to a common wire; if it has six, then two windings have a common wire by pairs and this common winding may be connected together. Just about anything can drive a unipolar stepper winding, such as transistors, driver chips, or MOSFETs, because current only flows in one direction.

Bipolar Stepper Motor

The bipolar stepper is named as such because two windings are located in the stepper, and each winding may be energized in two polarities, hence the term bipolar. The bipolar stepper requires an H-bridge driver for each winding because the current must flow in both directions. Bipolar steppers tend to be more powerful than an equivalent-weight unipolar stepper because with fewer windings needed, the windings used can take up more space. They are also usually larger, meaning more current will flow, and in DC motors, current equals power. This type of stepper has four wires, two per winding.

Universal Stepper Motor

This is an oddball in the stepper motor definitions. The universal stepper has eight wires coming out of it, two for each winding. These windings can be configured to form one of three configurations:

- Connect one lead of each winding together (making sure that the polarities are all the same) and use it as a unipolar stepper.
- Connect two pairs of windings together to form a series-winding bipolar stepper, again making sure of winding polarities.
- Connect two pairs of windings together to form a parallel-winding bipolar stepper; heed the same warning about polarities.

The series-winding bipolar is a little stronger than the parallel-winding bipolar, but the parallel-winding bipolar is a little faster. Pick your tradeoff. In any case, these are not common stepper motors; if you do get some, you can choose the configuration you want based on the driver you want to construct.

Now What? Where to Go from Here

I hope this chapter has made you more confident with the concept of modifying and adding electronics to your OOPic microcontroller so you can extend its capabilities and use it to control other devices. A microcontroller isn't very useful without sensors for feedback and the capability to control other electronic or mechanical devices. You should be able to read most schematics you find with ease and be able to build the circuits they show with a little practice.

I strongly recommend you pick up a copy of Don Lancaster's *TTL Cookbook*.[3] This is a great book to help you with your beginning skills in electronics. Of course, I also recommend my book *Building Robot Drive Trains* as a comprehensive compendium of movement and feedback for robots and the robot builder.

Finally, because I couldn't bear to write a chapter without some code or objects to talk about, oConverter and oStepper objects have been introduced. Both are excellent examples of how a high-level object can make our programming jobs much easier and even fun.

---

3. Published by Howard W. Sams & Co., a division of Macmillan, Inc., 4300 West 62nd Street, Indianapolis, Indiana 46268.

9

OOPIC I2C AND DISTRIBUTED DDE PROGRAMMING (PROJECT: REMOTE CONTROL)

At the core of the OOPic's communications strategy is the Phillips® *Inter-IC* (I2C) serial communications protocol. I2C is a synchronous serial method that uses a bidirectional data line and a clock line, so it is frequently called a two-wire communications protocol. For more information on the Phillips I2C specification, see their web site at www.phillips-logic.com/i2c.

The I2C protocol is a complex specification, so it won't be explained in all of its gory details here; that would take a whole book alone. Instead I'm going to discuss how the OOPic implements the specification and the two OOPic I2C buses, the *local bus* and the *network bus,* and the OOPic objects that access devices on these two buses. Your project for this chapter is *Dynamic Data Exchange Link* (DDELink) related; you'll connect a variable resistor to an OOPic and have that OOPic communicate with another one over DDELink to change the actions of a servo motor configured on the remote OOPic.

I2C: What Is It and How Does It Work on the OOPic?

First, most of this section consists of background information on the I2C protocols. You really don't need to know it if you are using I2C devices that have example OOPic code. The OOPic handles all the complex interfacing between the microcontroller and I2C devices. The following information is useful for those who are breaking new ground with new devices that have not been documented with the OOPic before.

Some basic information will be useful to know when interfacing I2C devices to the OOPic. The I2C bus is known as a multimaster bus. This means more than one master can exist on the bus, but only one can be in control at any given time. If you're having trouble visualizing the bus concept, an example of a single master and multiple slaves is shown in Figure 9-1.

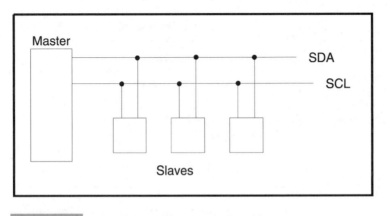

Figure 9-1 Master/slave I2C bus relationship

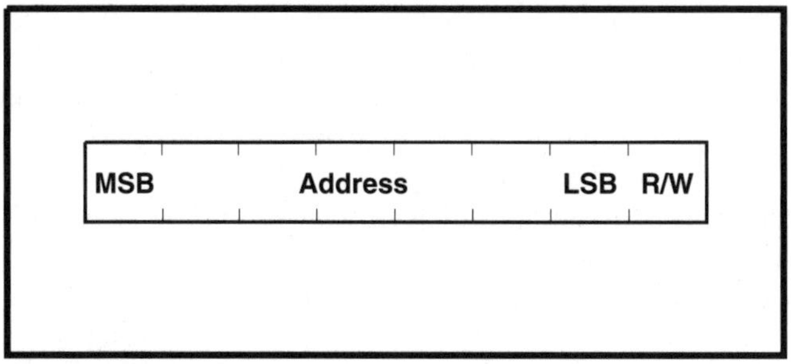

Figure 9-2 Phillips I2C 7-bit address specification

Several addressing schemes are used with I2C, but the only ones supported by the OOPic use the 7-bit addressing scheme. This is the default I2C addressing method, and it can access up to 127 devices, minus a few special addresses.[1] Figure 9-2 shows the addressing format for 7-bit I2C communications. It includes seven address bits, some of which are often hard-coded into the device and some are available as address pins on the IC, such as the A0-A2 pins on the 24LC64 *electrically erasable programmable read-only memory* (EEPROM) chip.

Bit 0 of the address byte is the R/W bit, which determines whether the I2C communication is a read (1) or a write (0). This address is simply the bus address of the device; it isn't the address of any particular location in the device. Subsequent bytes written to the device after the address byte has been written determine what happens next, which depends on the information the device requires.

Only the addressing scheme is specified in the I2C protocol, not the internal operation of the device. As long as the I2C device that you are using follows the Phillips specification, it works with the OOPic oI2C or oEEPROM objects. The I2C protocol is somewhat of a register-based one. It expects that an address be selected that is written to or read from and that you may either automatically increment to the next address or not by setting bits in certain device registers. These special registers are detailed in the documentation for your device. You'll need to read that documentation carefully to know what registers to access.

Some I2C devices out there do not follow the specification and use a custom addressing scheme; these will be problematic. Carefully read the manufacturer's product manuals before you buy one. If its addressing format doesn't look like Figure 9-2, don't get it; it won't work properly with the OOPic. Certain devices use a 10-bit addressing scheme, and the OOPic does not support the 10-bit addressing format.

Some devices, such as the Devantech® CMPS03 compass module, have hard-coded addresses, whereas other devices, such as the 24LC64 EEPROM, have their upper 4 bits hard coded and their lower 3 bits defined by pins that are either grounded or set high when the device is installed on the board. Check the documentation for your device carefully for its address.

1. I won't get into these. If you are curious, go to the Phillips site and search for "The I2C Bus Specification Version 2.1" for full details.

Specific rules have been established about how to write to and read from I2C devices. Fortunately for you, these rules are encapsulated in the oI2C, oEEPROM, and oDDELink objects. You only need to know which lines connect to which *input/output* (I/O) port on the OOPic to interface your I2C device. This is the great advantage of using a higher-level language like that in the OOPic , as opposed to programming your PICMicro in assembler. The gory details are hidden and you can concentrate on getting the job done, not building hardware and decoding communications protocols.

However, because you might be interested in the details, Figures 9-3 and 9-4 show the actual I2C communications format layout for 7-bit addressing. Checking these against a device you want to use will tell you if the device will work with the OOPic, because these are the formats the OOPic will use with the oI2C device to talk to it. This is the standard

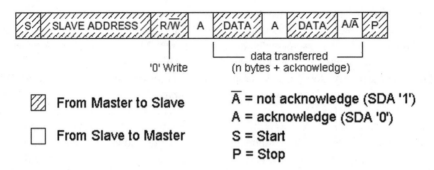

A Master Transmitter addressing a slave receiver with 7-bit address.

Figure 9-3 Master write and slave read 7-bit address sequence

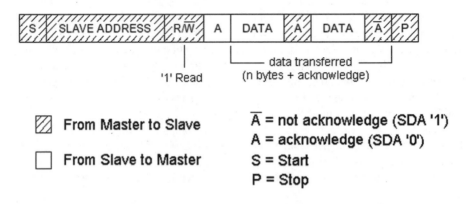

A Master reads a Slave immediately with 7-bit address.

Figure 9-4 Master read and slave write 7-bit address sequence

way of implementing a one-way single- or multiple-byte write to a 7-bit I2C device. If the OOPic is reading one or more bytes from an I2C device, this is the sequence of activity.

The I2C bus is a synchronous serial bus, so it is not as fast to write and read devices connected to it as it is to write and read the internal RAM of the OOPic processor. The great advantage is that it only takes two I/O lines to talk to dozens of coprocessors that can extend the capabilities of the OOPic controller.

An I2C Toolkit, with VB6 source code, can be downloaded from the OOPic software download page. It is also available on the CD-ROM that comes with this book. When working with the I2C tools, the OOPic programming cable is used as an I2C cable, so no additional hardware is required.

THE OOPIC LOCAL I2C BUS

The OOPic local I2C bus is used for programming the OOPic serial EEPROM, which is an I2C EEPROM, and the oI2C and oEEPROM objects also use this bus. The OOPic programming cable plugs into it to write your programs into the I2C EEPROM on the OOPic board. This bus can also be used to interface other I2C boards and chips to your OOPic for specific purposes like SONAR, temperature sensing, and a host of other possibilities listed in the appendices. You use the oI2C object to access these I2C coprocessors from the OOPic. Chapter 1, "OOPic Family Values," shows where to connect your special I2C devices to the various OOPic boards and access this I2C bus.

The OOPic S boards also have a second socket labeled *E1* where you can place your own I2C EEPROM to store and retrieve data. The Microchip® 24LCXX series of I2C EEP-ROMs and others that are fully compatible with them are supported in this socket. You cannot use a 24LC16 EEPROM in this socket, however, because it has no address set bits and it conflicts with the EEPROM loaded into the E0 socket.

To access the E1 socket's EEPROM and the OOPic default EEPROM, you must use the oEEPROM object. The OOPic *Integrated Development Environment* (IDE) stores your program in the default EEPROM from address 0 on up. If you are careful and know what is in your EEPROM, you can use the space at the "top" of the EEPROM's memory to store your own data you want to retain between power cycling your OOPic board. Use caution; if you don't do your math correctly, your program will clobber the OOPic program that is to be running!

Attaching I2C Devices to the Local I2C Bus

Although the I2C network is defined as a multiple master bus, the OOPic implementation is such that the OOPic is the master, always. Because of that, a pull-up resistor is located on the SDA (data) line and not on the SCL (clock) line. The OOPic is *always* driving the SCL line, except during programming, and at that time the programming cable pulls the OOPic reset line low so that it won't drive the SCL line during the programming cycle from the OOPic IDE.

For that reason, you don't need to use a pull-up on the SCL line on the OOPic Local I2C bus unless you have a cable on the bus that is longer than 6 inches (15 centimeters). If you need a resistor on the SCL line, place it near the device at the other end of the cable from the OOPic. Look on your OOPic board; if you have an OOPic I, chances are you only have

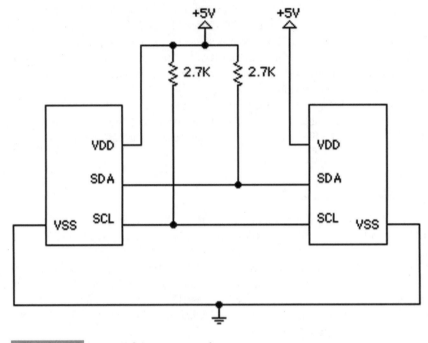

Figure 9-5 Local I2C bus connections

a 4.7K (yellow-purple-red-gold) pull-up on your SDA line. For best conditions, it has been determined that 2.7K (red-purple-red-gold) is the best pull-up resistor value. This resistor is located immediately to the left of the big 40-pin (PICMicro) chip.

If you need to upgrade your resistor, you have two options. Either remove the old ones and put in the new one (recommended only for experienced hands with soldering irons) or solder a 4.7K resistor on top of the one loaded in the board. Chapter 8, "Resistors in Series and Parallel," tells you why this is a viable solution.[2]

It is simple to add I2C devices to the OOPic local I2C bus. Figure 9-5 shows you how to do this. Pin numbers are not included on the generic I2C device because they will be different with every device and board you use. What is important is that you know which signals to connect where.

Note that the +5V from the OOPic is being used to run the additional I2C device. Only make this connection if you are using a *very* low current device or you have upgraded your OOPic voltage regulator to the more robust 1-amp version on the OOPic S board. The OOPic R has a 5V I/O power line you can use with no worries. The OOPic C cannot power anything else, so make sure you use your own 5V power line in that case. You'll probably be powering your OOPic C from that other 5V line anyway.

---

2. Go on; look back at chapter 8 where you passed over the section that had the math in it.

THE OOPIC NETWORK I2C BUS

The second bus is the OOPic network bus and is used by the oDDELink object to communicate to other OOPic microcontrollers. It is also used by the OOPic IDE parallel port debugger to get real-time object information. Chapter 1 explains which connectors are used for this bus on the various OOPic boards. The OOPic parallel port cable is plugged into this OOPic network bus when you are using the parallel port IDE debugger. All OOPic controllers in the network are daisy chained[3] using this network. Only other OOPic processors and the OOPic IDE debugger cable use the *network* I2C bus.

Connections to the network I2C bus are simplified by the fact that *only* other OOPics and the IDE parallel port debugging cable are supported on it. You need to use 2.7K pull-up resistors on the SCL and SDA lines for the network I2C bus; they aren't there on any OOPic board. These resistors are unnecessary only when networking to a PC, and the pull-ups are incorporated into the programming/debugging cable. Figure 9-6 shows the proper layout of the cabling and resistors for networking OOPics together.

It's simple to add these resistors to the network I2C bus. You can solder them right on the OOPic S board. On the OOPic R, they need to be installed in the cable if you are networking two OOPic R boards together, and of course, they are already installed if you are networking to an OOPic S board. The OOPic C also needs them added to the bus at some location. Figure 9-7 shows where the pull-up resistors can be installed on the OOPic S board.

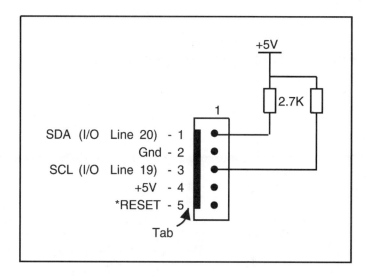

Figure 9-6 Networking OOPic requirements

3. *Daisy chained* means you can just keep adding new units to the end of the chain until you run out of address space to add any more.

Figure 9-7 Network I2C pull-up resistors

Make sure to attach the other side of the resistors to one of the +5V pads located on the OOPic board. Chapter 1 describes the part numbers you can use to make an OOPic networking cable. Don't connect the 5V line between two OOPic boards; all the other lines are just connected straight through.

Using I2C Devices on the Local I2C Bus

A wide variety of I2C devices and boards are available that will augment the capabilities of your OOPic controller. A simple Google search for I2C will find you enough hits to keep you awake every night for weeks. My experiments will show you how to interface a Devantech CMPS03 electronic compass board and how to store and retrieve data from an EEPROM loaded into the E1 socket on the OOPic S board. That should whet your appetite and provide you with enough information to interface any device that implements the standard I2C interface specification.

You've already seen how to connect the hardware to the OOPic local I2C bus. The OOPic objects that will talk to that hardware are oI2C and oEEPROM.

OI2C (ALL OOPIC FIRMWARE VERSIONS)

 The oI2C object has been with the OOPic since the first release. Because the communications bus of choice for the OOPic is the I2C, it makes sense that an I2C object should carry out communications with other I2C devices.

The OOPic oI2C object can communicate with only 7-bit addressable I2C devices. The *Mode* property defines how the oI2C device talks to the internal address registers of the I2C devices, not the chip-addressing format. The oI2C memory size is 5 bytes and its properties are outlined in Table 9-1.

This gets complicated if you don't have example code to work from, so I'll supply some examples. Also, some explanation will make this much clearer to those who want to be able to access devices no one has used before on the OOPic.

TABLE 9-1 OI2C PROPERTIES

Location (byte/word)	In Mode 0, this is a 16-bit address usually used in EEPROMs that follow the I2C address byte.
	In Mode 1, this is the 8-bit register address usually used in devices that are register based.
	Not used in Mode 2.
Mode (nibble)	0 (cv23Bit) means the I2C uses Location as the two address bytes that follow the device address byte.
	1 (cv10Bit) means I2C uses Location as the single address byte that follows the device address byte, usually to select a register.
	2 (cv7Bit) means the I2C uses the 7-bit I2C address mode for simple 1-byte writes or reads, usually for device commands.
Node (byte)	The I2C address of the device (refer to Figure 9-2).
NoInc (bit, flag)	0 (cvFalse) means the Location property is incremented each time the Value property is read or written.
	1 (cvTrue) means the Location property is not incremented each time the Value property is read or written.
	This is only used with Mode 0 or 1.
String (string)	Although listed, this property is not supported in oI2C.
Value (byte/word)	Width at 0 means bytes are read from or written to device.
	Width at 1 means words are read from or written to device.
Width (bit)	0 (cv8Bit) means read and write 8 bits at a time (1 byte).
	1 (cv16Bit) means read and write 16 bits at a time (2 bytes).

The cv10Bit and cv23Bit modes are somewhat misleading. The oI2C object only works with 7-bit addressable I2C devices; however, the oI2C properties handle the internal addressing modes of I2C devices that use 16-bit location addresses (such as EEPROMs) and those that use 8-bit location addresses (such as almost everything else). In fact, you will often use more than one Mode setting for a single device, depending on how you access it. Some more examples are provided later on, but for right now, you will learn what the I2C objects do and how they read and write to a device.

Although many I2C devices have an automatic address-incrementing capability, the OOPic I2C object does not use this mode. The oI2C object writes the device select address and then the register byte (in cv10Bit mode) or the address location (cv23Bit mode). Then it restarts the I2C sequence with a read of either a byte or 2 bytes depending on the Width property setting. This means the setting of the *NoInc* property determines any auto-incrementing of the address within the I2C device. The OOPic tracks the address it is writing to within the oI2C object and rewrites it every time, instead of relying upon the device to increment its own address. Each byte that is written to and read from the I2C device is done a single byte or word at a time via random access. Each I2C device may be different in how it handles addressing internally, but usually when you write an address out to a device, whether it is a 16-bit address or an 8-bit register address, that address is remembered, and the next read from that device returns the data that the address references. Let's do some experiments and see some examples of what I mean.

EXPERIMENTS WITH I2C DEVICES #A1: THE DEVANTECH CMPS03 ELECTRONIC COMPASS

I2C is complex and not the topic of this book, so I'm going to keep this discussion rooted solidly to what the OOPic can do. Here you're going to read directional settings from the Devantech CMPS03 electronic compass. It is an I2C device that is command and register oriented, meaning that the first byte that follows the I2C address of the device is the address that selects a register or a command that implements an activity.

Interfacing the CMPS03 with the OOPic
The first thing you need to do is connect the device *to* the local I2C bus. Figure 9-8 shows the pinout of the programming connector, which is one method of connecting to this I2C bus. This connector is the same on the OOPic S and R boards; the OOPic C has the local I2C bus signals on pads on the carrier chip. On the OOPic S board, you can also get local SDA and SCL lines on pins 1 and 3 respectively of the 40-pin connector. On all boards, these lines are usually denoted as LSDA and LSCL.

The Devantech CMPS03 board has nine pins on it, whereas for I2C interfacing only four are of interest: +5V, Ground, SDA, and SCL. The others are explained in the documentation for that device. Before using this board in a project, you must calibrate it. The easiest way to do that is to attach a button on pin 6 (Calibrate) and follow the manual procedure in the documentation. Figure 9-9 shows the pinout of the CMPS03 compass board.

To communicate with the CMPS03 board, you first send the control byte, which includes the I2C address of the board, with the R/W bit at 0. Then you send the address of the register you want to read from or write to. Make sure that the oI2C Width property is set to cv8Bit

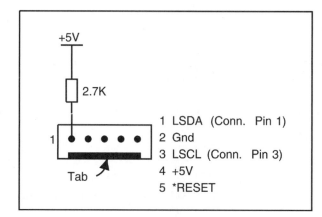

Figure 9-8 Local I2C bus connector

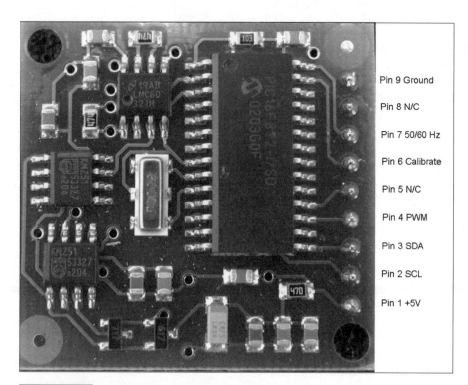

Figure 9-9 Devantech CMPS03 electronic compass

when you do this. The Mode property should be set to cv10Bit as well. If you are reading a 16-bit word back (to get the full 360-degree number, for instance), change the Width property to cv16Bit for that read. Figure 9-10 shows graphically what must happen to communicate with the CMPS03 board.

9

OOPIC I2C AND DISTRIBUTED DDE

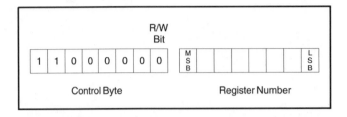

Figure 9-10 CMPS03 addressing format

You can interface any I2C device that uses the same read/write format as this board in the same way, and Figure 9-10 shows the addressing format only. OOPic handles the write/read format and you don't need to worry about it other than setting the Width property correctly for either 8-bit or 16-bit data.

Programming for the Devantech CMPS03 Device

Listing 9-1 displays the program I used to talk to my CMPS03. It works like magic and is simple to use. This program is written for the B.X.X firmware, and the A.2.X firmware version is on the CD-ROM as A2CMPS03.osc. It's not printed here in order to save space and clutter.

```
'B2CMPS03.osc
'Chapter 9 Devantech CMPS03 and oI2C Example
'Demonstrates using the oI2C object to work with
'I2C devices on the OOPic Local I2C bus.
'
'Just because I like to see what is happening, I've
'left the LCD attached to display the results.
'Remember to check your LCD's RS and E line connections!
'The Devantech CMPS03 needs to be calibrated before
'you will get meaningful numbers - Don't forget that.
'Make sure your RS and E lines are properly configured
'for the LCD!
'
'For OOPic II B.2.X firmware
'
'Copyright Dennis Clark 2003
'Permission is granted to use this code anyway you like
'as long as you say you got it from me.

Dim LCD as New oLCD
Dim compass as New oI2C

Sub Main()

    Dim ID as byte
    Dim bearing as word

    setupLCD
```

```
compass.width = cv8Bit              'Reading 8-bit registers
compass.node = 96                   'Address= 1100000
compass.Mode = cv10Bit              'One register address byte
compass.NoInc = cvTrue              'Don't increment address
```

'We're reading a single byte from the compass here. The
'OOPic will automatically place the byte value in the
'LSB of the word variable for us. The printData routine
'can handle any width of data sent to it, from a bit to
'a word, automatically.

```
compass.Location = 0                'ID byte register
ID = compass
LCD.Locate(1,0)
LCD.String = "ID"
printData(ID)                       'All that displayed it.
```

'The ID is a single byte. We now want to get heading
'information on a full 360 degrees of compass, so we
'want to use a 16-bit-wide data read.

```
compass.width = cv16Bit             'Getting 2 bytes now
compass.Location = 2                '0-3599 (359.9 degrees)
LCD.locate(1,4)                     'Setup on LCD for number
LCD.String = "bearing "
do
    bearing = compass               'Get the reading
    LCD.locate(1,12)
    LCD.String="     "              'clear the line
    LCD.Locate(1,12)
    printData(bearing)              'Print out the reading
    OOPic.delay = 100
loop

End Sub

Sub setupLCD()
```

'Traditionally, an LCD is used in 4-bit mode to save on
'I/O lines. These will be the upper 4 bits (4-7) on
'the LCD module. Typically, we tie the R/W line low so
'that all operations are writes. The RS line chooses
'whether we are doing a command or data write and the
'E line is the command latch to tell the LCD to use
'the information present on the data lines.

```
LCD.IOLineRS = 27                   'RS line to the LCD module
LCD.IOLineE = 26                    'E line to the LCD module
LCD.IOGroup = 3                     'I/O lines 28-31 (pins 32-26)
LCD.Nibble = 1                      'Upper nibble
LCD.Operate = cvTrue

OOPic.delay = 3                     'wait for LCD to come up
LCD.Init                            'Perform the icky initialization
LCD.Clear                           'Clear the screen
LCD.Locate(0,0)                     '
LCD.String = "OOPic CMPS03"
OOPic.delay = 100                   'Wait a little

End Sub
```

(Continued)

9

OOPIC I2C AND DISTRIBUTED DDE

```
Sub printData(wt As word)
'Print out a word, byte, nibble, or bit value to LCD.
'This simply outputs one character at a time by finding the
'digit in the 10,000's place (65,535 is highest number) and
'works down to the 1's place.  No leading zeros are output.

Dim wTemp as word
Dim bTemp as byte
Dim mTemp as word

bTemp = 0
mTemp = 0

wTemp = 10000
While wTemp > 0
    mTemp = wt/wTemp
    if ((mTemp > 0) OR (bTemp = 1) OR (wTemp = 1)) then
        wt = wt - mTemp * wTemp
        bTemp = 1                       'Now print trailing zeros
        LCD.Value = mTemp + 48       'convert to ASCII character
    End If
    if wTemp = 1 then
        wTemp   = 0
    else
        wTemp = wTemp /10
    End If
wend

End Sub
```

Listing 9-1

I greatly dislike leading zeros, which is why my code always includes the *printData()* subroutine. It prints numbers out to the LCD without leading zeros. Note too that this subroutine does not use any object memory, only variable memory, so it won't hurt your VC space either. I've formatted my output for a 2×16 LCD; that's why the display is so terse.

OEEPROM (ALL OOPIC FIRMWARE VERSIONS)

The oEEPROM object is dedicated to accessing any I2C device that is addressed like the Microchip 24LCXXX I2C EEPROMs. Its memory size is 5 bytes and its properties are outlined in Table 9-2.

TABLE 9-2   OEEPROM METHODS

Data (Value1, Value2 . . .)	Stores data into EEPROM above a program's maximum used address location. Bytes, words, and quoted strings are supported as data.

The oEEPROM object is identical to the oI2C object already given. It is identical in all respects, with one addition. The oEEPROM object has the *Data* method. This method is much like the basic *Data* statements that hold data to be read back serially in the same order in which they appear in the *Data* statement. It's actually more flexible than that, because you can *fast-forward* through the data stored if you know the offsets and formats of that data. A means of doing this is shown in the following section. The key to using the *Data* method safely is that it always stores its data in the program EEPROM, even if the object you are using with the *Data* method is addressing another EEPROM.

The *Data* method is the only way to directly store string literals in the oEEPROM object. As with the oI2C object, the *String* property, although listed in the manual, is not supported, and bad things will happen if you try to use it.

EXPERIMENTS WITH I2C DEVICES #A2: 24LC32 EEPROM IN THE E1 SOCKET

If you are using your OOPic as a data-logging device, you will want to save that data where it won't disappear after you turn off the power. An EEPROM is an ideal choice for such data storage. The 24LC32 is a 4KB I2C EEPROM that fits perfectly into the OOPic S board's E1 socket. No extra hardware, no extra wiring—how much simpler can you get?

Two programs are discussed here, and they both use the OOPic oEEPROM object. This object is identical in all respects to the oI2C, but with one addition, the *Data* method, which can be used to create data storage space simply in the program EEPROM.

Interfacing the 24LC32 EEPROM to the OOPic

No wiring is required for this experiment. The OOPic S board already has a socket in it labeled E1 (no, that isn't the address; it's short for EEPROM 1). That socket has its address pins (see Figure 9-11) wired at a different address than the default program EEPROM. You need to use an EEPROM that is larger than the 24LC16 in order for the address lines on the

Figure 9-11 24LC32 EEPROM pinout

9

OOPIC I2C AND DISTRIBUTED DDE

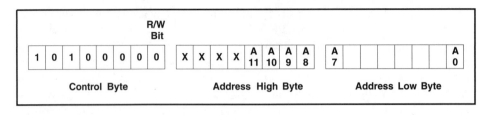

Figure 9-12 24LC32 addressing format

chip to be active. With the 24LC16 and lower-capacity chips, the A2-A0 address selects lines on the chip that are not connected, and any address that selects the main OOPic EEP-ROM also selects this one, resulting in corrupted data on the bus and a nonfunctional OOPic. If you are using an OOPic R or C, you have to connect the EEPROM using the local I2C bus connector as shown in Chapter 1 for the various platforms. This project focuses on the OOPic S board.

To communicate with the 24LC32 EEPROM, you first send the control byte, which includes the I2C address of the board, with the R/W bit at 0. Then you send the address bytes (2 of them, high-order one first) whose location you want to read from or write to. The Mode property must be set to cv23Bit because this is the mode that oEEPROM uses to send out two address bytes to an I2C device. In general, you will be writing and reading single bytes from the EEPROM, so set the Width property to be cv8Bit. Sometimes you'll be using 2-byte words, which requires Width to be set to cv16Bit.

Figure 9-12 shows graphically what must happen to communicate with the 24LC32 EEP-ROM. You can interface any I2C device that uses the same read/write format as this device in the same way. The figure shows only the device-addressing format. The data write/read format is handled by the OOPic objects and you don't need to worry about that part other than to make sure you have the Width property set correctly for the data size.

Programming for the 24LC32 EEPROM #1: Simple Data Storage and Retrieval

Listing 9-2 is the first of the EEPROM examples. It performs simple data storage and retrieval on the A.2.X firmware to show you how simple it is to use the oEEPROM object. It also shows the proper way to store and retrieve information from an EEPROM.

As you can see, you can store both bytes and words with the oEEPROM object. You just have to know which you are doing so you can set the object to the proper Width property before you write out or read your data. I used the *Str$()* function in this program because a compiler bug prevents a word data type from being defined in the A.X.X firmware options in the 5.01 release of the IDE. You can't store strings with the oEEPROM by using the String property; this is not supported and will crash the OOPic if you try it. To be able to store a string in an EEPROM, you need to store the ASCII equivalent bytes and convert them back to strings with the *chr$()* function, or you use the *Data* method, as my next experiment will show.

```
'A2I2CEEP.osc
'Chapter 9 24LC32 EEPROM and oI2C Example
'Demonstrates using the oEEPROM object to work with
'I2C devices in the OOPic E1 EEPROM socket.
'Just because I like to see what is happening, I've
'left the LCD attached to display the results.
'Remember to check your LCD's RS and E line connections!
'
'For OOPic I A.2.X firmware
'
'Copyright Dennis Clark 2003
'Permission is granted to use this code anyway you like
'as long as you say you got it from me.

'LCD objects
Dim LCD as New oDataStrobe
Dim nibs as New oDio4
Dim RS as New oDio1
Dim E as New oDio1

Dim Edata as new oEEPROM

Sub Main()

    setupLCD

    Edata.width = cv8Bit              'Reading 8-bit data values
    Edata.node = cvE1                 'Address= 1010100
    Edata.Mode = cv23Bit              '2-byte addressing
    Edata.NoInc = cvFalse             'increment address

    Edata.Location = 0                'First address location

    'start sending data out
    Edata = 32
    Edata = 129

    Edata.width = cv16Bit             'Store words now
    Edata = 2500

    'Now read it all back and display what we got

    Edata.Width = cv8Bit
    Edata.Location = 0                'Reset address
    LCDLocate(1,0)
    LCD.String = str$(Edata)
    LCD.String = str$(Edata)

    Edata.Width = cv16Bit             'Get 16-bit words back
    LCD.String = str$(Edata)

End Sub

Sub setupLCD()
[Deleted for brevity, full source is on the CD-ROM]
End Sub

Sub LCDInit()
```

(Continued)

9

OOPIC I2C AND DISTRIBUTED DDE

```
[Deleted for brevity, full source is on the CD-ROM]
End Sub

Sub LCDClear()
[Deleted for brevity, full source is on the CD-ROM]
End Sub

Sub LCDLocate(row as byte ,colm as byte)
[Deleted for brevity, full source is on the CD-ROM]
End Sub
```

Listing 9-2

Programming for the 24LC32 EEPROM #2: String Data Storage and Retrieval

This example is for the OOPic II. Here strings are stored using the *Data* method of the oEEPROM object. This is a simple matter, and Listing 9-3 shows that all you need to know is the starting address of the data storage to use any number of strings. Why would you want to do this? If you have a user interface that asks a lot of questions or displays a lot of text that is predefined, you need a place to store these prompts and text. Listing 9-3 shows how to accomplish this.

```
'B2Text.osc
'Chapter 9 24LC32 EEPROM and oI2C Example
'Demonstrates using the oEEPROM object to work with
'I2C devices in the OOPic E0 EEPROM (program) socket.
'This code demonstrates saving text strings in the
'EEPROM to use later as prompts in your programs.
'Just because I like to see what is happening, I've
'left the LCD attached to display the results.
'Remember to check your LCD's RS and E line connections!
'
'For OOPic I B.2.X firmware
'
'Copyright Dennis Clark 2003
'Permission is granted to use this code anyway you like
'as long as you say you got it from me.

'LCD object
Dim LCD as New oLCD
Dim Edata as new oEEPROM

Sub Main()

    Dim n as byte
    Dim slen as byte
    Dim loc as word

    setupLCD
```

```
Edata.width = cv8Bit              'Reading 8-bit data values
Edata.node = cvE0                 'Address= 1010000
Edata.Mode = cv23Bit              '2-byte addressing
Edata.NoInc = cvTrue              'Don't increment address

'start sending data out and saving the string start locations.

Edata.Data(10,"Testing...")       'Save list entry
loc = Edata.Location              'Save beginning of list
Edata.Data(15,"The second one.")  'next list entry
Edata.Data(14,"From me to you")
Edata.Data(99)                    'Flag for end of list

'Since we know where the first data starts, we can get to any
'string in the list by just counting through them and discarding
'that which we don't want.  Now let's read our strings back and
'display them.

Edata.NoInc = cvFalse             'Now we want to increment
Edata.Location = loc              'address of first data
slen = Edata
while slen <> 99                  'Look for end of list flag
    LCD.Locate(0,14)              'Place string length here
    printData(slen)
    LCD.Locate(1,0)              'Where to print our string
    LCD.String="            "    'clear the line
    LCD.Locate(1,0)              'Where to print our string
    for n = 1 to slen
        LCD = Edata              'print out our string
    next n
    slen = Edata                 'get length of next string
    OOPic.delay = 200
wend
```

'Now let's reset and choose a string that we want from the middle
'of the list. Let's say we want the second string only.

```
Edata.NoInc = cvFalse             'Now we want to increment
Edata.Location = loc              'address of first data
slen = Edata                      'length of first string

LCD.Locate(1,0)
LCD.String="            "         'clear the line
LCD.Locate(1,0)                   'Where to print our string

Edata.Location = Edata.Location + slen 'fast forward one string.
slen = Edata
for n = 1 to slen
    LCD = Edata
next n
```

'We can flip through the list in this fashion to any line stored,
'and all we have to do is add the string length of each string to
'the location to get to the next one. We don't even need to know
'the exact location of any string but the first one.

(Continued)

9

OOPIC I2C AND DISTRIBUTED DDE

```
End Sub

Sub setupLCD()
[Deleted for brevity, full source is on the CD-ROM]
End Sub

Sub printData(wt As word)
[Deleted for brevity, full source is on the CD-ROM]
End Sub
```

Listing 3-1

This program shows that you can store text in an EEPROM easily with an OOPic. The *Data* method is aimed at users who don't have a separate EEPROM. Because it always starts at the first address past the current program, the *Data* method always uses the program EEP-ROM, even if the object that the method is attached to is addressing the EEPROM in the E1 socket.

Another limitation of the *Data* method is that although it stores both bytes and words of data, it does *not* automatically switch the Width property of the oEEPROM object, so you must save all your word values in homogeneous *Data* method calls. You should also know that the section of the EEPROM you are reading is saved as words in order to get the correct data back out. Strings and bytes are both bytes of data, so they can be freely mixed in the same *Data* method call. You just have to know what is where when you read it back.

I2C ADDRESSING SUMMARY

I2C addressing can be confusing, so it bears repeating how to use the OOPic to access I2C devices. Here are the salient points:

- If you are writing commands to an I2C device, or saving a value in a register that has previously been assigned, use the cv7Bit Mode property. This does not send any address bytes beyond what is required to select the device. You do not use the Location property when doing this.
- If you are writing data to, or reading data from, a register that is specified in the Location property, use the cv10Bit Mode property (as shown in Figure 9-10). This sends out the device address byte and a byte defining a register location. It then restarts and either writes or reads a byte of data. If you have Width set to cv16Bit, you will write or read a word (two bytes) of data.
- If you are writing data to, or reading data from, a 16-bit address, such as an EEPROM, set Mode to cv23Bit (as shown in Figure 9-12). This writes out the device address byte and then 2 bytes of the internal address. It then restarts and either writes or reads a byte (or word) of data. You must remember to set Width as cv8Bit or cv16Bit, depending upon whether you are accessing byte or word data..
- You can switch Location, Mode, and Width freely between I2C accesses to the same device.

Using DDELink on the Network I2C Bus

As stated earlier, DDELink stands for Dynamic Data Exchange Link. OOPic boards and chips may exchange data at any time they choose as long as the source and target objects are linked to DDELink objects in their respective OOPics.

The network bus supports the use of only other OOPic devices and the parallel port cable. You can network up to 127 other OOPic devices, in other words, as many as you want. The process of networking OOPic devices together sounds complicated but isn't really. The OOPic is designed to network using I2C, and all the objects are designed to be accessed through a DDELink.

ODDELINK (ALL OOPIC FIRMWARE VERSIONS)

 This object networks OOPic devices together for the run-time exchange of data. Only the Default or Value data properties can be configured as data exchange items. The oDDELink memory size is 8 bytes and its properties are outlined in Table 9-3.

The key information to remember is that *only* the Value or Default Property may be transferred between a master and a slave that are DDELinked. An oDDELink can be configured to be a master and a slave at the same time. If the oDDELink is transmitting data, it uses the Input link; if it is receiving data, it uses the Output link. A master can either retrieve data from another OOPic DDELinked object or transmit data to another OOPic DDELinked object. Likewise, a slave may either send or receive data. It depends on how the oDDELink is configured. In either case, an oDDELink object can be configured as both a master and a slave at the same time.

This is the sequence by which you set up a DDELink master/slave relationship:

1. Define your *virtual circuits* (VCs) for both the master and slave OOPics before you start coding. This is the first step in the design of any program. Make sure you work through your use cases so that you fully understand what it is you want your programs to be doing.

2. Write and compile your slave OOPic program, as shown in Listing 9-4. You must do this first so that you can find the address of the DDE object the Master OOPic will talk to. The quickest way to do this is to use the *View* menu and pull down to *Opp Codes* so that you are seeing the OOPic program file output in the right pane. Look for the DDE entry, as shown in Figure 9-13. To the immediate left is the object address, in this case, 41.

3. Write and compile your master OOPic program, as shown in Listing 9-5. Use the address you found in step 2 as the *Location* property in the master DDE object. Use the OOPic.node number of your slave OOPic code, as shown in Listing 9-4, as the Node number of the OOPic to contact.

Simple, isn't it? It's probably the easiest intermicrocontroller communications link you will find anywhere. Let's look at an example now.

9

OOPIC I2C AND DISTRIBUTED DDE

TABLE 9-3 ODDELINK PROPERTIES

Address (number pointer)	All objects have an Address property, which is referred to by a master's Location property.
Direction (bit, flag)	0 (cvSend) means data is copied from the local object specified by the Input property to the remote object specified by the Node and Location properties.
	1 (cvReceive) means data is copied from the remote object specified by the Node and Location properties to the local object specified by the Output property. This defines only the master's actions.
Input (number pointer)	A link to a local object whose default property value is sent to the remote object's default property.
Location (byte)	The Address property of the remote slave object.
Node (byte)	The Node property of the remote OOPic. (0 means inactive.)
Operate (bit, flag)	Must be 1 (cvTrue) for this object to be active.
Output (number pointer)	A link to a local object whose default property value is updated with the data received from the remote object.
Sync (bit, flag)	Sends or retrieves data to or from the remote object just once. This bit is cleared when a sync is completed (default property).
Transmitting (bit, flag)	Is 1 when this OOPic is transferring data; it is 0 otherwise.

```
----:<@Dim-Begin>:
----:<@Dim#1-Begin>:
0020:041 '->DDE
0021:001 'ArraySize
0022:008 'ObjSize
0023:007 'ObjID:oDdelink
0024:135 'Dim
----:<@Dim#1-Exit>:
----:<@Dim#2-Begin>:
0025:049 '->finger
0026:001 'ArraySize
0027:006 'ObjSize
0028:136 'Get1Value
0029:163 'ObjID:oServox
0030:135 'Dim
----:<@Dim#2-Exit>:
0031:251 'Command Set2
0032:021 'SetDimDone.
```

Figure 9-13 Getting slave DDELink address value

Project: Remote Control

In this project, you first construct a master program that transmits the value of a variable resistor (potentiometer, or pot for short) to another OOPic where a servo will be moved according to the value. Because that is too simple by itself, I've added a bit more. The slave OOPic (the one with the servo) will watch to see if a value is out of the normal range for a hobby servo. Then it will send a flag back to the master OOPic telling it so. This means that the DDELink object in the master program will have a slave mode that receives the "out of range" error flag, and the slave program will have its DDELink object configured to also have a master mode to send the "out of range" flag back to.

HARDWARE LAYOUT

Unlike some previous projects, this one has a simple hardware interface. Figure 9-14 shows where the hardware devices are connected to the OOPic.

Again, you'll notice various capacitors being used in a smoothing or bypass mode to try to keep the servo from chattering. These work reasonably well, but some additional software antichatter work will also need to be done. Keep all your wires short. If your DDE network wires are longer than a few inches, twist the ground and +5V lines around the SCL and SDA lines for noise reduction purposes. Insulated and shielded CAT5 cabling is a good choice for this task. Don't route the DDE network wires near any other digital or power

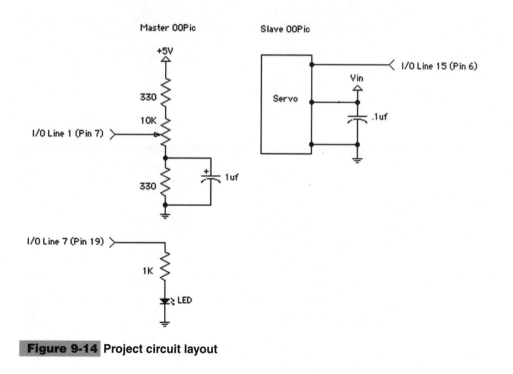

Figure 9-14 Project circuit layout

wires. Keep your *analog-to-digital* (A2D) lines short, because the hardware A2D will pick up ground loop currents and other analog noise if you give it any opening to do so. If you aren't careful, your project will chatter[1] because of the electronic noise the wires will pick up from your work bench.

DESIGN THE VC

I like programs that are 100 percent VC, so both of these programs are. Everything is linked and automatic. I've used some new objects with these programs, and I explain why I used them.

oChanged(O)(C) (B.X.X and Later Firmware Versions)

This object compares one input value with its output value, toggles the *Changed* property if the values differ, and then updates the Output property. When using the *oChangedO* version, no Output link exists, but instead an 8-bit *Value* property is compared and updated. Finally, when using an *oChangedC* or *oChangedOC* variant, the comparison only occurs when the *ClockIn* property is toggled. This is a useful processing object when you only want something to happen if a change occurs, not constantly. The oChanged memory size is 3, 4(oChangedO), and 5(oChangedC). Its properties are outlined in Table 9-4.

TABLE 9-4 OCHANGED(O) PROPERTIES	
Changed (bit, flag)	Is set to 1 if the Input property changed during the last cycle of the object list loop. It is 0 otherwise.
ClockIn (bit, flag)	Links to a flag that clocks the object into action for one object list loop cycle.
Direction (bit, flag)	0 if the Input property is greater than the last value saved. 1 if the Input property is less than the last value saved.
Input (number pointer)	Links to the value being watched for change. It can be any number, bit, nibble, byte, or whatever.
InvertC (bit, flag)	0 (cvFalse) means the compare and copy function is performed when the Flag property that the ClockIn property links to transitions from 0 to 1. 1 (cvTrue) means the compare and copy function is performed when the Flag property that the ClockIn property links to transitions from 1 to 0.
	(continued)

1. Chatter means that your data values will bounce around and not be stable.

Operate (bit, flag)	Must be 1 (cvTrue) for this object to be active (the default property).
Output (number pointer)	Links to the object holding the value that was last read from the Input property (oChanged and oChangedC).
Value (byte)	The 8-bit value of the last input read when not using the Output property (oChangedO and oChangedOC).

This object is being used in the project so that the DDELink transfer only occurs when the A2DX changes. This reduces the bandwidth taken up by DDE transfers for the servo. Hopefully, this also reduces the likelihood of jitter.

oOneShot (All OOPic Firmware Versions)

 A one-shot is a digital device that takes an input, and no matter how long that input is at a 1 level, only a short pulse comes from its output. You can think of this as a form of debounce object. Often you only want a signal to be sent once and not continuously for any single transition of a property. This is how to do it. The oOneShot object does not reset or enable another 1 transition pulse to occur until its input has returned to 0 for at least one object list loop cycle. The oOneShot memory size is 3 bytes and its properties are displayed in Table 9-5.

You can use this object just like an oWire object; it has much the same functionality as oWire but will only briefly output the logic level, instead of as a steady state.

TABLE 9-5   OONESHOT PROPERTIES

Input (flag pointer)	Links to the flag that will be the trigger level for the one shot.
InvertIn (bit, flag)	1 inverts the logic of the input, so that a 0 level is seen as the trigger. 0 means normal operation, and a 1 level is required to trigger the one shot.
InvertOut (bit)	Inverts the logic of the trigger before storing the result of the operation and linking to the Output property.
Operate (bit, flag)	Must be 1 (cvTrue) for this object to be active (default property).
Output (flag pointer)	Links to the flag that gets the one-shot pulse.
Result (bit, flag)	The result of the one-shot operation as a value, not a link.

This object is used in the project so that the DDELink is only synched once when a change occurs, rather than continuously. This reduces the bandwidth taken up by the DDE activity.

Figure 9-15 shows the VC devised for the master OOPic, and Figure 9-16 illustrates the slave OOPic VC. All choices that were made have been designed to minimize the amount of time taken up talking over the DDELink bus. In this manner, messages are less likely to be lost because the bus is busy when a message is sent.

The VC for the master OOPic program is a bit more complex than many. The whole thing starts when *interval*'s clock pulse, which happens every 85 milliseconds, triggers a comparison in the test object. If the current value of *pot* is different from the last one read and stored in *test.Value,* the Changed property triggers a *DDE.Sync,* and the oA2DX Value property is sent to the slave OOPic. When a 1 comes in through the receive (slave) side of the DDE, this turns *LED* on. When this happens, you start a clock with the timeout object whose output is inverted. At the end of one cycle of timeout, a 0 pulse is loaded into LED, which turns the LED and the oClock object off. In this way, the LED stays on for about a second and then resets so that another OutOfRange indicator can be seen if one should occur.

This VC for the slave program, shown in Figure 9.16, is a bit less convoluted. When a message comes in through DDE, it is given to *finger* to move the servo. If the *OutOfRange* property is set, this signal is linked to an oBit object, *OOR,* which in turn is linked to the Input property of DDE. You cannot link anything but a Value or Default property to an oDDELink object, which is why the intermediate oBit is used to hold the OutOfRange value. At

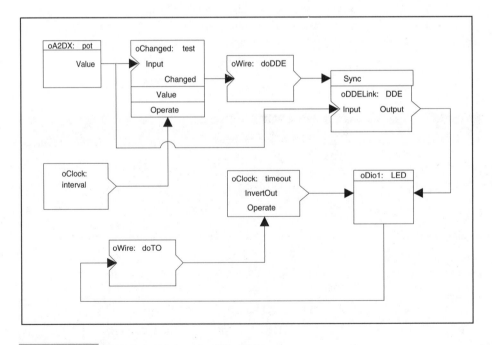

Figure 9-15 The project's master VC design

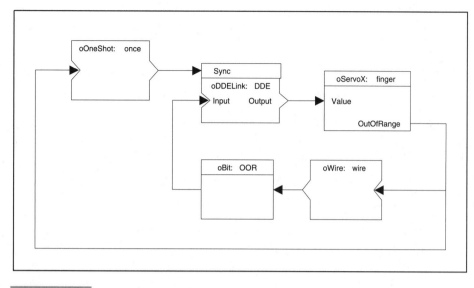

Figure 9-16 The project's slave VC design

the same time OOR is given to DDE, the oOneShot object *once* is signaled to trigger the *DDE.Sync* property to send the message back to the other OOPic. A one shot is used here so that only one out of range message is guaranteed to be sent for a single operation.

WRITE THE CODE

Now we get to the good part. These VCs are a little complicated, and this is done to maximize the efficiency of the network node transfers. In order to show that a master DDELink can also be a slave DDELink, both sides of this discourse, the master and the slave, have been designed to be both send and receive capable. Getting lots of value from oDDELink objects is a good idea, because they are expensive memory-wise. Listing 9-4 is the master OOPic program, the one that handles the potentiometer, and Listing 9-5 is the slave program that handles the servo.

A couple of things should be noted here. An oClock has been used as a form of free-running one shot. When the LED gets set because of an out of range error, this enables the Operate property on the timeout object. This object has its output inverted, so while the clock is counting, the output is high, which leaves the LED turned on. When the end of the timer occurs (about a second later), the output pulses low for one object list loop cycle and the LED is turned off, which also turns the oClock object timeout off. Listing 9-5 shows the use of a DDE slave OOPic.

9

OOPIC I2C AND DISTRIBUTED DDE

```
'DDEMaster.osc
'Provides the Master side of the DDELink project for Chapter 9
'This program demonstrates the use of a DDE Master OOPic in
'a DDELink network.
'
'For OOPic II B.2.X firmware
'
'Copyright Dennis Clark 2003
'Permission is granted to use this code anyway you like
'as long as you say you got it from me.

Dim DDE as New oDDELink
Dim pot as New oA2DX              'Uses URCP values
Dim test as New oChangedO        'Only send if changed
Dim doDDE as New oOneShot        'For linking Transmitting property
Dim interval as New oClock       'update timer
Dim LED as New oDio1             'Out of Range indicator
Dim timeout as New oClock        'resets the LED to 0
Dim doTO as New oWire            'Connect up the timeout

Sub Main()

    OOPic.node = 1

    'Set up the pot here
    pot.IOLine = 1
    pot.Center = 0
    pot.Limit = cvTrue           'Limit to +/- 127
    pot.Operate = cvTrue

    'This helps remove some of the servo jitter by limiting
    'how often a compare is made.
    interval.Rate = 232          'every 85 ms do update.
    interval.Operate = cvTrue
    interval.Output.Link(test.operate)

    test.Input.Link(pot)         'Set up the pot
    test.Operate = cvTrue

    doDDE.Input.Link(test.Changed)   'Only send a new value if
    doDDE.Output.Link(DDE.Sync)      'There was a change.
    doDDE.Operate = cvTrue

    DDE.Location = 41            'Address of slave object
    DDE.Node = 2                 'Slave OOPic node address
    DDE.Direction = cvSend       'Sending data out
    DDE.Input.Link(pot.Value)    'Send new reading
    DDE.Operate = cvTrue

    LED.IOLine = 7              'Set up Out of Range indicator
    LED.Direction = cvOutput

    'This will reset the LED to 0 after about a second.  When
    'the LED is cleared, the timeout object will be stopped.
    doTO.Input.Link(LED)
    doTO.Output.Link(timeout.Operate)
    doTO.Operate = cvTrue

    'This operates as a form of delayed reaction; otherwise, the
    'LED would be set the first time an Out of Range is received
```

```
                'and would never get unset - Not very useful.
                timeout.Mode = 1              'Do a single pulse each cycle
                timeout.Output.Link(LED)     'clear the LED
                timeout.InvertOut = cvTrue   'go low after timeout
                timeout.Operate = cvTrue

                DDE.Output.Link(LED)          'Get back data as a slave from this.

End Sub
```

Listing 9-4

```
'DDEslave.osc
'This is the slave side program for Chapter 9 Projects.
'This program demonstrates the use of a DDE Slave OOPic
'in a DDE network.
'
'For OOPic II B.2.X firmware
'
'Copyright Dennis Clark 2003
'Permission is granted to use this code anyway you like
'as long as you say you got it from me.

Dim DDE as New oDDELink
Dim finger as New oServoX              'Uses URCP values
Dim wire as New oWire                  'DDE sync .
Dim OOR as New oBit                    'A bit holder object
Dim once as New oOneShot               'DDE sync trigger.

Sub Main()

    OOPic.node = 2

    'This object will get the data from the Master OOPic
    finger.IOLine = 15
    finger.Center = 0
    finger.Operate = cvTrue

    DDE.Output.Link(finger)
    DDE.Operate = cvTrue

    'Now let's make a master object to send back to the other
    'OOPic.  This tells the other OOPic that a value sent
    'was out of range of the servo being used.  We need to
    'use an oOneShot object so that the DDELink won't
    'constantly transmit and tie up the bus.

    wire.Input.Link(finger.OutOfRange)  'Save out of range bit
    wire.Output.Link(OOR)
    wire.Operate = cvTrue
```

(Continued)

9

OOPIC I2C AND DISTRIBUTED DDE

```
'Only talk to the other OOPic if OutOfRange is detected
once.Input.Link(finger.OutOfRange)
once.Output.Link(DDE.sync)
once.Operate = cvTrue

DDE.Input.Link(OOR)
DDE.Direction = cvOutput
DDE.Location = 41                      'Same address over there.
DDE.Node = 1                           'Other OOPic node address
```

Listing 9-5

An alternative to DDELink Networking

Some folks have found that DDELink networking can be too slow for their uses. The DDELink code must properly update pointers, flags and other data in both OOPic master and slave processors before it completes the DDELink task. This can take a lot of time depending upon the objects involved. There is another way to read and write data between two OOPics by using the OOPic I2C *local* bus on the master OOPic and the OOPic I2C network bus on the slave OOPic. This is a little complicated, so read closely, this is the process.

Let us assume that you have an oI2C object defined called *noopic* and you have given your target (slave) OOPic an I2C address using the *OOPic.Node* property. You then set noopic.Node equal to the target OOPic's node address that you gave it. An OOPic looks just like a 256 byte memory on the I2C local bus. You then connect the OOPic I2C local bus on the master OOPic to the network I2C bus on the slave OOPic.

Use one of the processes given in Chapter 3 "Linking Properties by Using the oRam object" to find the address of the object and the property within that object. Add 128 to that address since the object data is kept in RAM bank 1 of the PICMicro processor.

To read a byte from the network port of another OOPic you:

```
noopic.width = cv8bit    ' You can only read or write bytes
noopic.Mode = cv10bit    ' Use a single register address
noopic.NoInc = cvFalse   ' You can use autoincrement with a read

noopic.Location = (objAddress +128) ' The beginning of the object
```

To read the third byte of an object whose address is 41 using auto-increment:

```
noopic.Location = (41+2+128)
theByte = noopic
theByte2 = noopic  'Gets the fourth byte, and so on.
```

You can also use random addressing and not use the auto increment feature.

To write to an object (again at address 41 as seen in the .omp file)

```
noopic.width = cv8Bit   'again, only bytes
noopic.Mode = cv23bit   'You need the address and a MASK for a write.
noopic.NoInc = cvTrue   'you can ONLY write bytes

noopic.Location = (objAddress + 128)*256 + &H00  ' nothing is masked
```

For example, to write to oStepper's rate property, which is only 7 bits:

```
noopic.Location = (41+128+2)*256+&H80   'mask off bit 7, which is
                                        'the direction bit
noopic = 55
```

The mask value is thus:
Each corresponding bit marked with a '0' will be written
Each corresponding bit marked with a '1' will NOT be written

There is no mask to use when reading.

Whew, now all you have to do is find the location within the object of all of the properties. Which you can do by just reading the object values out and fiddling with bits. Beware, some aren't there, like oDio direction and value bits, those are in their respective PIC registers, others like the oTimer values or PWM prescale and period properties are also in PIC registers, not in memory.

I have One final caveat to share. Because this procedure can only read and write bytes, you cannot guarantee that a word read in two steps will be synchronized. That is, if you read the value of a word that is updated from the oTimer object, one of the bytes of the value may have changed and may not be the one that should go with the first byte you read. This is one of the housekeeping functions that oDDELink handles, it makes sure that you get synchronized data when you read a word from an object.

Now What? Where to Go from Here

This chapter should open your eyes to the near-infinite possibilities of networking OOPic controllers and other I2C devices. The programming is simple and the range of applications is huge.

The key to using new I2C devices with the OOPic is in understanding what the device wants you to do to communicate with it. The OOPic operates with any 7-bit addressable I2C device. Carefully read the documentation to see how the device's internal implementation wants to be addressed. The OOPic has a mode that will allow this addressing.

9

OOPIC I2C AND DISTRIBUTED DDE

In the case of this chapter's project, the remote control of servos, you should see that the capability to have one controller send detailed data to another over a network can make some projects simple. Take, for example, a distributed control system throughout a house. Your main unit can control slave devices in every room. If you want to change how your program does something, you only need to change the program on your master controller because it handles the logic for all the other controllers; they just follow orders. Those features make upgrades simple.

Further signal conditioning can be done in software to reduce servo jitter with this project. Try using an oMath object to remove the *least significant bit* (LSB) of the A2DX value so that minor changes there don't get transmitted out to the remote servo, but this can cause the servo to get *jerky* with a whole bit value missing. You might even try an averaging scheme or a more sophisticated means of gaining a *deadband,* a region around which changes are ignored, in order to reduce servo noise by including the oCompare object that enables a *Fuzzyness* property. Still another option is to use code to provide a *sliding window* of averaged values that can be systematically fed to the remote servo.

On occasion you'll notice that the OutOfRange LED doesn't light when you think it should. Why would this be so? It is because a VC cannot implement a "memory" of an event. When the slave servo detects an out of range condition and signals for that to be sent back, if the master is currently transmitting a new value to the slave, the DDE object is busy and that event transmission will most likely be dropped. See if you can think of a way around this problem within the VC without using program code.

You don't need to have multiple DDELink objects if you don't mind using code instead of pure VC logic. The master can send a message to a slave detailing which object it wants to write to or read from. The slave code can then reconfigure the DDELink object to the correct object and send back a message to the master that it is ready. Then the master can send the data down. Of course, you could simply use the DDELink object as a "telephone" between the master and slave, and have the slave do all the work to read or write to the objects. One way will work better than the other depending on your application. The OOPic enables complex logic to be implemented in a VC, and it's fun to play with them as well!

OOPIC ROBOTICS AND URCP (PROJECT: A ROBOT THAT TOES THE LINE)

Every new firmware version of the *Object-Oriented Programmable integrated circuit* (OOPic) that has been released comes closer to achieving a robotic controller. The B.X.X firmware has added dozens of new high-level objects designed specifically for robots. The release of the OOPic R board has capped the pyramid. This board is fully robot enabled with special connectors and a tiny size that makes it ideal for mini-sumo or line-follower competitions where the size of the robot is important.

Along with the B.X.X firmware came Savage Innovation's proposal for the *Uniform Robotic Control Protocol* (URCP). I'll explain what URCP is all about first, and then I'll list the various special robotics-oriented objects and how they relate to each other and to URCP. Finally, I'll discuss a line-following robot that will be a high scorer in any competition you enter and that is built of simple and inexpensive components that can be purchased off the shelf and built in an hour. I'm so pleased with this little guy that you will find a QuickTime video on the CD-ROM of the robot cruising through the Acroname (www.acroname.com) line-follower competition course in less than a minute. As you can tell, I think this is very cool. Cool is a technical word in the robotics research labs by the way.

URCP, The Uniform Robotic Control Protocol

The URCP is an open standard proposed by Savage Innovations as a means by which a robot may be controlled and guided; it also enables a robot to respond to instructions that are only 1 byte wide and of a uniform format. The 1-byte objective has several advantages: You use less memory-storing and passing control variables, you transfer instructions via serial links faster, and the commands can be evaluated faster.

A standard was created so that manufacturers of robotics sensors and accessories can all be designed to work together on any platform that uses the standard. Imagine what the Internet would be like if no standard means of communicating existed among computers and if every computer company used its own language. Now imagine how easy it would be to work with robots if they all used a common interconnect protocol like URCP?

We'll see what the future holds for URCP; it's a simple standard and no one wants any licensing fees to use it. Hopefully the robotics component manufacturers will adopt URCP or a future variant of it.

Currently, URCP is defined for three categories of control:

■ Heading and direction
■ Speed
■ Distance

Let's break down the concept by these three categories and see what they mean and how the OOPic uses them.

URCP HEADING

URCP heading information is conveyed using a single signed byte with the range of -128 to $+127$. 0 is straight ahead and -128 is straight backwards. This means that the circle of the robot's motion is broken into 256 discrete positions, instead of the circle's usual 360 (degrees). The negative numbers are to the left, and the positive numbers are to the right. Figure 10-1 shows a graphical representation of URCP heading information. This discrete 256-step heading representation is called *binary radians*, or *brads* for short.

One thing that should be immediately obvious is that the numbers will wrap nicely within a byte. This means that if your robot is moving at a heading of 64 (full right, assuming that forward is 0) and you give an instruction to change direction of -128, the resulting number will be -64 with no interpretation. If you were using 360 degrees and you subtracted 180 degrees from 90 degrees, you would get a result of -90 degrees, which would then have to be translated to 270 degrees to put you back on the standard circle. This doesn't seem like much, but try this sometime in your robot and see how much easier it is to interpret numbers that add and subtract nicely without needing to be massaged into a meaningful number for your robot. What is meaningful to us is completely different for your robot; the robot deals with concrete data and finite, discrete steps.

The URCP direction is primarily intended to be a relative reference for your robot with the 0 value meaning straight ahead. You may also choose to make the 0 mark the absolute direction north when dealing with compasses.

Of course, the most obvious advantage of using 256 steps for a circle is that it fits into a single byte of data; 360 degrees would take 2 bytes. You lose about 30 percent of your trigonometric resolution but gain a great deal in terms of speed of execution and 50 percent

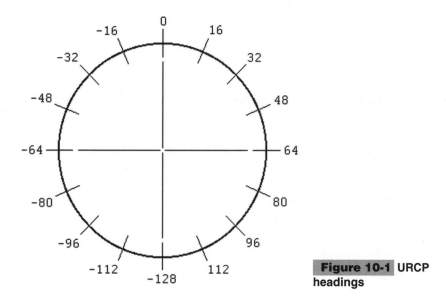

Figure 10-1 URCP headings

less memory space is used. Finally, it's simple, efficient, and easy to understand programming tasks with URCP.

URCP SPEED

URCP speed is rather simple: +127 means full forward, and −128 means full reverse. Therefore, in 1 byte you have both motor speed and direction. The programming interface is also simple: Subtract to go less forward and add to go more forward. Obviously, this means that lots of subtraction ends up going in reverse, and lots of adding ends up going forward. This makes a speed integration formula simple, and vector-speed[1] algorithms are easy to program. URCP speed is not a reference to the speed of your robot, but rather a representation of how much of its maximum capability to use in its motors.

URCP DISTANCE

URCP distance is the OOPic's consistent method of distance measurement. With SONAR, your usual measurement is a number of microseconds, and with Sharp *infrared* (IR) range finders, the measurement is volts. You then have to convert these to a meaningful distance value before your program can use the data to make a decision. By using the URCP distance-measuring objects, this is done for you by converting the time or voltage values into a measurement based upon a 64-step-per-foot metric. This means the maximum range that can be reported is 4 feet (about 1.24 meters). The OOPic manual says that this is enough to define the personal space around a robot. I guess you could say it is a definition of the *high-interest* zone for the robot's reaction. Currently, only the oIRRange object uses URCP distance values. This object is designed for the Sharp GP2D02 range finder, which has a maximum range of about 31 inches (80 centimeters).

The URCP standard of 64 steps per foot was not planned to be an English unit of measure. It just turned out that the division of steps coincided with the 1-foot unit coincidentally.

OOPic II Robotics Objects

Okay, let's dive right into the fun stuff: the robotics-specific objects that appear in the OOPic B.X.X firmware versions. Many of these objects are simply wrappers around generic objects with certain defaults already set. These wrappers save the programmer configuration time and make a program easier to understand and modify. I'll note such wrapper objects when they appear.

The robotics objects can be broken down into four basic categories:

- *Motion objects* that control DC motors, hobby servos, and stepper motors
- *Direction objects* that interpret compasses, proximity detectors, and bumpers

---

1. A vector includes both speed and direction. In this case, the direction is only forward and reverse.

- *Range-finding objects* that interpret IR range-finding and SONAR readings
- *Navigation-processing objects* that integrate direction and speed settings before sending data to motor control objects

Later sections of this chapter discuss similarities among the objects that are simply variations or wrappers around a common proto-object. This chapter does not cover the details of each of these objects, because far too many of them exist to talk about them all. Instead, each of the objects and how they would be used are discussed. In the "Experiments with OOPic Robotics Objects" section, I'll pick one from each category and show how it is used. Finally, the project for this chapter ties all the pieces together and includes a navigation object as well. Some of these objects are URCP and some are not. You'll find that the URCP-based objects interface in an elegant manner. Let's look at the robotics objects now.

OOPIC MOTOR OBJECTS

I'll list all the objects in this class of hardware objects and only go in depth about one or two of them. Some of these objects aren't going to be strictly robotics related, but they are highly related to the class of robotics objects and form the core around which specific robotics objects are built. Most will be hardware objects, but a few processing objects are tightly linked to motion control and belong here. I'll not deal specifically with the servo or stepper objects here, because those were covered in detail in Chapter 5, "Analog-to-Digital and Hobby Servos (Project: Push My Finger)," and Chapter 8, "OOPic Interfacing and Electronics (and Steppers and Seven-Segment LEDs)," respectively.

I'll start out with the proto-objects and then list those wrapper classes built around them. One oddball object is in this category: the *oQEncode* object, which, strictly speaking, does not control a motor, but rather reports the actions of a motor.

DC Motors

All the motors in this chapter are, strictly speaking, DC motors, but it's tedious to be absolutely specific and say, "Brushed DC motors without attached drive trains *and* controller circuitry." Therefore, I'll distinguish these DC motors by saying they are all the DC motors that aren't steppers or hobby servos. Now I've been very specific.

The prototype object for all DC motors is the *oPWM* object. Around the oPWM object are wrapped oDIO1 bits for *pulse width modulation* (PWM), direction, and brake lines to the various controller boards and chips. Two hardware PWM channels are located on the OOPic processor: IOLine17 and IOLine18. The keys to understanding PWM on the DC motor objects are the properties *Prescale*, *Period*, and *Value*. Table 10-1 shows the *Prescale* property values to get the supported divides for PWM on the two OOPic PWM channels.

PWM controls the speed of a DC motor by varying the *on* time of the motor. The most common way is to change the *duty cycle* of a fixed frequency signal; the duty cycle is the ratio of *on* to *off* in an oscillating waveform. The nature of the DC motor's windings is such that the motor does not see a chopped-up signal, but instead sees a varying level of DC voltage proportional to the *on* percentage of the duty cycle. The higher the voltage (higher duty cycle), the faster the motor runs. Figure 10-2 shows an example of 10, 50, and 90 percent

TABLE 10-1

VALUE	MEANING
0	No divide occurs; the base PWM frequency is 5 MHz.
1	Divide by 4; the base PWM frequency is 1.25 MHz.
2	Divide by 16; the base PWM frequency is 312.5 KHz.
3	Same as 2.

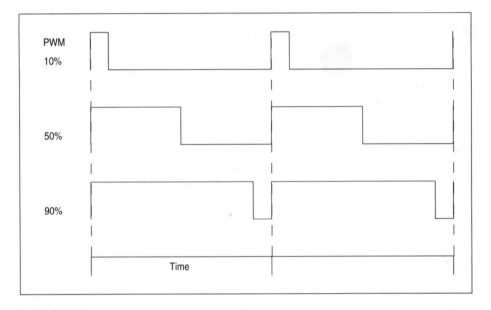

Figure 10-2 PWM duty cycle examples

duty cycles, which would translate to 10 percent of the maximum voltage up to 90 percent of the maximum voltage.

I told you all that so you can understand this next bit. To set the PWM duty cycle, which represents the percentage of the maximum voltage applied to your motor, use the Period and Value properties. The period of a waveform is the inverse of the frequency. The Prescale property sets the divide down to the PWM hardware from the system 5 MHz clock. To further divide down to a more usable PWM frequency for dealing with DC motors, the OOPic uses the Period property.

The process works like this: At the beginning of every cycle, the oPWM object sets the PWM line to 1 and a clock is started whose total number of tics is equal to the Period property setting. This clock is incremented at the Prescale frequency and checked against the

Value property. If the count is equal to Value, the PWM output line is set to 0. When the count is equal to the Period property, the clock is cleared, and the PWM line is set to 1 again. This means that the Period property sets the final PWM frequency, and at the same time it sets the PWM *resolution*. In a nutshell, resolution is how many pieces the pie can be divided into. In essence, it is the number of choices you get. With the oPWM object, resolution and frequency are inversely related. If you want a higher PWM frequency, you'll have to settle for a lower-speed selection resolution. Two examples follow.

If you choose Prescale to be 2, giving a 312.5 KHz frequency, but you really want a 20 KHz PWM frequency (which is a good frequency for high-quality DC motors), set the Period property to 312.5/20, which is 15.625 and can be rounded up to 16. This means that you will have 16 speed settings to choose from. Therefore, a Value setting of 8 means that the motor will run at 50 percent of maximum voltage, which roughly translates to half speed.

More than one way exists to get 20 KHz though. If you choose Prescale to be 1 (1.25 MHz clock), your Period property can be set to 63, which is a much better resolution. Again, if you leave Prescale at 0, a Period property equal to 250 is also 20 KHz.

Again, if you choose Prescale to be 2 and want a 2 KHz PWM frequency (which is a good choice for lower-quality DC motors), set the Period property to be 312.5/2, which is 156.25, rounded to 156. This gives you 156 speed settings you can use in the Value property to select your PWM duty cycle. Therefore, a Value property setting of 16 will be a speed setting of roughly 10 percent.

Something else also needs to be mentioned here. Because the largest number the PWM module can take is 255 and the largest number a URCP object can have is 127, the actual number of steps you get is only half that of the Period property. This is because 127 needs to reach 255, so its value is doubled to do so. This doubling of the Value property occurs no matter what the Period property is set to. Therefore, you will only get half the number of steps (or resolution) the Period property is set to. The following are some examples:

- If Period equals 255, then -128 to $+127$ or only 127 steps are really available.
- If Period equals 16, then -8 to $+8$ is your range.

This keeps the functionality of the object consistent no matter what the Period property setting is.

Take care never to set the Value property to a value higher than half the Period property. In such a case, the motor will be full on because a Value counter match will never set the PWM line to 0.

The following objects are built around the *oPWM* DC motor controller, and all of them use URCP:

- **oDCMotor**   This object is designed to work with the LMD18200 H-bridge, which uses a single direction line, a brake line, and a PWM line.
- **oDCMotorWZ**   This object is identical to oDCMotor. Just don't bother with the *IOLineB* property, because this board doesn't have a brake.
- **oDCMotorMT**   This object is identical to oDCMotor except that the PWM signal is inverted to work with this motor controller board.
- **oDCMotor2**   This object is designed to work with the L293, L298, and TI754410 H-bridges that have two direction pins and a PWM line.

The *oMotorMind* object is based on a serial port and sends serial commands to the Solutions Cubed™ MotorMind DC motor controller boards. This device is not URCP aware, so see the instruction manual for the correct commands to send it.

The last DC motor control object is *oPWMX*, which enables you to send DC motor PWM signals on any *input/output* (I/O) line of the OOPic. Its frequency is fixed at a little over 500 Hz. This is a low PWM value but is perfectly adequate for crude DC motors such as those found in portable drills. Your resolution is only 15 steps, which again is adequate for most tasks.

Stepper Motors

The specifics of steppers are covered in Chapter 8, so this section does not delve deeply into them, but a short description of what each object does is in order.

Two basic stepper motor objects exist: *oStepper* and *oStepperSP*. The *oStepperL* and *oStepperSPL* objects are also available, but I don't recommend that you use them. They were created for a specific purpose that is so obscure you will never need them. (If you just *have* to know, send me an email sometime and I'll try to explain them to you.)

The distinction between the two basic stepper motor objects is simple. The oStepper object enables you to set the number of steps to take and the direction via *virtual circuits* (VCs), but not the speed (Chapter 3, "OOPic Object Standard Properties," shows you how to get around that barrier). oStepperSP enables you to use URCP to set the speed of the stepper when in continuous run mode. When you are in specified-steps mode, you set the number of steps. This means you can use oStepperSP just like a DC brushed motor with URCP controls. The URCP speed is set by linking to the Value property, and the number of steps is set by linking to the *Steps* property.

Hobby Servo Motors

The specifics of hobby servos are covered in Chapter 5, so only the basic differences among the objects are described here. The four hobby servo objects are as follows:

- **oServo**   Is the standard servo controller.
- **oServoX**   Uses URCP values from −128 to +127 to position a servo.
- **oServoSP1**   Uses URCP values from −128 to +127 with 0 sending no pulse.
- **oServoSP2**   Uses URCP values from −28 to +127 to modify the rate at which full-motion servo refresh pulses are sent. It does not vary the pulse width, just repetition.
- **oServoSE**   The oddball in the bucket, this uses an I/O line to send serial information out to a Scott Edwards® *Serial Servo Controller II* (SSCII) serial servo controller board.

The basic servo object is *oServo* and it serves as the prototype for the rest. This object is not URCP but uses absolute servo values. One thing to remember is that although the OOPic has a 64-step resolution for servo values between the normal servo limits, if you use a higher value, you will go *well* outside a standard servo limit. If you're using a normal hobby servo, it *will* be damaged. User beware!

The three defined URCP servo objects are *oServoX, oServoSP1,* and *oServoSP2.* The first two objects are functionally identical except for one item. The oServoX object outputs the centered servo pulse (usually 1.5 milliseconds) at setting 0. oServoSP1, on the other hand, outputs no pulse at all at setting 0. This is because oServoSP1 is intended to be the object for hacked servos that are being used as drive motors for wheeled robots. You've seen how difficult it is to find the exact center location of a servo such that a 0 value will cause the wheels to stay still. Well, oServoSP1 fixes all that by simply not putting out a pulse at 0, guaranteeing that the wheels won't move.

The oServoSP2 object is an experiment to see if the speed of a robot could be effectively changed by altering not the width of the control pulse, but the repetition rate of a full-movement pulse being sent to a servo. The general principle of operation is based on the knowledge that a servo gets weaker as the pulse repetition rate to it slows down. Although this actually works (I've tried it), the movement is so jerky and slow that I don't recommend using it.

Finally, the Scott Edwards SSCII controller object, oServoSE, is a serial-communications-based object that sends 2400-baud 8N1 serial commands to an SSCII board to control servos. You can connect this object to any I/O pin of the OOPic. Refer to Chapter 5 for details on its usage.

OOPIC DIRECTIONAL OBJECTS

Two basic kinds of directional objects are available: those that report the direction you are going, like a compass, and those that report the detection of an external influence, like bumpers and proximity detectors. In fact, the oCompassDN object is the same object as the oBumper4 object (and the oJoystick object too for that matter). The former deals with absolute headings, and the latter deals with absolute locations on the robot. Finally, oTracker reports on the location of a black line on a white background, which is useful for line-follower robots and is a topic for this chapter's project.

Compass Heading Objects

The following is the list of the compass-type objects in the OOPic II firmware. They all return their values as URCP heading readings.

- **oCompassDN**   Is a compass object that uses the Dinsmore™ 1490-type digital compass module. This module shows the eight cardinal compass points as a combination of four open-collector[2] digital outputs. You'll be seeing this one later on in this section.
- **oCompassVX**   Controls the Vector™ V2X electronic compass. This device uses a clocked data-in and data-out line, and the OOPic II hides all that confusion from you.

---

2. This means you'll need pull-ups on the outputs of the compass signal lines.

Proximity Detectors

Both *IR proximity detectors* (IRPDs) and bumpers fall into the proximity detectors category. Why bumpers? Because, in my opinion, running into something is the closest proximity you can get to an object. The following is a list of proximity detectors that return URCP values, except for the IRDP objects, which are not URCP:

- **oBumper4**   This object looks at and reports on the contact of four bumpers: front, back, left, and right. The contact is reported as a URCP number where 0 is straight ahead.
- **oBumper8**   This object is really two oBumper4 objects, one skewed 1/8 of a turn from the other. This gives you contact information on front (left, right, and center), rear (left, right, and center), left center, and right center. When you look at the bit patterns, you'll see those two distinct oBumper4 objects in there.
- **oIRPD1**   (Not URCP) Since this object has a single detection output, URCP doesn't make much sense. Therefore, it simply reports a 0 or 1 for undetected and detected, respectively.
- **oIRPD2**   (Not URCP) This object is designed to use the Lynxmotion™ IRPD module, which is capable of detecting left, right, or straightahead obstacles. The left and right detection bits can be linked directly to other objects in a VC.
- **oTracker**   Designed as a simplified interface to a sophisticated line-tracker robot, this object needs either four or three digital line sensors to compute a URCP heading. The heading is then handed off to another object such as oServoX or oNavCon, which will control your robot's steering. This oTracker object is used in this chapter's project.

The oBumper objects may seem difficult to comprhehend at first, but a simple means exists for understanding them. Look at Figure 10-1 again, the URCP heading layout. When you assign the I/O lines to the oBumper4 object, start by making bit 0 the 0 position. Then, in clockwise order, bit 1 is the 64 position, bit 2 is the -128 position, and bit 3 is the -64 position. When two adjacent bumpers are depressed, the number reported is the average of the two bumpers. In the case of right rear, it does not average -128 and $+64$. It averages $+128$ and $+64$ (which makes more sense) to get $+96$. When you assign the I/O lines to oBumper8, it is the same; bit 0 is the 0 position. Going around the circle in a clockwise direction, bit 4 is -128, or rear. Again, when two, three, or even four adjacent bumpers are depressed, the values are averaged. Again, when the right-side bumpers are averaged with the rear bumper, 128 is used, not -128. When averaging values the left side, -128 is used.

OOPIC RANGE-FINDING OBJECTS

Range-finding objects are a class of objects that locate the distance to an obstruction or external object. Some of them are SONAR and some are infrared in nature. The SONAR objects all use the OOPic 16-bit timer represented in the oTimer object, and the IR range finder uses the oA2D channel objects to work its magic. Only the oIRRange object restricts the URCP range values. The SONAR objects have much more range than the 4-foot URCP range limit. The OOPic range-finding objects are as follows:

- **oIRRange** This object is designed to use the Sharp GP2D12 analog IR range finder. It uses an *analog-to-digital* (A2D) port for its input. The maximum returned value from oIRRange is 128, for a maximum of 2 feet in 64 steps per foot resolution using URCP. An example of this is provided later in this section.
- **oSonarDV** This object is a wrapper around an oTimer functionality and uses a Devantech SRF04 SONAR board. It can connect to any I/O lines, and Chapter 6, "OOPic Timers, Clocks, LCDs, and SONAR (Project: SONAR Ping)," explains how to use oSonarDV.
- **oSonarPL** This object wraps oTimer to control a Polaroid™ 6500 module. This module has a much greater range than the Devantech unit, and this object has a maximum range of 32,768 in URCP 64-steps-per-foot readings.

OOPIC NAVIGATION OBJECTS

This object interprets the outputs of directional sensors and decides how to update motors' inputs in order to move a robot.

The *oNavCon* object takes heading and speed information, integrates the two pieces of data, and feeds motor-speed information to two DC motors or hacked hobby servos to steer your robot. This object is used in the line-tracker robot project at the end of this chapter.

Experiments with OOPic Robotics Objects

To give you a feel for the operation of some of these objects, I've included a few experiments for you to try. Each illustrates some useful robot functionality and gives you an idea of what you can accomplish with relative ease.

EXPERIMENT: CONTROLLING TWO DC MOTORS WITH AN L298 H-BRIDGE

Examples are the best way to explain a complicated topic. This experiment shows how to configure two *oDCMotor2* objects to control an HVW Technology L298 dual-motor controller board. This is an inexpensive board that works with voltages down to 5V or 6V and currents up to 2 amps. Figure 10-3 is a picture of this dual-motor controller board, which is perfect for a small robot. Figure 10-4 shows how to configure the I/O lines from the OOPic to the controller board. Use this board with the Period property set to 255 and the Prescale property set to 2. It might not really handle high PWM frequencies unless you replace the kit diodes with diodes similar to 1N5821 or SB550, which are high-speed diodes. I use this

Figure 10-3 HVW Technology L298 dual-motor driver board

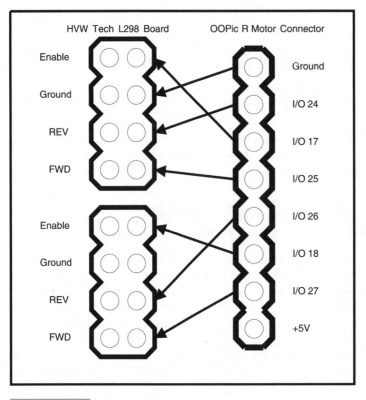

Figure 10-4 OOPic and HVW Technology L298 board connections

board in one of my mini-sumo robots to control a Tamiya™ dual-motor gearbox, and it works quite well. The board comes as a kit that you have to solder together.

oDCMotor2 (B.2.X Firmware and Later)

 All the oDCMotor objects (except oMotorMind) use the oDCMotor2 template for properties. Subtle differences can be found among the oDCMotor objects, but they are similar enough to oDCMotor2 that you can use this discussion to help you configure them. This object is specific to an L293, L298, or TI754410 H-bridge chip. Refer to the previous discussion to understand how to configure your PWM frequency and duty cycle. The oDCMotor2 memory size is 5 bytes and its properties are outlined in Table 10-2.

TABLE 10-2 ODCMOTOR2 PROPERTIES	
Brake (bit, flag)	When this bit is set, the object sets all the control lines to the *brake* condition. This means that both motor outputs are at the same state and the PWM line is not cycling. *Mode* and *InvertOutB* define the operation of this state. Check your motor driver documentation for how to get an active brake; then configure this object to match it.
Direction (bit, flag)	If the *Unsigned* property equals 0, this bit reflects the motor direction.
	0 is used for positive Value settings.
	1 is used for negative Value settings.
	If *Unsigned* equals 1, you set or clear this bit to determine motor direction.
InvertOutB (bit)	Enables you to change the output level of the *Brake* functionality.
	0 means both motor lines are logic 1.
	1 means both motor lines are logic 0.
InvertOutD (bit)	When set to 1, the motor control line logic is reversed. This is useful if you can't unsolder and swap your *IOLine1* and *IOLinet2* control lines to change this logic.
	0: When Direction equals 0, IOLine1 equals 1 and IOLine2 equals 0, and vice versa.
	1: When Direction equals 0, IOLine1 equals 0 and IOLine2 equals 1, and vice versa.
IOLine1 (byte)	One of the two direction control lines to the H-bridge.
IOLine2 (byte)	The other direction line to the H-bridge.

(continued)

TABLE 10-2   ODCMOTOR2 PROPERTIES *(Continued)*

IOLineP (bit)	The I/O line to use for the PWM signal (connected to enable on L298).
Mode (bit)	Determines how the PWM line is set when the Brake mode is asserted.
	0 (cvOff): When Brake equals 1, PWM output is 0.
	1 (cvOn): When Brake equals 1, PWM output is 1.
	If you choose incorrectly, don't worry; you may never notice. One way causes the motor to stop immediately, and the other gradually brings the motor to a stop. .
Operate (bit, flag)	The bit must be 1 in order for the object to respond to input.
Period (byte)	The number of clock tics that define the PWM time period. (Refer to previous section.)
PreScale (nibble)	Sets the clock divider for the PWM hardware. (Refer to previous section.)
	0 means the base frequency is 5 MHz.
	1 means the base frequency is 1.25 MHz (system clock divided by 4).
	2 means the base frequency is 312.5 KHz (system clock divided by 16).
	3 is the same as 2.
Unsigned (bit)	0 means the Value range is -128 to $+127$.
	1 means the Value range is 0 to 255.
Value (byte)	This sets the on time for the PWM signal. (Refer to previous section.)

IOLine1 and IOLine2 are the two signals that connect to the H-bridge direction lines. They are also known by other names, such as A1 and A2. It can bend your brain to try to figure which way a motor will turn when you attach *this* wire to *that* signal line, so I just connect the wires, and if the motor turns the wrong way for the setting I want, I swap them. IOLineP is the PWM line and typically connects to the *enable* pin of the chip or board you are using.

Listing 10-1 shows how I configured the OOPic R board to control these motors. The OOPic R has a convenient 8-pin DC motor control connector, which is why I chose it for this project. You can just as easily configure this experiment using an OOPic II or II+. You just have to make a different set of cables.

```
'dualDCM2.osc
'Chapter 10 DC Motors experiment.
'Demonstrates the use of oDCMotor2 to operate an L298
'dual H-bridge for driving two DC motors at two PWM
'frequencies.
'
'For OOPic II B.2.X firmware
'
'Copyright Dennis Clark 2003
'Permission is granted to use this code anyway you like
'as long as you say you got it from me.

Dim Lmotor as New oDCMotor2
Dim Rmotor as New oDCMotor2
Dim n as New obyte

Sub Main()

    Lmotor.IOLineP = 17          'PWM output
    Lmotor.IOLine1 = 24          'Direction pin
    Lmotor.IOLine2 = 25          'Other direction pin
    Lmotor.Prescale = 2          '312 KHz base rate
    Lmotor.Period = 255          '1.2 KHz rate
    Lmotor.Operate = cvTrue

    Rmotor.IOLineP = 18          'PWM output
    Rmotor.IOLine1 = 26          'Direction pin
    Rmotor.IOLine2 = 27          'Other direction pin
    Rmotor.Prescale = 2          '312 KHz base rate
    Rmotor.Period = 255          '1.2 KHz rate
    Rmotor.Operate = cvTrue

    'OOPic for/next loops do not like negative numbers, so
    'here is a way to send negative numbers to objects.

    for n = 0 to 128             'Ramp down in one direction
        Lmotor = n - 128
        Rmotor = n - 128
        OOPic.delay = 5
    next n

    for n = 0 to 127             'Now ramp back up
        Lmotor = n
        Rmotor = n
        OOPic.delay = 5
    next n

    Lmotor = 0
    Rmotor = 0

    'Now let's change the PWM frequency from 1.22 KHz to
    '20 KHz, with about the same number of steps.  What did you
    'notice that was different?

    Rmotor.PreScale = 0
    Rmotor.Period = 250
    Lmotor.PreScale = 0
    Lmotor.Period = 250
```

(Continued)

```
        for n = 0 to 128              'Ramp down in one direction
            Lmotor = n - 128
            Rmotor = n - 128
            OOPic.delay = 5
        next n

        for n = 0 to 128              'Now ramp back up
            Lmotor = n
            Rmotor = n
            OOPic.delay = 5
        next n

        'Hit the brakes.
        Lmotor.Brake = cvTrue
        Rmotor.Brake = cvTrue

        Do:Loop

End Sub
```

Listing 10-1

This experiment has been designed simply to show you what to expect when using the URCP DC motor objects. I made these motors first with 1.22 KHz PWM and then with a 20 KHz PWM frequency. What did you notice that was different? I'll give you a hint: 1 KHz is well within the human hearing range. How did your motors do at the higher frequency? If you have low-quality motors, you'll find that it takes a higher-percentage PWM to get the motor to start turning. Why? The reasons can be complicated. A simple one is that the armature-bearing resistance is higher on low-quality motors. Winding inductance can be another reason, but that is another story. Regardless, I connected my Escap gearhead motors to the circuit and at 20 KHz they started turning at low PWM values. Those are $50 motors, so the lesson is you get what you pay for.

EXPERIMENT #2: RANGE FINDING WITH THE SHARP GP2D12 IR RANGE FINDER

This experiment shows you what to expect when using the oIRRange object to get range information.

oIRRange (B.X.X Firmware and Later)

This object gives you readings from a Sharp GP2D12 analog IR range finder device. The GP2D12 has a range from 10 to 80 centimeters (4 to 31 inches), and the oIRRange object reports to a maximum range of 24 inches (URCP 128). Thus, 7 inches of range capability are lost, but you also lose the hassle of figuring out what the range is. The GP2D12 does not report a linear analog voltage change as the distance increases; the OOPic oIRRange object handles this by using a lookup table and reporting the distance in a linear fashion. Because of this nonlinearity, some variation occurs in the results that different units report. You can use the Center property to adjust

TABLE 10-3   OIRRANGE PROPERTIES

Center (byte)	Enables you to adjust the center point (-128 to $+127$) of your ranging unit (refer to previous info).
IOLine (nibble)	I/O lines 1 through 7 can be used. Remember, the OOPic II and II+ offer seven potential A2D lines each, whereas the OOPic I only has four.
Operate (bit, flag)	Must be set to 1 for this object to respond to input.
Value (byte)	The distance to the nearest object in URCP format; 0 to 128 is the range.

Figure 10-5 Sharp GP2D12 IR ranging module

your unit until the results are 128 when nothing falls within 24 inches of the sensor's range. The oIRRange memory size is 5 bytes and its properties are outlined in Table 10-3.

It's fairly easy to tweak your IR ranger to get proper values. Mine came in as a 20 when nothing was in front of it. To get it to report range values, I connected my trusty *liquid crystal display* (LCD) display that I've so often used throughout this book. Figure 10-5 shows a picture of the Sharp GP2D12 ranging module.

I connected my Sharp IR ranger to A2D line 7, just to show that I could, on my OOPic II board. I didn't use the OOPic R board for this experiment because that board has the three LED/button I/O lines taking up the last three A2D lines, and I wanted to try using a different one. This also allowed me to make a connector that used the $+5$V, ground, and I/O line 7 on the 40-pin connector with a single 3-pin plug since those pins were adjacent to each other on this connector. Clever, huh? Figure 10-6 shows my I/O line and power connections for the LCD and GP2D12 IR ranger with an OOPic II board.

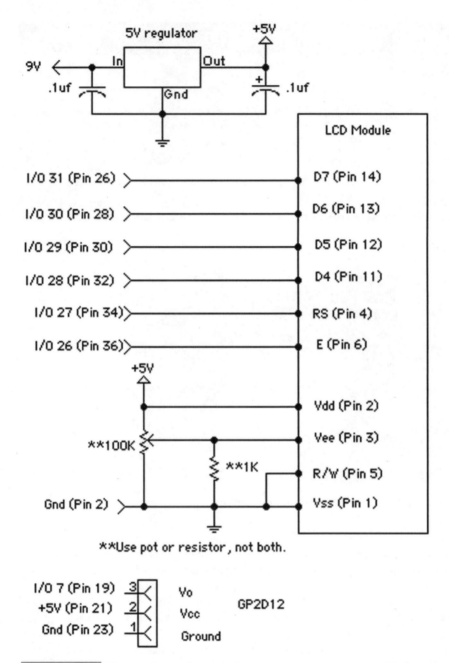

**Use pot or resistor, not both.

Figure 10-6 IR ranger experiment setup

Listing 10-2 is the code used to configure and read the GP2D12.

Notice that I didn't have to tweak the Center property much to bring it in line. Still, I found that my range values were off; they were short by about 10 percent, which was easy to fudge to get a reasonably close measure. If you are metric by nature, remember that 2.54 centimeters are in an inch, and 12 inches are in a foot, which gives 30.5 centimeters per foot. You can then convert URCP range values to centimeters by just dividing by 2, and that'll be as close as the accuracy of the device can muster anyway.

```
'ranger.osc
'Chapter 10 Sharp IR Ranger experiment.
'Demonstrates the use of the oIRRange object and tweaks
'it to show proper range readings.
'
'For OOPic II B.2.X firmware
'
'Copyright Dennis Clark 2003
'Permission is granted to use this code anyway you like
'as long as you say you got it from me.

Dim range as New oIRRange
Dim LCD as New oLCD
Dim n as byte

Sub Main()

    OOPic.node = 1

    LCDsetup                            'Setup LCD

    'Configure the ranger
    range.IOLine = 7                    'Yup, B.X.X has 7 A2D lines.
    range.Center = 7                    'Not too far off.
    range.Operate = cvTrue

    'Now take readings and translate them to feet.  I'm using
    'feet because inches don't round out well.  Also print
    'the URCP number as a reference point.  I added a "fudge"
    'factor in because I read a little short, about 10%.
    do
        LCD.Locate(1,0)
        LCD.String = "                 "
        LCD.Locate(1,0)
        n = range + (n/10)              'mine reads a little short
        printData(n/64)
        LCD.String = "."
        printData((n Mod 64) * 10 /64)
        LCD.String = " feet"
        LCD.Locate(1,10)
        printData(n)
        OOPic.delay = 100
    loop

End Sub
```

(Continued)

```
Sub LCDsetup()
[Text deleted for brevity, full source is on the CD-ROM]
End Sub

Sub printData(wt As byte)
[Text deleted for brevity, full source is on the CD-ROM]
End Sub
```

Listing 10-2

EXPERIMENT #3: COMPASS DIRECTIONS WITH THE DINSMORE 1490

Eventually you are going to want your robot to know which direction it is going. The compass is the natural device to tell you. This experiment uses a low cost compass that has low resolution.

oCompassDN (B.X.X Firmware and Later)

The Dinsmore 1490 compass module is an inexpensive compass that reports the compass headings it detects using the eight cardinal points: N, NE, E, SE, S, SW, W, and NW. The four cardinal points are straightforward, just N, E, S, and W. The other four directions are combinations of two adjacent points. The needle of the compass in this package is encased in a highly viscous fluid that dampens its movements and also makes it slow to respond. The needle position is detected by four Hall-effect sensors positioned at 90-degree intervals around the case. Hall Effect devices sense magnetic fields, in this device they sense the presence of the compass needle, which is magnetized. The direction north is arbitrarily chosen and the other directions are determined from that spot. The oCompassDN memory size is 4 bytes and its properties are listed in Table 10-4.

The hardware is simple to assemble, just tedious. The Dinsmore 1490 compass has 12 leads coming out of it. The outputs of each of the Hall-effect sensors are open collector, so a pull-up resistor is needed. If you use the OOPic's I/O lines 8 through 15 for your required nibble, you can set OOPic. Pull-up to 1 and use the OOPic internal pull-up resistors instead of your own external ones. A picture of the device is shown in Figure 10-7. It has the diameter of a nickel and is about an inch tall. A 90-degree swing takes approximately 2.5 to 3.5 seconds for the compass to complete, with no overshoot. Unlike the other electronic compasses, the 1490 module stands a 12-degree tilt and suffers little accuracy disturbance. Of course, you only have a 45-degree resolution to start with, but it is good enough for tweaking dead reckoning work.

TABLE 10-4   OCOMPASSDN PROPERTIES

IOGroup (nibble)	0 represents I/O lines 1 through 7.
	1 represents I/O lines 8 through 15.
	2 represents I/O lines 16 through 23.
	3 represents I/O lines 24 through 31.
Nibble (bit)	0 means the lower 4 bits of the selected I/O group are used.
	1 means the upper 4 bits of the selected I/O group are used.
Operate (bit, flag)	Must be set to 1 for this object to respond to input.
Unsigned (bit)	0 means signed Value numbers are allowed.
	1 means only unsigned Value numbers are allowed.
Value (byte)	URCP directional value returned −128 to +127.
	I used the signed values and modified my *printData()* function to handle negative numbers.

Figure 10-7 Dinsmore 1490 digital compass module

The compass module has no fixed north position. When you wire it up, you can choose which direction you want to call north by how the signals are interpreted. Figure 10-8 shows how I connected my module to the OOPic for oCompassDN configuration to get a 0 value for north, −64 for west, 64 for east, and −128 for south. Note that the pins are placed counterclockwise around the can. The pinout in Figure 10-8 is shown from the top of the part and the A, B, C, and D denote which output is connected to which OOPic I/O line on an OOPic II board. Listing 10-3 shows the code used to set up oCompassDN with the Dinsmore 1490 compass, and it also is used to configure the hardware.

This program prints out the direction the compass is pointing every second. It handles both negative and positive numbers and the slow cycle rate nicely matches the slow cycle time of the compass itself. This is an inexpensive part; it is crude, with only a 45-degree resolution, but it can handle small inclines up to 12 degrees without a loss of accuracy. The printed URCP values match those in Figure 10-1 exactly if you consider 0 north. This is a fun project with lots of potential uses for robotics.

Project: Line-Tracker Robot

This project is a little different from previous projects. Here I will give you the location to find a simple and inexpensive robot platform, show you how to mount an OOPic R board to it, explain how to wire up the necessary sensors, and finally show you how to program your robot to be a successful line-following competitor. I have decided to use a commercial robot platform because it's a pain to have to build everything from scratch, and this platform is so inexpensive, easy to assemble, and sturdy that I just couldn't resist it.

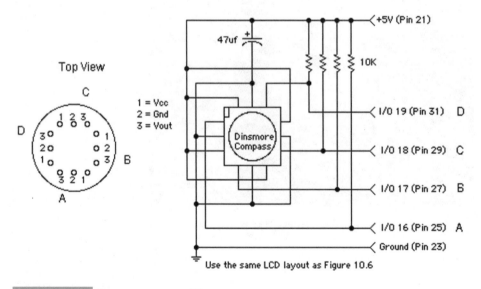

Figure 10-8 Compass connections

```
'compassdn.osc
'Chapter 10 Dinsmore 1490 compass experiment.
'Demonstrates the use of the oCompassDN object used
'with the Dinsmore 1490 compass and how to configure
'the hardware.
'
'For OOPic II B.2.X firmware
'
'Copyright Dennis Clark 2003
'Permission is granted to use this code anyway you like
'as long as you say you got it from me.

Dim compass as New oCompassDN
Dim LCD as New oLCD

Sub Main()

    OOPic.node = 1

    LCDsetup                          'Setup LCD

    'Set up the compass.  How we connect the outputs of the
    'compass is completely arbitrary, just so long as we go
    'in order around the case counterclockwise.  This layout
    'depends upon the order going counterclockwise in order
    'from I/O line 16 to I/O line 19.  Turn the module until
    'you find the side that gives a 0 for north.
    compass.IOGroup = 2               'I/O 16-24
    compass.Nibble = 0                'I/O 16-19
    compass.Operate = cvTrue

    'Now we point it around and see what numbers come back.
    'In this way, we find the orientation such that 0 is
    'north, which is arbitrary, but sounds good.
    do
        LCD.Locate(1,0)
        LCD.String = "                "
        LCD.Locate(1,0)
        printData(compass)
        OOPic.delay = 100
    loop


End Sub

Sub LCDsetup()
[Deleted for brevity, full text is on the CD-ROM]
End Sub

Sub printData(wt As byte)
    'Print out a byte, nibble, or bit value to LCD.
    'This simply outputs one character at a time by finding the
    'digit in the 100 place (255 is highest number) and
    'works down to the 1's place.  No leading zeros are output.

    Dim wTemp as byte
    Dim bTemp as byte
    Dim mTemp as byte
```

(Continued)

```
bTemp = 0
mTemp = 0

if (wt AND &H80) > 1 Then              'negative number
    wt = NOT wt + 1                    'convert to positive
    LCD.String = "-"                   'note negative
End if

wTemp = 100
While wTemp > 0
    mTemp = wt/wTemp
    if ((mTemp > 0) OR (bTemp = 1) OR (wTemp = 1)) then
        wt = wt - mTemp * wTemp
        bTemp = 1                      'Now print trailing zeros
        LCD.Value = mTemp + 48         'convert to ASCII character
    End If
    if wTemp = 1 then
        wTemp   = 0
    else
        wTemp = wTemp /10
    End If
wend

End Sub
```

Listing 10-3

In fact, this project has worked out so well that you will find a QuickTime movie of my robot traversing the Acroname line-following competition course on the CD-ROM accompanying this book. It takes less than a minute to finish the course and could be faster if you drive the servos with a higher voltage than the 6V that I used.

THE ROBOT PLATFORM

I wanted an inexpensive and easy-to-assemble robot for this project so that I wouldn't have to dedicate a whole chapter or even a whole book to constructing one. I found just such a robot at www.junun.org/MarkIII/Store.jsp.

I normally don't promote any single shop over another, but this place has the best little robot kit I've seen. It's clever in its mechanical design where many pieces rely upon each other to create a simple but durable robot. Here is what I bought to build my line-follower robot. All prices are in U.S. dollars as of March 1, 2003:

1 Mark III Chassis kit	$ 9.50
3 QRB1134 IR photo-reflectors	$ 5.25
1 Injection molded wheels	$ 6.00
2 High-torque ball-bearing servos	
(any hackable Futaba splined servo will do)	$ 20.50
Total cost:	$ 41.25

Just add your OOPic R robotics board and you are good to go. Figure 10-9 shows what my Mark III OOPic R robot looks like fully built for this project. You can't see the colors, but it's jet black with stylish bright yellow wheels. Because the OOPic R board screw holes do not match the uprights on the Mark III robot base, I mounted a 1/8-inch Lexan™ layer to 1the chassis and then mounted my OOPic R board to that panel. It took me 15 minutes to make that little kludge using a hacksaw, drill, and pliers.

You could simply buy the full OOPic II+ Mark III kit and get the same robot a little cheaper, but you wouldn't have the OOPic R's nifty robotics connectors and you'd need to remap where all the I/O lines came to headers on the Mark III board.

The last finicky phase of the robot chassis construction is to mount the IR reflector sensors on the front of the robot so they can detect a black line effectively. The QRB1134 sensors are not small, so you're limited in how close you can get them side by side. Figure 10-10 shows how I mounted and "aimed" the outer sensors. The Mark III robot is designed as an entry-level mini-sumo robot frame, so its sensors are mounted at the extreme edges of the front wedge. I drilled two holes so I could mount those edge sensors as close to the center sensor as I could. I then adjusted them all so that none of them comes in contact with the surface of the mat. If they do touch the Acroname line follower course mat, the robot could jam up or skip over some areas of higher friction.

Figure 10-9 Mark III OOPic R line-following robot

Figure 10-10 Mounting the line sensors

ELECTRONICS HARDWARE LAYOUT

You can plug the servos into the OOPic R board at I/O lines 28 and 31. I have used a jumper between the *MP* and *SP* power lines on the *Optional Power* connector and moved *S4* of the power jumpers to the *SP* setting. This allows me to power the OOPic R board and the servos from the same battery that was plugged into the main *Power* connector. See Chapter 1, "OOPic Family Values," for a full description of what I am talking about. I could have also powered the servos by plugging in another battery just for them, but this robot doesn't have a whole lot of space for more battery packs, so I chose to look elegant instead, and use a single battery for everything.

Next I created a simple protoboard that holds the resistors for the QRB1134 sensors. This small board also makes all the power and ground lines common so that out of the 12 wires that go into it for the sensors, only 5 are run to the OOPic R board: 2 power and 3 sensor outputs. Figure 10-11 shows where everything is connected to everything else and what I/O lines are used on the OOPic R board.

Did you notice I'm connecting the outputs of analog sensors to OOPic R digital input lines? This is because I carefully set the bias on the photo-transistors of the QRB1134 sensors such that they read nearly 0 when they are over a white reflecting surface. They read from 3V to nearly 5V when over the black line. This is even in the worst lighting situations (when there's bright lights around). The OOPic sees a logic 1 at any voltage over 0.8V and a logic 0 under that level. This means I didn't need to build any special A2D comparator circuit to use with the oTracker hardware object that only takes digital inputs. I could simply plug the sensors in. I've verified that they work great in this application.

After building the sensor board, I wrapped it in electrical tape to insulate it and stuffed it between the Lexan plate and the robot chassis with only the wires I needed going to the OOPic R board. Finally, I used some 22-gauge *insulation displacement connectors* (IDCs) as the connectors on my wires and plugged them into the OOPic R board. I recommend this to anyone that builds robots. A good connector is essential and a bad one can sink you on competition day.

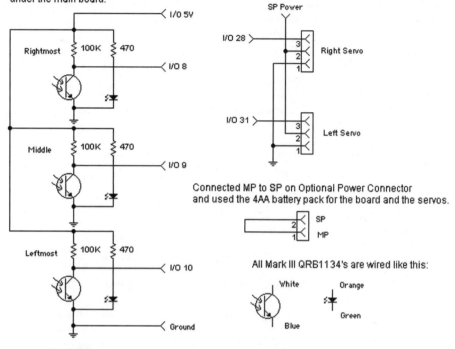

These are all on another PC board stuffed under the main board.

Connected bank 4's power jumper to the upper position.

Connected MP to SP on Optional Power Connector and used the 4AA battery pack for the board and the servos.

All Mark III QRB1134's are wired like this:

Figure 10-11 Mark III OOPic R hardware connections

DESIGN THE VCS

Ah, back to good old project planning again after all this time hacking code for experiments. It feels good to know where you're going before you start. An ancient Chinese saying goes something like this: "A man with no destination will never end his journey." The robot hobbyist saying is similar: "If the designer does not know what his robot is supposed to do, it will never be completed." I harp on this all the time out there on the robotics boards, and I'll do it again here. The designer *must* keep three things in mind when building a robot. These three items are related and inseparable. If you change one, you'll have to change the other two as well. They are

- The robot's task
- The robot's intended environment
- The robot's sensors

These three are so tightly coupled you should say them all in the same breath. Fortunately, you've already defined all three objectives, making your robot easier to design and allowing you to tweak it until it's perfect. Those objectives are as follows:

- Follow a 3/4-inch-wide black line on a white mat.
- Create the white vinyl mat with a 3/4-inch wide black line on it. Some gray areas should be there to be checked with your sensors.
- The sensors are three QRB1134 IR photo-reflectors mounted 3/4-inch on-center (center of the sensors are 3/4 of an inch from each other) from each other front and center on the robot.

Okay, you're ready to design. I like using 100-percent VC designs when I can, which this project does. Three things must be accomplished:

- Look for the black line and get the values to decide which way to turn.
- Translate the turn data into motor control data.
- Use motor control data to change the speed of the hacked servos.

The OOPic B.2.X firmware has specific objects for all three of those needs:

- oTracker is used to track the black line and send back directional change data.
- oNavCon (and oNavConI) translate directions into motor commands.
- oServoSP1 is used to control the hacked servos.

Let's look at these objects in detail to see why I chose them.

oTracker (B.X.X Firmware and Later)

 This object is specifically designed to use digital sensor input in the form of *go/no-go* binary inputs to determine the location of a black line on a white background. The oTracker object then determines the direction in which the robot needs to go, and the object formats a URCP reading that describes how much it needs to turn to do so. Changes come in the ranges $+/-8$, $+/-16$, $+/-24$, and $+/-32$. The farther the sensor reading is from being centered, the higher the directional change value is.

Normally, this object requires four sensor inputs in the form of right, middle right, middle left, and far left to get a full directional change resolution, but a Width option enables a three-sensor mode to be used successfully, which is the mode used for this project. The oTracker memory size is 5 bytes and its properties are shown in Table 10-5.

Two properties need some explanation; the first is Width. This object expects to have 4 sensors to detect a line and uses 4 bits to determine the next direction that needs to be taken. In some cases, room for four isn't available, as in this project. In that case, you set Width to 1 and oTracker does the following:

1. Bit 0 *least significant bit* [LSB] is left alone.
2. Bit 2 is shifted over to bit 3's *most significant bit* [MSB] position.
3. Bit 1 is copied to bit 2's position.
4. Bit 3's input is ignored.

TABLE 10-5   OTRACKER PROPERTIES

Center (bit, flag)	1 signifies that the robot is centered on the line.
Direction (bit, flag)	0 means the last movement of the line was toward the LSB sensor.
	1 means the last movement of the line was toward the MSB sensor.
InvertIn (bit)	0 indicates a high-logic input means the line is under the sensor.
	1 indicates a low-logic input means the line is under the sensor.
IOGroup (nibble)	0 represents I/O lines 1 through 7.
	1 represents I/O lines 8 through 15.
	2 represents I/O lines 16 through 23.
	3 represents I/O lines 24 through 31.
Mode (bit)	0 means that, if no line is detected, you can assume the line is far outside either the left or right sides, depending on the last direction. Value is set to either -32 or 32 depending on the Direction property.
	1 means that, if no line is detected, you can assume the line is between the center two sensors. Value is set to 0 and Center is set to 1.
Moved (bit, flag)	If set to 1, the line has moved since the last time Moved was set to 0. This simply means something has changed.
Nibble (bit)	0 means use the lower four bits of the selected IOGroup.
	1 means use the upper four bits of the selected IOGroup.
Operate (bit, flag)	Must be set to 1 for this object to respond to input.
OutOfRange(bit, flag)	When this bit equals 1, the line is not under any sensor.
Unsigned (bit)	When this bit equals 1, don't use URCP; Value is unsigned.
Value (byte)	The URCP (or unsigned) directional data (default property).
Width (bit)	0 means use four sensors for line detection.
	1 means use three sensors for line detection.

Basically, this assumes that the center two sensors are always the same value. The object can then treat this sensor data in the same way as a 4-bit input. The only difference is that you get the $+/-8$, 24, and 32 values, and not the $+/-16$ value. It works great.

Now, just what do you see in Value and the other outputs when the line changes position under the sensors? Table 10-6 explains this. An X in a sensor position means that the black line is seen under that sensor. An $<$ or $>$ means that the direction bit is set in either the MSB or LSB direction, respectively. The sensor pattern is MSB to LSB, and four sensors are always assumed to be in use.

TABLE 10-6   INTERPRETING OTRACKER'S RESULTS

SENSORS	MODE	SIGNED VALUE	UNSIGNED VALUE	CENTER	OUTOFRANGE
---->	1	−32	4	0	1
---X	0 or 1	−24	5	0	0
--XX	0 or 1	−16	6	0	0
--X-	0 or 1	−8	7	0	0
-XXX	0 or 1	−8	7	0	0
-XX-	0 or 1	0	8	1	0
XXXX	0 or 1	0	8	1	0
----	0	0	8	1	1
XXX-	0 or 1	8	9	0	0
-X--	0 or 1	8	9	0	0
XX--	0 or 1	16	10	0	0
X---	0 or 1	24	11	0	0
<----	1	32	12	0	0
All other	0 or 1		No change		

You can see the places where the three sensor readings have holes in the directional information. Only one set exists: the $^+/-16$ slot.

oNavCon(I,C) (B.X.X Firmware and Later)

This processing object is designed to take turn information (like oTracker) and convert it into motor control speed information for two motors driving a robot in differential drive mode. Basically, oNavCon takes the value of the object linked by the Input1 property (the speed value) and adds and subtracts the value of the object linked by the Input2 property (the turn value). The resulting two numbers (the two motor's speed values) are then stored in the objects linked by the Output1 and Output2 properties where Output1 equals Input1 + Input2, and Output2 equals Input1 − Input2.

In the case of oNavConI, the Input2 (turn value) is replaced with an 8-bit Value property so no external object needs to be linked to this object. In the case of using hacked hobby servos, if you use oNavConI, the reverse value needed to handle a motor that is flipped over (because it's on the other side of the robot) is automatically handled. It is confusing, but it works, and you'll see the code that does this later on. For those inevitable unmatched motors, the Center property enables you to dial in a bias to one side to offset that drift. The

TABLE 10-7   ONAVCON(I,C) PROPERTIES

Center (byte)	Added to Input2, which is used as the turn value to offset a motor inequality bias.
ClockIn (flag pointer)	Links to a flag used to clock this object (oNavConC and oNavConIC only).
Input1 (byte pointer)	Links to a Value property used as the robot speed.
Input2 (byte pointer)	Links to a Value property used as the turn value.
InvertC (bit)	Inverts the active logic sense of the ClockIn flag when it equals 1 (oNavConC and oNavConIC only).
Limit (bit)	When equal to 1, the motor values are limited to -128 and $+127$.
Operate (bit, flag)	Must be 1 for this object to respond to input (default property for oNavCon and oNavConC).
Output1 (byte pointer)	Links to the motor Value input for one side of the robot.
Output2 (byte pointer)	Links to the motor Value input for the other side of the robot.
Value (byte)	Used by oNavConI as the turn value (default property for oNavConI and oNavConIC).

oNavCon(I,C) memory size is 7 bytes (8 bytes for clocked [C] versions) and its properties are shown in Table 10-7.

This processing object does some simple math from linked inputs and sends the results to linked outputs. That's it; you don't have many details to worry about. Your number one issue is deciding which motor is Output1 and which is Output2. Make sure you choose such that the positive values make your motor go forward and the negative ones make it go backward. Everything else will be taken care of.

oServoSP1 (B.2.X Firmware and Later)

This object is covered in detail in Chapter 5, so I'll not repeat information here. The reason I chose this servo object is that it is URCP, so I can link it to oNavCon, and when the 0 position is used, the OOPic sends no servo pulses out, which guarantees that the wheel won't spin.

I found that, after hacking, my two servos didn't have a correct center position. I created a quick program for each of them and found the values I needed to center them by outputting pulse values until the servo stopped. I then divided those numbers by 2 and used them as the offsets for the servo. They worked great. The reason I'm being so careful is that I need the servo to spin in one direction with positive URCP speed values and to spin the other way with negative ones. I had to have a centered servo for that to work.

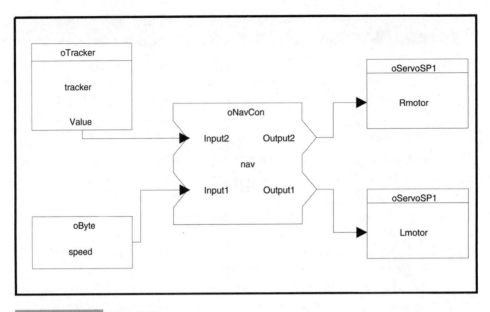

Figure 10-12 Robot VC

I explained all that so that you could actually design the VC for this robot. It's just too cool to do the whole robot activity using a single VC. Figure 10-12 shows the VC layout I've chosen to use.

WRITE THE CODE

Here's the part you've all been waiting for. In fact, you've probably skipped the first parts of this chapter just to get to this section. For shame! Go back and read it! I didn't write it for you to ignore it, you know.[3]

Listing 10-4 shows the initial program I wrote to handle line-following using the OOPic R robot and it works perfectly. You should now pop the CD-ROM into your computer and look at the QuickTime movie I made showing it breezing through the Acroname line-following competition. I've shown this robot running this course to everyone I know, and most of them aren't robot crazy like I am, so now they dodge me when they see me coming with that gleam in my eye.

Did you notice the dummy servo object in the code? This was put in because a bug is currently in the 5.01 OOPic compiler that does not recognize Refresh as a valid property. Thus, I wasted a few bytes, put in a dummy servo object, and set the Refresh property to 1. It doesn't matter that I don't even set an I/O line for that servo; the point is to set that refresh bit so that the OOPic servo system doubles the refresh rate of the servos. This doubled refresh rate makes them stronger, which translates to higher speeds and more power (for you sumo wrestlers out there).

3. Okay, I'm kidding. You bought the book. You can decide what to do with it.

```
'linetrak.osc
'Chapter 10 robotics and URCP project.
'Demonstrates the basics of URCP heading and direction
'information and how to build a simple line-following
'robot that works really, really well.
'
'For OOPic II B.2.X firmware
'
'Copyright Dennis Clark 2003
'Permission is granted to use this code anyway you like
'as long as you say you got it from me.

Dim tracker as New oTracker
Dim nav as New oNavCon
Dim Lmotor as New oServoSP1
Dim Rmotor as New oServoSP1
Dim speed as New oByte
Dim dummy as New oServo

Sub Main()

    OOPic.node = 1

    'I'm using an analog signal on these I/O lines from the
    'QRB1134 IR proximity sensor.  The circuit that I chose
    'gives a swing of almost 3V when the sensor is over a
    'black line and about 0.0V over white and near white.
    'This robot only had room for three QRB1134 IR photo-
    'detectors up front, and the Width property when set to 1
    'tells oTracker that.  It then drops bit 4, shifts
    'bit 3 to that position, and makes bit 2 be the center
    '2 bits; this reduces your fine motor tweaking but
    'gives fully symetrical output just as if you had 4
    'detectors.
    tracker.IOGroup = 1              'I/O 8-15
    tracker.Nibble = 0              'narrowed to I/O 8-11
    tracker.Width = 1              'Only use 3 sensors
    tracker.Operate = cvTrue

    'Here's the magic. This object will translate speed
    'and heading into servo control values to steer the
    'robot.  I guessed which servos went where at first, and
    'then looked at what it did and modified links to get
    'the correct directions.  I set the initial speed
    'at 32 so that at the extremes the wheels would stop.
    'Because oTracker will output a zero when the robot is
    'directly on the line, this will be added to one servo
    'and subtracted from the other. Since the motor is
    'reversed on one side, I did an InvertOut on the side
    'whose (+) values made the servo go backwards.
    speed = 32                      'Go Forward full speed
    nav.Input1.Link(speed)
    nav.Input2.Link(tracker)        'heading information
    nav.Limit = cvTrue              'don't exceed -128 or 127
    nav.Output1.Link(Lmotor)        'differential for left
    nav.Output2.Link(Rmotor)        'differential for right
    nav.Operate = cvTrue
```

(Continued)

```
'I'm using oServoSP1 because a 0 value there means the
'pulses to the servos stop guaranteeing that the servo
'won't be spinning.  Since one servo is the mirror image
'of the other, you need to reverse the output to that
'one so the same value will make both go the same
'direction.  The center property adjusts so + goes one way
'and - goes the other for each servo.
Lmotor.IOLine = 31                      'Left servo
Lmotor .center = 3
Lmotor = 0
Lmotor.Operate = cvTrue

Rmotor.IOLine = 28                      'Right servo
Rmotor.InvertOut = cvTrue              'This one went backwards
Rmotor.center = 5
Rmotor = 0
Rmotor.Operate = cvTrue

'This dummy is here just to boost the refresh rate.  A
'bug in the current compiler doesn't allow Refresh in
'the oServoSP1, and it should be there.
dummy.refresh = cvTrue

End Sub
```

Listing 10-4

When I looked at the oNavConI object, I noticed that by making the turn input the constant added to one speed and subtracted from another, $+/-32$ would be sent to the servos if the speed is set to 0 and the Value (turn) property is set to 32. This effectively mirrors the off-side servo without having to invert that servo's signal. The reason I chose 32 as the turn value is because oTracker has a maximum turn request value of 32. This would stop a wheel on one side and make the other side go full blast, and smaller numbers have smaller effects. I tried this experiment, and it worked great. Listing 10-5 shows the modified line-follower robot code using these side effects.

Just for coolness' sake, I added code to sense when the robot has reached the end of the course so that it could beep twice and then stop. That might impress the judges in case of a tie in a competition. It's worth a try anyway.

```
'linetrak2.osc
'Chapter 10 robotics and URCP project.
'Demonstrates the basics of URCP heading and direction
'information and how to build a simple line-following
'robot that works really, really well.  This one,
'however, uses side effects of oNavConI to simplify the
'program and reduce the object space needed slightly.
'As a final treat, this program will sense when the
'robot has finished the course and will beep and stop.
'
'For OOPic II B.2.X firmware
```

```
'Copyright Dennis Clark 2003
'Permission is granted to use this code anyway you like
'as long as you say you got it from me.

Dim tracker as New oTracker
Dim nav as New oNavConI
Dim Lmotor as New oServoSP1
Dim Rmotor as New oServoSP1
Dim dummy as New oServo
Dim beep as New oFreq
Dim count as Byte

Sub Main()

    OOPic.node = 1

    'I'm using an analog signal on these I/O lines from the
    'QRB1134 IR proximity sensor.  The circuit that I chose
    'gives a swing of almost 3V when the sensor is over a
    'black line and about 0.0V over white.  This robot only
    'had room for three detectors up front; the Width

    'property when set to 1 tells oTracker that.  It then
    'drops bit 4, shifts bit 3 to that position, and makes
    'bit 2 be the center two bits; this reduces your fine
    'motor tweaking but gives fully symetrical output just
    'as if you had four detectors.
    tracker.IOGroup = 1                  'I/O 8-15
    tracker.Nibble = 0                   'narrowed to I/O 8-11
    tracker.Width = 1                    'Only use 3 sensors
    tracker.Operate = cvTrue

    'Here's the magic; this object will translate speed
    'and heading into servo control values to steer the
    'robot.  I guessed which servos went where at first, and
    'then looked at what it did and modified links to get
    'the correct directions.  I set the initial direction
    'to 32 so that at the extremes the wheels would stop.
    'Because oTracker will output a 0 when the robot is
    'directly on the line, 32 will be added to one servo
    'and subtracted from the other, giving an automatic
    'value pulse reversal, so no InvertOut is needed on any
    'servo.
    nav.Value = 32                       'Go Forward full
    nav.Input1.Link(tracker)             'heading information
    nav.Limit = cvTrue                   'don't exceed -128 or 127
    nav.Output1.Link(Lmotor)             'differential for left
    nav.Output2.Link(Rmotor)             'differential for right
    nav.Operate = cvTrue

    'I'm using oServoSP1 because a 0 value there means the
    'pulses to the servos stop, guaranteeing that the servo
    'won't be spinning.  Since one servo is the mirror image
    'of the other, you need to reverse the output to that
    'one so the same value will make both go the same
    'direction.  Center property adjusts so + goes one way
```

(Continued)

```
'and - goes the other for each servo.
Lmotor.IOLine = 31                    'Left servo
Lmotor .center = 3
Lmotor = 0
Lmotor.Operate = cvTrue

Rmotor.IOLine = 28                    'Right servo
Rmotor.center = 5
Rmotor = 0
Rmotor.Operate = cvTrue

'This dummy is here just to boost the refresh rate. A
'bug in the current compiler doesn't allow Refresh in
'the oServoSP1, and it should be there.
dummy.refresh = cvTrue

beep.Value = 59853                    '440 A note
count = 0

'Let's be cool now.  The robot will detect when it has
'completed the course.  This count of 40 was what I
'came up with as a safe number of "out of range" readings
'to denote the end of the course.  Too low a number
'would result in the robot quitting during a long,
'sharp corner; too high and it would never quit, always
'finding a line before the timeout.  Of course, this
'number is highly dependent on the speed of your robot!
do
    if (tracker.OutOfRange = 1) then
        count = count + 1               'got an out of range
    else
        count = 0                       'must be consecutive
    end if
    if count > 40 then                  'beep a little bit and
        beep.operate = cvTrue           'then stop the robot.
        OOPic.delay = 20                'makes a nice startup
        beep.operate = cvFalse          'button too.
        OOPic.delay= 20
        Beep.operate = cvTrue
        OOPic.delay = 50
        beep.operate = cvFalse
        OOPic.Operate = cvFalse
    end if
loop

End Sub
```

Listing 10-5

Now What? Where to Go from Here

I dedicated this chapter to URCP and robotics. This is a *huge* topic area, and believe me, it was tough to tune it down so it didn't take up the whole book (that book will come later, trust me). The new OOPic II URCP robotics-specific objects are easy to use and incredibly powerful. You can build up several VCs and have each of them implement some specific behavior in your robot. Then all behaviors run at all times, which means you never have to worry about a piece of code missing something.

The biggest problem you will have is implementing some kind of priority handler that determines which VCs can take control of output devices, such as motors. This is not impossible; in fact, it's not really all that difficult. You can use oCompare objects to check the priority assigned to a VC and only let the highest-priority one get through.

How do you assign a priority to a VC? Simple, by controlling when you define (Dim) it in your program. You can have the oCompare compare the addresses of the output objects and only let the higher address (or lower one—take your pick) take control of the motors. Once again, I leave that design as an exercise for the reader. I know I'm going to work on it. Maybe we'll meet at a competition some day.

OOPIC R SERIAL CONTROL PROTOCOL (SCP)

This chapter is dedicated to the brand-new capabilities of the OOPic R board. This is the first OOPic programmed over the serial port, which liberates it from the mysterious errors found on the various PC parallel port hardware configurations. Not much is known about the *Serial Control Protocol* (SCP) right now because the protocol is quite new, and although it is used to program the OOPic R through the serial port by the *Integrated Development Environment* (IDE), the documentation is too sketchy for us to use it in our programming experiments.

Here you will find all the SCP commands, formats, and procedures you need to start experimenting with SCP. I've even included some simple programs written in HotPaw™ Basic[1] for the PalmOS™ *personal digital assistant* (PDA) to interface to an OOPic R board. The bundled CD-ROM includes a promotional installation demo of HotPaw Basic for your PalmOS PDA, so you can load it all up and get everything in this chapter running right from the CD-ROM. If you like it, I encourage you to spend the tiny fee for the program and support tthe author.

This chapter contains both OOPic programs and PalmOS programs that show what can be done with SCP, a Palm PDA, and the OOPic R (or OOPic C or II+) in robotics and distributed control.

What Is SCP?

SCP is the new serial-port-based communications protocol for use with the OOPic II+, C, and R variants. It enables program downloading, debugging, and external communications through a standard RS-232C port on your computer.

The OOPic IDE no longer talks directly only to the *electrically erasable programmable read-only memory* (EEPROM) to download a program to the OOPic B.2.X+ firmware-equipped OOPic controllers. Now it can talk to the OOPic *operating system* (OS) via special serial commands. This means you can program other serial-port-equipped devices to interact directly with OOPic objects and even download new programs without using the OOPic IDE running on your PC.

SCP opens the OOPic hardware up to much more sophisticated and diverse interactions with distributed robotics, control hardware, and control software by using a remote PC, Pocket PC, Palm Pilot, or any other serial-port-equipped device. In fact, Savage Innovations had the PalmOS in mind when SCP was developed because the default serial port speed is 9600 baud, which is the default PalmOS serial port speed as well. In addition, the serially attached device can be used as a terminal to the OOPic, providing a form of user interface through the touch screen and keyboard. The SCP system has also taken special steps to enable the OOPic serial connection to be done with a wireless radio link by handshaking all commands and not requiring any special shorted pins or hardware flow control.

---

1. An inexpensive Basic interpreter that runs on the Palm OS (www.hotpaw.com/rhn/hotpaw).

The rest of this chapter deals with the hardware and system requirements for the SCP system, as well as the interface protocol and how to use it.

SCP Hardware and Firmware Requirements

Only a select few OOPic variants use the SCP communications protocol, and these devices restrict the use of some of the *input/output* (I/O) lines on the OOPic thus equipped. The following OOPic controllers are equipped with the firmware to use SCP communications:

- **OOPic R** This board even has the DB9 serial connector installed. SCP is usable right away; nothing needs to be added.
- **OOPic C** This board is the size of a large airmail stamp and requires a carrier board with a DB9 connector on it. The OOPic C does the RS-232-level translation already.
- **OOPic II+** This board has the correct firmware and PICMicro chip but lacks the RS-232-level translation hardware and a DB9 serial port connector. Chapter 1, "OOPic Family Values," and Chapter 7; "OOPic Events, Keypads, and Serial I/O (Project: A Mini-terminal)," give you plans for adding the needed circuits to facilitate serial communications.

OOPic firmware versions B.2.X+ are required to use SCP. The B.2.X+ firmware only runs on the PICMicro 16F877 chip as well. With this firmware, the serial port's I/O lines 22 and 23 are restricted for use as *only* serial ports. Initially, those I/O lines are also owned by SCP; however, an escape sequence can be issued that forces SCP to relinquish control of the serial port and "lurk" as a monitor in the background, waiting for another escape code and valid node address to bring it back in control. Although the SCP system is lurking in the background, the serial port can be used for normal serial communications. SCP immediately takes or releases control of the serial port upon detection of the following escape codes:

- *Take control: \0,* which is Escape 0 in ASCII encoding
- *Release control: \A,* which is Escape A in ASCII encoding

This enables your other programs to use the serial port if you so desire. Even if you issue the *Release Control* escape code, I/O lines 22 and 23 are forced to remain as serial ports; they cannot be reassigned as digital I/O lines.

You need to be running OOPic IDE software version 5.01 or later to use SCP (the serial port) to program and debug your OOPic R, C, or II+. The install for this IDE is included with the bundled CD-ROM.

Using SCP

SCP is a completely new set of instructions whose syntax is terse. It is intended to provide a fast means for accessing specific address locations for bits, bytes, words, and data. The OOPic is all about addresses and the specific data stored in them. Now, instead of using *Inter-IC* (I2C) to access those locations, you're going to use SCP. Unlike the I2C-based access, however, you can take control of the OOPic to single-step instructions and even call subprograms and functions.

This section serves as a complete list of the SCP commands and an introduction to using SCP to access program objects within a running OOPic program. You can even download whole programs using SCP, and I'll provide an example of how to do this as well, but first you'll need to see the command set.

THE SCP COMMAND SET

The first thing you must know and internalize is that SCP is *not* a programming language. It is an interface protocol similar to the commands used to access your hard disk or graphics card. As such, it may not be simple or intuitive to learn and use. With practice, you will begin to understand how it is formatted and become proficient. Just remember it's supposed to be lean, fast, and of high functionality in a small space. Those are *not* the keywords for "easy to understand."

Now onward with the commands. In the following sections, the character denoting the SCP command is listed first, and then the function of that command is given, followed by the description of what the command does.

\

Begin Escape Sequence

The Escape character is the first character of a two-character Escape Sequence. The Escape character is followed with a single digit that represents the serial node of the OOPic to communicate with. SCP mode is entered when the received serial mode is either 0 or matches the OOPic.SNode property value. If the received serial node is not 0 and does not match the OOPic.SNode property, then SCP mode is turned off. The Escape character can be issued at any time.

0 to 9, A to F
Numbers

These are the values written out or read back from the OOPic B.2.X+ firmware. Remember that when writing directly to EEPROM or RAM, you *must* format the bytes in two-digit hexadecimal. However, you *must* use regular decimal when writing to default object memory.

G

Read Memory Type Register

This command reads the 8-bit Memory Type control register and sends its value out the serial port. A *g* is returned when the Read Memory Type command is completed. Regular decimal numbers are returned.

H

Store Memory Type

This command stores a value in the 8-bit Memory Type control register (see Table 11-1). Store Memory Type is preceded with the decimal number to be stored. An *h* is returned when Store Memory Type is completed. Regular decimal numbers are required.

TABLE 11-1 MEMORY TYPE REGISTER		
BIT(S)	**USE**	**DESCRIPTION**
7	Autoincrement address	0 means that the 16-bit Memory Address control register autoincrements each time the memory is read or written. 1 means the address is unchanged when memory is accessed.
6	Memory Type Bit—C	0 means internal memory is accessed. 1 means a local I2C port is used to access external memory.
5, 4	Memory Type Bit—B, A	Internal memory. 11: N/A. 10: read/write internal EEPROM. 01: read/write RAM. 00: read/write object's default property. Local I2C port addressing. 11: N/A. 10: Use 7-bit addressing, no internal address register. 01: Use 10-bit addressing, 8-bit internal address register. 00: Use 23-bit addressing, 16-bit internal address register.
3	Reserved	
2-0	Block Size	Number of bytes read or written. 000 reads 1 byte; 111 reads 8 bytes.

11

OOPIC R SERIAL CONTROL PROTOCOL

I
Read Memory Address Register

This command reads the 16-bit Memory Address control register and sends its value out the serial port. An *i* is returned when the Read Memory Address command is completed. Regular decimal numbers are returned, not hexadecimal.

J
Store Memory Address

This stores a value in the 16-bit Memory Address control register. Store Memory Address is preceded with the decimal number to be stored. A *j* is returned when Store Memory Address is completed. Regular decimal numbers are required.

K
Read Sub address Register

This command reads the 8-bit Subaddress control register and sends its value out the serial port. A *k* is returned when the Read Subaddress command is completed. Regular decimal numbers are returned.

L
Store Sub address

This command stores a value in the 8-bit Subaddress (address within the byte) control register. Table 11-2 shows the subaddress values when accessing RAM. When accessing I2C devices, the subaddress is the I2C address of the device (80L is the program EEPROM).

Store Subaddress is preceded with the decimal number to be stored. An *l* is returned when Store Subaddress is completed. Regular decimal numbers are required.

TABLE 11-2   SUBADDRESS VALUES

BITS(S)	USE	DESCRIPTION
7–6	Block number	One of the four blocks of RAM (see Table 11-3).
5–3	Bit shift value	The number of positions to left-shift the bits to match the bit position in RAM. 000: 0 bits, no shift. 111: 7 bits, maximum shift value.
2–0	Number of bits	The number of bits to access. 000: 8 bits. 001: 1 bit. 111: 7 bits.

TABLE 11-3   MEMORY BANK LOCATIONS

RAM				INTERNAL EEPROM	
BANK 0	BANK 1	BANK 2	BANK 3	BANK 0	BANK 1
32 bytes Objects 96 bytes	Operating System 128 bytes	40 bytes Variables 72 bytes 16 bytes	128 bytes	128 bytes	128 bytes

M
Read Memory

This command reads the memory specified by the control registers and sends its values formatted in two-character-per-digit hexadecimal numbers out the serial port unless you are reading the default object value memory. In that case, regular decimal numbers are returned.

An *m* is returned when the Read Memory command is completed. If accessing the local I2C bus and the program execution has not been stopped, Read Memory returns an asterisk (*). If accessing the local I2C bus and the addressed I2C device does not acknowledge, then Read Memory returns an exclamation point (!).

N
Store Memory

The Store Memory command stores value in the memory specified by the control registers. Store Memory is preceded with the value to be stored, formatted in two-character-per-digit hexadecimal, unless you are writing to the default object value memory. In that case, regular decimal numbers are required.

An *n* is returned when the Store Memory command is completed correctly. If you are accessing the local I2C bus and the program execution has been stopped, Store Memory returns an *s*. If accessing the local I2C bus and the program execution has not been stopped, Store Memory returns an asterisk (*). If accessing the local I2C bus and the addressed I2C device does not acknowledge, Store Memory returns an exclamation point (!).

O
Exit Single Step

This command exits Single Step Mode. An *o* is returned when the Exit Single Step command has completed.

P
Start Single Step

This command starts Single Step Mode. Start Single Step also steps through the first instruction. A *p* is returned when the Start Single Step command has completed.

Q
Query Buffer

Query Buffer returns the contents of the serial port's input buffer. A *q* is returned when this command has completed.

R
Clear Buffer

This command clears the contents of the serial port's input buffer. An *r* is returned when this command has completed.

S
Stop Program

Stop Program stops the program execution, which makes the local I2C bus available for use by the SCP module. An *s* is returned when Stop Program has completed. If no EEPROM is used in the E0 socket, Stop Program returns an exclamation point (!). If the program execution has already been stopped, Stop Program returns an asterisk (*).

> **Note:** Exiting SCP mode does not restart the program. Any active *virtual circuits* (VCs) are not affected by this command and will continue to operate.

T
Start Program

Start Program returns the local I2C bus to Program Execution mode, which results in the application program resuming from where it was paused. A *t* is returned when Start Program has completed. If the local I2C bus was already in Program Execution mode, this command has no effect.

U
Stop Acknowledge Commands

Stop Acknowledge Commands sets the other commands to silent mode which means that every command that doesn't return data will not respond with their completion character.

V
Start Acknowledge Commands

This sets the other commands to verbose mode where the return character is sent over the serial port after the commands complete. A *v* is returned when the Start Acknowledge Command has completed.

W
Reset

This command resets the OOPic. Upon resetting, the OOPic is no longer in Serial Control mode and no character is returned to confirm that the Reset command has completed.

X
Exit SCP Mode

This command exits the SCP mode. An *s* is returned when the Exit SCP Mode command has completed.

Y
Branch to Address

Branch to Address exits the SCP mode. A *y* is returned when the branch is started. No indication is made from SCP about whether the branch actually worked.

AN INTRODUCTION TO SCP PROGRAMMING

Now that you've seen the command set, you've got an inkling of what SCP can do. As usual, it is helpful to see some examples to see how to use the commands and interpret the results. These introductory examples are designed to help you with that, but they are *not* complete programs; they are snippets that would be used during an SCP session. I'm not even going to try to show everything the SCP can do. In fact, most of the time I'm going to stick to objects' default properties. This is the least likely way to cause a program to go haywire and the most easy-to-understand use of the application. However, the following includes some of the more esoteric things you can do with SCP as a template for future projects. Try it and see what works. The worst you can do is lock up your OOPic R, C, or II+ and have to do the "EEPROM not inserted" reset procedure detailed in Chapter 2, "The OOPic IDE and Compiler."

Entering and Exiting Serial Control Mode

These commands were shown earlier, but not explained. To take control of the serial port for SCP, use \0 or \n, where *n* is the OOPic.SNode set in your OOPic program.

This tells the SCP command system lurking in the background to start paying attention to SCP commands and acting on them. Now all numbers and letters are interpreted as SCP commands and parameters. You won't be able to use the serial port for anything else when you are in SCP command mode. To verify that you are indeed in SCP mode, send the *verbose* command V, which returns *v* unless you have used *U*, the *Stop Acknowledge* command, in which case it returns nothing.

To exit SCP mode, send an illegal escaped address such as \A. This tells SCP you don't want it to pay close attention anymore. SCP now is lurking in the background looking for an escaped address, such as \0 to tell it to pay attention again.

11

OOPIC R SERIAL CONTROL PROTOCOL

Setting the Object Address

Before you can read or write the default object property, you need to set the address. This is accomplished with the *J* command, preceded with the object address in decimal format. In fact, all object default property reads and writes use and return values in decimal format. For example, if you know that your oByte object is at address 120, you would set the address in SCP as \0120J. This is the set object address.

In reality, you don't have to use the \0, but it doesn't hurt. However, after you have done one \0 to address an OOPic, all you have to do is issue regular commands without the escape slash. 120J is the same because it sets the object address in the default (power-up) memory mode and replies with a *J* when completed, unless you have used *U*, the Stop Acknowledge command.

One difference must be pointed out. If you use a \0 without using a *V*, only commands that send data back will return a suffix response. The *j* or *n* returns won't happen unless and until you send a *V* (verbose) command.

Reading and Writing an Object's Default Property

The command for reading a memory location (no matter which mode or format it is in) is *M*. This command simply reads the address that the previous *J* command has set. Data returned from the OOPic is always be suffixed with an *m*. You'll have to strip that off to process the information numerically.

To write to the current memory address, use <n>N, which means that you precede the *N* with the data to be written, be it bit, byte, nibble, or word. For instance, let's say you are writing 100 to the oByte object whose address you just set. You would do it using 100N, after which SCP returns *n* unless you have used *U*, the Stop Acknowledge command. You can continue to write or read from the current address without having to continue sending the address first.

Resetting the OOPic

Sometimes you get all messed up and just need a quick reset to start at the beginning. Two commands will do this: *X* and *W*.

X is the *Reset and Stop* command. This resets the OOPic to the beginning, stopping the program and making sure no VCs are still running. It should return an *s* when it completes, but I've found that some terminal programs and serial input commands in HotPaw Basic won't see the *s*. They will see either nothing or a garbage character sequence. This command leaves SCP in control.

W is the *Reset and Run* command. This resets the OOPic to the beginning and starts the currently loaded program. This is the one to use if you've just loaded a new program or modified one already in memory. This command does not return anything; it does its work and turns SCP off (into lurk mode) silently.

S stops the currently running program (VCs stay in operation) and makes the OOPic local bus available for you to use via SCP. Using *T* restarts the program where it left off, so treat this as a "pause" in execution.

These are all the commands you need to know to start experimenting with OOPic SCP. Later I'll show you some more commands that can get you into even more trouble.

SCP Experiments

Now that I've frightened you with how complex SCP is, I will show you that it's actually simple to write SCP programs that deal with the OOPic B.2.X+ firmware and objects. My secret is to stick with the default mode and only read or write the default properties of the objects. This is the easiest way to use SCP, because it starts up and defaults to the OOPic object default property address mode. In fact, if you are going to write to an object, you should use this mode because it activates various object activities that set flags and take care of housekeeping chores.

To access any other object property, you need to know exactly where a bit or nibble is in an object and set several registers to read or write to that memory location. Those locations are not easy to find. As you saw in Chapter 3, "OOPic Object Standard Properties," digging that data from the OOPic's oRam object and the *Opp Codes* display of the view pane is tedious. It can be done, and then you can use the *Sub address* and *Memory Type* registers to go after specific bits and nibbles.

For these experiments, I decided to hack together another robot. It is basically the same one I showed you in Chapter 10, "OOPic Robotics and URCP (Project: A Robot That Toes the Line)," but it needed to be bigger to enable me to put my Palm M100 on it. I tried finding a way to put the Palm PDA on Mark III, but as you can see in Figure 11-1, that was unstable, unsatisfactory, and just plain ugly.

Thus, Klutzilla was born, a simple platform essentially the same as Mark III, but large enough to put bigger stuff on it. Klutzilla has an OOPic R board, two hacked hobby servos, an SRF04 SONAR unit, and a *Todays Toys and Technology* (TTT) *IR proximity detector* (IRPD) board (my own creation[2]). Of course, it also carries on it a Palm M100 PDA as the central brain that talks to the OOPic R board. While Klutzilla was slightly better than stacking everything on top of a tiny Mark III robot, it isn't the best. Figure 11.2 shows all the electronics on a very nice small and inexpensive platform called Scooterbot sold by BudgetRobotics.com. This is a $50 kit that comes with all the needed hardware, battery holder and two hacked RC servos. The kit goes together very easily in an hour, including the time that I took to drill 4 holes and make the boom for the SONAR board. They also sell a complete kit that includes the OOPic R board with all mounting holes drilled for it, or they sell just the shell with no motors or controllers. That ugly serial cable hides very nicely in between the upper and lower shelf of this kit. All versions of the robot carry that huge serial cable used to hotsync the Palm to a computer. I'm planning on getting another cable and hacking it down to use on the robot, leaving my main cable for the hotsync operation. You *do* want to save that hotsync cable because, trust me, you don't want to develop code using Graffiti™. You want to use your own text editor on your PC and then create a memo you download to your PDA when you do a sync.

---

2. Available on my web site www.techtoystoday.com for a modest price.

Figure 11-1 A really bad robot idea

Figure 11-2 Klutzilla, a slightly better robot idea

But first, let's try some SCP commands using that good old standby in your Windows machine, Hyperterm™. After that, we'll move on to much more complex and fun subjects.

SCP COMMANDS USING A TERMINAL EMULATOR

Before you dive into the world of Palm communications with the OOPic R, let's try some simple interfacing between the PC and the OOPic R by using Hyperterm to send and receive SCP instructions. This will allow you to get comfortable with SCP in a simple environment. Listing 11-1 is a simple *light-emitting diode* (LED) flasher program called *simple.osc* that you can download into the OOPic R. It will enable you to talk to the oDio1 object while you practice SCP.

Compile and download Listing 11-1, and then open Hyperterm while your serial cable is still connected to your OOPic. Make sure that Hyperterm is configured for 9600 baud, 8 bits, no parity, and 1 stop bit. Also check that it is directly connected to the COM port on which you program your OOPic R.

> **Note:** When you are typing commands in Hyperterm, the OOPic R does *not* echo the commands back. You won't see what you are typing unless you have *local echo* enabled.

Now let's do some SCP commands. Your board should be flashing the center (yellow) LED after the program has downloaded. This means that a program is running, which is important for your experiments.

Type in **\0V**. The \0 tells SCP to wake up and start paying attention to the serial port for SCP instructions. The V (verbose) then replies back with a *v*, saying the command was executed. If it doesn't, do the commands over again in case you did them wrong. Remember to always send SCP commands as capital letters.

Now type **S**. You will then get an *s*, and the LED stops flashing. You've just told SCP to stop the program from running.

The first object created is always at memory location 41, but if you don't believe me, you can look to find that address by reading the OOPic simple.omp file, which is shown in Listing 11-2. It has been edited a little to leave out the bits you aren't interested in. Notice that the omp file lists the addresses of all the objects in a program. This is the place to go to find these addresses. Remember that you'll use this information again later.

```
Dim LED as New oDio1

Sub Main()

    LED.IOLine = 6
    LED.Direction = cvOutput

    Do
        LED = OOPic.HZ1
    Loop
End Sub
```

Listing 11-1

11

OOPIC R SERIAL CONTROL PROTOCOL

```
Add Type          Name            ClE Size
--- ----------    ------------    --- ----------
041 oDio1         LED              0  001 Byte(s)
--- ----------    ------------    --- 000
041 Value         led.value
--- ----------    ------------    --- 000
041 Flag          led.nonzero
041 Flag          led.value
```

Listing 11-2

To read and write to the oDio1 object, you first need to set its address. Type **41J**, and you will get a *j*. Boring, but it tells you that SCP has set the address you told it to set.

Now type **M**. If your LED was on when you stopped the program, you will see *1m* returned on the screen. If your LED was off, you'll see *0m* instead. Type **M** a second time, and again, you'll get the same value. The address does not change, so you'll always be reading the same location.

Let's turn the LED on and off now. This shows that you have full control of an object remotely via SCP. Type **0N** to turn the LED off or type **1N** to turn it on. Or you can do both over and over again to make it blink, which is what you would do if you were controlling the object remotely. In each case, after you type the command, SCP replies with *n* to tell you it understood and acted upon your command.

Type **T**, and the LED starts blinking again because you told SCP to continue the program; it starts up right where it left off when you stopped it. If you wanted to have some fun and single step through commands, at this point you could type **P**, which takes the OOPic into single-step mode. When you get tired of sending P out, which executes a single EEPROM instruction each time. Then type **O**, which exits single-step mode and resumes the program flow.

Interfacing a PalmOS PDA with the OOPic R

Both the Palm PDA and the OOPic R have the same DB9 serial connector pinout. They'll both be female connectors, so you won't be able to connect them together. Therefore, you'll need what is commonly called a *null modem* adapter, where pins 2 and 3 (Rx and Tx) are swapped and pin 5 (ground) is the same. Figure 11-3 shows how you would wire this adapter. The pins are referenced as if you were looking at the back of the male DB9 connectors while they are plugged into your OOPic R and the PDA hotsync pigtail connectors.

Now you can continue on with the rest of the chapter, which deals with the PDA to OOPic R serial communications protocol.

Terminal Control with a Palm PDA

You can do the following experiment directly on your Palm if you have a terminal emulator on your PDA. I'm using Dicon 1.5, which is a freeware program for the Palm PDA that you can find at www.palmgear.com/software/. Enter dicon in the search window at the upper left and hit go! to find the software. It works great because it not only displays the ASCII code, but it also shows the hex code in a separate window. This enables you to see

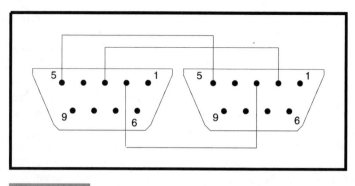

Figure 11-3 Null modem adapter between PDA and OOPic R

why some return codes look wrong. This book isn't about using a Palm PDA, so I won't go into detail about how to get the program into the PDA, but here is the short version:

1. Open your Palm desktop.
2. Click Hotsync, Install Handheld files.
3. Add the Dicon.prc file.
4. Hotsync your Palm to install the program.

Those are the baby steps, but you can see how simple it is to take control of your OOPic R with SCP commands. Now you'll move on to meatier topics such as programming your Palm PDA to take control.

INSTALLING THE HOTPAW BASIC SOFTWARE ON THE PALM PDA

I could write a whole book on the PalmOS PDA, but that isn't this book's topic. It took me one afternoon to go from zero knowledge to near expert on the Palm desktop, Hotsync, and the Palm PDA—to the level I needed at any rate. Go ahead and take your time to play with the Palm PDA so you understand how it works.

Here are the general requirements for using the HotPaw Basic interpreter on the Palm PDA. On the CD-ROM, under the *Palm* directory you will find two directories that have files you need to install. They are *in order of installation*:

- Palm\ybas138b0\mathlib11m\mathlib.prc
- Palm\ybas138b0\ybasdir\ybasic.prc

Use the file add and hotsync actions presented in the last section to install these files on your PalmOS PDA. You must install the *mathlib.prc* file before the *ybasic.prc* file, because it is a dependency file and the program won't work if you do it in the wrong order. You might also want to install an *examples* directory under *ybasdir*. Those are memo files, so open your Palm desktop and, using Windows Notepad and the copy and paste functions, create new memos with those files. You can't use the *add applications* method to create memos because those have to be created in the Palm Desktop program.

Get used to this method of creating memos. Trust me when I tell you that you don't want to do program development with a stylus on your PDA. You'll want your keyboard!

Note that all HotPaw programs are named something like #atipcalc.bas. That name is taken from the first line in the memo. Memos in the PalmOS don't have filenames; they are stored in a memo database that is searchable by the first line in the memo, which is why that first line is special.

You'll want to create a HotPaw category in your memos (or notes as they are sometimes called on the Palm desktop) for your HotPaw programs. You can create this category either on your PDA before a hotsync or in the Palm desktop. Use the Memo Pad and not the Notepad to edit or create new HotPaw programs.

The manual for HotPaw Basic is located at Palm\ybas138b0\ybasdir\quickref.txt and is in plaintext format. You'll also want to look at some example code, which can be found at Palm\ybas138b0\ybasdir\yb_tutorial.txt. HotPaw is called *ybasic* when it is installed on your PDA. Just tap on it to start the interpreter.

I created a program category named *programming* on my PDA and placed my terminal emulator and HotPaw interpreter there. You don't have to do that, but it does make it easier to separate your various tasks on your PDA when you give them different categories. To start a PalmOS program, just tap on its screen icon. When HotPaw is starts, it lists all the memos that start with the pound sign (#a) and end with *.bas*. I created a category of memos called *HotPaw* to hold my programs. In the upper-right corner of the screen is a pull-down menu you can use to select which category of programs you will look at. To run a program, select the memo you want with the stylus and tap the Run button.

Read the HotPaw manual for details on writing programs, but you should be aware of a few facts:

- All commands are lowercase.
- Variable and subroutine names are case insensitive.
- Although the programs have no line numbers, HotPaw assigns them a line number, starting with 1. To avoid confusion while you are debugging, don't leave any blank lines in your program; that makes it easier to count lines to find the one with the syntax error.
- All HotPaw memos must start with the pound sign (#) and end with .bas (this is the first line).
- A Memo Pad memo has a limit of 4000 characters, but you can use the *#include* pragma to pull in another memo that executes in-line like a macro.
- All variables are global and no dynamic linking takes place, so recursion is not supported.
- You can use *gosub* or *call* to call subprograms.
- A function may return a value by assigning that value to the name of the subprogram. For example, sub doit(this) . . . doit=that . . . end sub . . . returns the value of *that* like a normal function call (a = doit(5)). You cannot return a string.
- The system has a few bugs. The one I found is that the $<>$ compare does not always work, but fortunately there is always a workaround for that.

■ Remember to delete the old code from the PDA before you hotsync new code into it. If you don't, you'll have dozens of memos with the same name. Sometimes you want this; sometimes you don't.

Okay, now you've got a PDA, HotPaw Basic, a serial Hotsync cable, and your OOPic R. You're ready to go with some example programs to control your OOPic R via SCP from your PDA.

CONTROLLING THE OOPIC R WITH A PALMOS PDA AND HOTPAW BASIC

Now the real fun begins. From here on out, you'll be controlling your OOPic R with a Palm PDA. In my case, I'm using a Palm M100 (because I got it really cheap on eBay and it has a serial connector, not *universal serial bus* (USB), which makes this the perfect package). The M100 only has about 2MB of memory in it, but because I'm only writing robot code there, I've got *plenty* of RAM to work with. You can use any Palm PDA that is as new or newer than the Palm III. The main issue is that you need to have PalmOS version 3.1 or newer to run HotPaw Basic. This program does not work with Visor™ PDAs because the Visor does not output RS-232-level signals; it outputs *time to live* (TTL)-level serial data. You'd need to put in an RS-232-level translator circuit. Check the documentation and with your local gurus about any other particular PalmOS-based PDA to see what they send out their serial port.

Listing 11-3 simply creates a collection of objects that are used as input and output devices on a robot. SONAR and IRPD input objects are used, as well as LED and servo motor output objects. The program doesn't do anything but sample the SRF04 SONAR board periodically; it relies upon the PDA to control the robot.

I could have created a VC for the SONAR system so that a program doesn't need to sample the data, but I wanted to have all objects quiet when I stopped the program via the SCP command. If that VC had been made, the SONAR would still be running even after the program stopped. The hobby servos on my robot are oriented with the servo output spline toward the front of the robot. If you build yours the same way, the motors will go forward with positive URCP values and backward with negative ones. If you built yours differently, you'll have to reverse all the numbers to control the robot movement.

Because no soldering was done at all in the creation of this robot, you can see which devices were connected where by just reading the code in Listing 11-3. Everything on this robot was off the shelf and simply required making the proper cable. I used an oDio4 object for the IRPD so I could read the 2 bits from my IRPD board in a single read without having to do 2 reads and then mathematically combine the 2 bits into a single number. The upper 2 bits of the oDio4 object were tied low with jumpers so that they would always read 0. It is a bad idea to leave an input line floating. Like the Mark III robot, I have jumpered SP to MP on the OOPic R optional power connector (see Chapter 1, "OOPic R Hardware Details") so I can run the robot on a single set of batteries for the motors as the OOPic R board. Here I'm using a 6-cell, 7.2V AA *nickel cadmium* (NiCd) battery pack for everything.

```
'Klutzilla.osc
'Chapter 11 Klutzilla robot code for use with the SCP
'programs.
'Demonstrates SCP control of a robot.
'
'For OOPic II B.2.X+ firmware
'
'Copyright Dennis Clark 2003
'Permission is granted to use this code anyway you like
'as long as you say you got it from me.

Dim SONAR as New oSonarDV
Dim Lservo as New oServoSP1
Dim Rservo as New oServoSP1
Dim IRPD as New oDio4
Dim LED as New oDio1

Dim dummy as New oByte

Sub Main()

    OOPic.SNode = 1              'Use SNode for serial debug

    OOPic.delay = 100           'Wait a little

    'Configure the SONAR device
    SONAR.IOLineP = 9           'Trigger to SRF04 board
    SONAR.IOLineE = 8           'Echo back from SRF04
    SONAR.Operate = cvFalse

    dummy = 222                 'Just to experiment with.

    'Configure our simple LED
    LED.IOLine = 6              'Middle one on OOPic R board
    LED.Direction = cvOutput
    LED = 0

    'Configure our servos
    Rservo.IOLine = 28
    Rservo.InvertOut = cvTrue
    Rservo.Operate = cvTrue
    Rservo = 0

    Lservo.IOLine = 31
    Lservo.Operate = cvTrue
    Lservo = 0

    'Configure the IRPD
    IRPD.IOGroup = 1            'I/O Lines 12-15
    IRPD.Nibble = 1            'Upper bits
    IRPD.Direction = cvInput

    do
        SONAR.Operate = 0
        SONAR.Operate = 1       'Trigger one reading

        While SONAR.Received = cvFalse   'wait for echo
        Wend
```

```
        OOPic.delay = 50

    loop

End Sub
```

Listing 11-3

Listing 11-4 is the HotPaw Basic program I'm using to show you how nicely SCP commands from the PDA control the OOPic R. This isn't really a robot program. It's just a test section that performs simple SCP and HotPaw functions to read the SONAR and turn the LED you've configured on and off.

```
'[Item A] Open the serial port, 9600 baud 8N1
open "com1:",9600 as #5 else bolo

'[Item B] Send SCP to turn on LED
print #5,"\059J1N";

'[Item C] Send SCP to read SONAR value
print#5,"\041JM";

'Delay long enough for serial data
'to get sent out.
for x=1 to 50
next x

'[Item D] Get the number of bytes sent
s=fn serial(5)

'[Item E] Read number back, take off the "m"
a$=get$(#5,s-1)
a = val(a$)
a = a/64
'[Item F] Display the SONAR result
print "SONAR reads "; a;" feet"

'Turn off the LED
print #5,"\059J0N";

'Close serial port and exit.
close #5
goto sdone
bolo:
print "error on open"
sdone:
end
run
```

Listing 11-4

In the listing, Item A shows how you open the serial port for the PDA. The serial port is *always* #5. You can use a variety of baud rates, but the OOPic defaults to 9600, so that is what is being used here. Item B shows you a combined escape, address set, and write data command that turns the LED on the OOPic R board on. When you send multiple commands in quick succession, SCP does not return a handshake code unless a *V* is sent or the last command sent results in data being sent back, which will append an *m* to the data. Item C shows a full escape, address, and read command, which results in the SONAR data with an m appended to it being sent back.

I then wait a little while for all the data to be sent to the PDA and Item D checks to see how many characters were sent. Item E retrieves the data from the PDA's buffer (which is 1024 bytes long) except for the last character, which is the *m* you don't want, and calculates feet from that data. The undeclared variable is, by default, a floating point number, which Item F displays in a pop-up window. After you click the OK button to dismiss the window, the program turns the LED off and exits.

Because both the OOPic and the PDA are programmed in Basic (that was nice of me, wasn't it?), you should see some similarities in the syntax. HotPaw Basic is built from an older style of nonobject-oriented Basic, which those of you used to Microsoft's Qbasic should recognize. Now let's make a more elaborate program on the PDA.

The following HotPaw program lets you use your stylus to change motor speeds, turn the LED on and off, and read the IRPD and SONAR values on demand. The current readings and settings are displayed, and the robot will do as it is commanded to do. Listing 11-5 is the HotPaw program I wrote to do this.

```
#scpCntrl.bas
'clear Palm Screen
draw -1

'create our action buttons
form btn 130,20,20,12,"Set",1
form btn 130,40,20,12,"Set",1
form btn 130,60,20,12,"Set",1
form btn 130,80,20,12,"Clr",1
form btn 130,100,20,12,"Read",1
form btn 130,120,20,12,"Read",1

'Place our Value fields
form fld 90,20,20,12,"0",1
form fld 90,40,20,12,"0",1
form fld 90,70,20,12,"0",1
form fld 90,100,20,12,"0",1
form fld 90,120,20,12,"0",1

'Place address fields
form fld 60,20,20,12,"47",1
form fld 60,40,20,12,"53",1
form fld 60,70,20,12,"59",1
form fld 60,100,20,12,"127",1
form fld 60,120,20,12,"41",1

'Name our lines
draw "Left servo",1,20
draw "Right servo",1,40
```

```
draw "LED",1,70
draw "IRPD",1,100
draw "SONAR",1,120

'Open serial port and start SCP

open "com1:",9600 as #5 else bolo
cmd$="\0V"
print #5,cmd$;
for x = 1 to 50
next x
s = fn serial(5)
rec$ = get$(#5,s)
draw ">"+rec$,130,150

while 1

    '[Item A]Look for a button (screen) press.
    a$ = input$(1)
    k=asc(a$)-13

    if k < 7 then
        'These are the LED set/clear pair
        if k = 3 then s$(2) = "1"
        if k = 4 then s$(2) = "0"
        draw s$(2),90,70
        if k > 3 then k = k -1

        'Create SCP command strings
        addr$ = s$(4+k)
        value$ = s$(k-1)
        if k < 4 then
            'Data out commands
            cmd$="V"+addr$+"J"+value$+"N"
        else
            'Data returned commands
            cmd$="U"+addr$+"JM"
        endif
        draw "sending "+"               ",50,150
        draw "sending "+cmd$,50,150
    endif

    'Now do commands and set fields as needed
    print #5,cmd$;

    'Wait long enough to get all the data
    for x = 1 to 50
    next x

    '[Item B]Get returned data or status and update screen
    s = fn serial(5)
    rec$ = ""
    if k < 4 then
        'These are output only
        if s > 0 then rec$ = get$(#5,s)
        draw "          ",130,150
        draw ">"+rec$,130,150
    else
        'These receive data from OOPic
```

(Continued)

11

OOPIC R SERIAL CONTROL PROTOCOL

```
            rec$ = left$(get$(#5,s),(s-1))
            draw "        ",130,150
            s$(k-1) = rec$
            if k = 4 then
                draw "        ",90,100
                draw s$(3),90,100
            elseif k = 5 then
                draw "         ",90,120
                draw s$(4),90,120
            endif
        endif
wend

'put this here as an exit point
'Close serial port and exit.
close #5
goto sdone
bolo:
print "serial port won't open"
sdone:
end
run
```

Listing 11-5

This program is considerably more complex at first glance. It sets up a PalmOS *form* that enables us to put buttons, graphics, and data fields onscreen. It uses a simple database that hides the details behind instructions to draw buttons and display data fields. The most complex part of the program is Item A, which builds the command string to send out, based on the button pressed and the field it may access, and Item B, which interprets the data (if any) coming back.

The key to understanding this section is that buttons 3 and 4 both deal with field 2 (the LED). The variable k holds the number of the buttons pushed. The first button defined is 1, but the first field defined is 0. I wish that had been more standardized, but that is what it is. The first four buttons, which write data out, have their return codes placed in the lower right of the screen to the right of the greater than ($>$) character. The U in front of the read commands suppresses the j memory set return value to make parsing the number easy, but because I want the return values to show up on the data output commands, those are prefaced with a V for verbose. If you see an asterisk (*) or an exclamation point (!) displayed down there, then the program could not communicate properly with the OOPic objects. Figure 11-4 is a screenshot of this program on my M100 PDA.

A *lot* of commands, functions, and operations can be done with HotPaw Basic. I *highly* recommend that you spend an afternoon reading the manual, which is quite short, and become familiar with it. I've barely scratched the surface of its capabilities here.

Again, I found the addresses of the objects by looking at the klutzilla.omp file, which is shown in Listing 11-6. This file is pretty useful overall as it gives the addresses of various flags, properties, and, most importantly, the Value or default property. It also gives the sizes of the various objects as well.

The addresses found in this file are hard-coded into the application for the PDA; however, I set them as fields in the form, so they can be changed by tapping on the numbers and

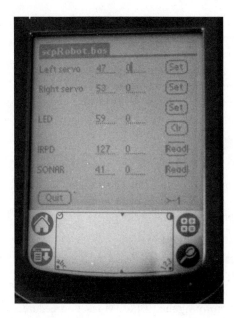

Figure 11-4 scpCntrl.bas
PDA screen

Add	Type	Name	ClE	Size
041	oSonardv	SONAR	0	006 Byte(s)
047	oServosp1	Lservo	0	006 Byte(s)
053	oServosp1	Rservo	0	006 Byte(s)
127	oDio4	IRPD	0	001 Byte(s)
059	oDio1	LED	0	001 Byte(s)
125	oByte	dummy	0	002 Byte(s)
---	---	---	---	000
041	Value	sonar.value		
047	Value	lservo.value		
053	Value	rservo.value		
127	Value	irpd.value		
059	Value	led.value		
125	Value	dummy.value		
---	---	---	---	000
042	Flag	sonar.operate		
044	Flag	sonar.received		
046	Flag	sonar.timeout		
043	Flag	sonar.transmitting		
048	Flag	lservo.operate		
052	Flag	lservo.invertout		
054	Flag	rservo.operate		
058	Flag	rservo.invertout		
127	Flag	irpd.nibble		
059	Flag	led.nonzero		
059	Flag	led.value		
126	Flag	dummy.msb		
125	Flag	dummy.nonzero		

Listing 11-6

writing in new ones while the program is running. This program is particularly useful when looking for and setting the center point of the servos. You can enter a new value and just tap the *Set* button for that object. You can also find the values for the SONAR and IRPD by changing something and tapping the Read button for the respective object. Play around with the code and see if you don't think of new and interesting projects of your own.

And speaking of new and interesting projects, how would you like to be able to program your OOPic R board from your PDA? Read on and I'll show you how.

PROGRAMMING THE OOPIC R WITH A PALMOS PDA

SCP is not only capable of interacting with your programs, but it talks directly to the OOPic object and, as such, can program the PDA as well. In fact, when you are using the serial port IDE programming cable, you are using SCP to program your OOPic.

Resetting the OOPic R If It's Hung Up

You can really shoot yourself in the foot here if you aren't careful, but no catastrophe is without a solution. If you find that you've hung up your OOPic R board with a poorly chosen or hacked program, you can get it back by performing the EEPROM reset procedure:

1. Turn off power to the OOPic R.
2. Remove the EEPROM from its socket.
3. Turn on the OOPic R.
4. Try to program the OOPic R with a correct program.
5. When the IDE says, "Insert the EEPROM," plug the EEPROM back in.
6. Download the good program.

This always works. Fortunately, you can *hot* plug (plug and unplug something when it is powered on) the EEPROM, or you really would be sunk when you got a bad program in the board!

Getting the Download Sequence

Let's go back to the first program in this chapter where you blink the LED on the OOPic R board. This is a simple and small program that shows how to collect the byte codes and dialog needed to create a downloader from the PDA to the OOPic R (or OOPic C or II+). Follow the next steps to create a listing of the SCP dialog needed:

1. Compile the program simple.osc as you did before.
2. Open the Communications Control window by going to View, Comm Control. If it has text in it, close the window and open it again.
3. Download the program to the OOPic R. The Communications Control window will have new text in it that looks like Figure 11-5.
4. Copy that text from the Communications Session window and paste it into Windows Notepad.
5. Separate out all the lines so that each line is a single command. See Listing 11-7 for what this file looks like when you are done.

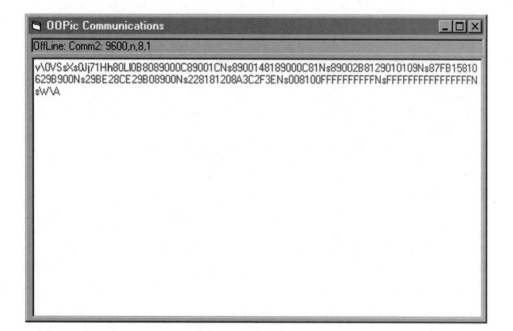

Figure 11-5 Communications Control window after a download

```
\0V
S
X
0J
71H
80L
0B8089000C89001CN
8900148189000C81N
89002B8129010109N
87FB15810629B900N
29BE28CE29B08900N
228181208A3C2F3EN
008100FFFFFFFFFFN
FFFFFFFFFFFFFFFFN
W
\A
```

Listing 11-7

After you clean up the output shown in Figure 11-5, which involves basically placing all the individual commands on their own lines and removing the SCP replies (the lowercase letters), you end up with Listing 11-7, which you can convert into a download program.

Everything you need is here. Here is what the OOPic R has been told to do:

- **\0V** Alerts the OOPic R that you're talking SCP and puts it in verbose mode.
- **S** Stops the currently running program.
- **X** Resets the OOPic R and halts all VC and program code.
- **0J** Sets the address register to the beginning.
- **71H** Selects the local I2C bus for 8-byte transfers, and auto-increments the address (see Table 11-1).
- **80L** The I2C address is 80, which is the program EEPROM (see Table 11-2).
- **...N** Eight-byte blocks of program code written into I2C EEPROM.
- **W** Resets the OOPic and starts the program running.
- **\A** Puts SCP into Lurk mode.

You just have to add the program statements to send the data and commands, and record the replies coming back so that you know it worked. Listing 11-8 shows the program I wrote to download the program to the OOPic R. This also works on the OOPic C because it has the same three LEDs on it. Primitive, yes, but you don't want me to do *all* your work for you, do you?

```
#dloadpn.bas
'Program for use on Palm PDAs to download a simple LED flasher to
'the OOPic R board.

'Open the serial port, 9600 baud 8N1
open "com1:",9600 as #5 else bolo
'Start download conversation
comm$ =""
print #5,"\0V";
gosub getStat()

'Set addresses and modes
print #5,"S";
gosub getStat()
print #5,"X";
gosub getStat()
print #5,"0J";
gosub getStat()
print #5,"71H";
gosub getStat()
print #5,"80L";
gosub getStat()

'Now download the program
print #5,"0B8089000C89001CN";
gosub getStat()
print #5,"8900148189000C81N";
gosub getStat()
print #5,"89002B8129010109N";
gosub getStat()
print #5,"87FB15810629B900N";
gosub getStat()
print #5,"29BE28CE29B08900N";
gosub getStat()
print #5,"228181208A3C2F3EN";
gosub getStat()
```

```
print #5,"008100FFFFFFFFFFN";
gosub getStat()
print #5,"FFFFFFFFFFFFFFFFN";
gosub getStat()

'Now reset OOPic and exit SCP
print #5,"W";
'This getstat() won't get anything,
'but it will use a little time.
gosub getStat()
print #5,"\A";

'Print out session results
print comm$

'Close serial port and exit.
close #5
goto sdone

sub getStat()
    'Wait for results to come back
    for x = 1 to 50
    next x
    'Then get them
    s = fn serial(5)
    if s > 0 then
        comm$ = comm$ +get$(#5,s) + " "
    endif
return

bolo:
print "Serial Port error on open"
sdone:
end
run
```

Listing 11-8

11

OOPIC R SERIAL CONTROL PROTOCOL

Did you notice that in Listing 11-8 I used *gosub* to call subroutines and in Listing 11-5 I used *call*? I did that just to show that you can use the syntax you are most comfortable with. HotPaw seems to support just about all of them.

You can do some obvious enhancements. For instance, you can use the PalmOS memo database commands in HotPaw Basic to look up and read a memo that contains commands like those in Listing 11-7. That would be a lot simpler than hand generating a download code like this one, but I leave that as an exercise for you. Perhaps you'll find something like that on my web page one of these days.

A SIMPLE PALM-PDA-CONTROLLED ROBOT

The chapter wouldn't be complete without a simple robot project that made Klutzilla truly autonomous, so here it is. This robot (using the klutzilla.osc program on the OOPic R) will wander a room randomly and when confronted with imminent collision (IRPD discovered) will look around (with SONAR) to find the least obstructed path to take. Because it has only dead reckoning to determine how far it turned, it may be a bit hit or miss in its selection. Maybe I should have put the compass on here (hint, hint). The point behind this experiment is not to make a really robust robot, but to simply display the potential for the PDA to control the robot effectively.

Three basic behaviors are to be used here: *Wander, Avoid,* and *Escape.* Wander roams the room, Avoid tries to avoid running into things, and Escape looks for the best way out of a problem. It's a good platform on which to build more elaborate behavior in the future. Listing 11-9 shows this new HotPaw program for your robot.

```
#irobot.bas
'Set up constants and variables.
'Set wander, avoid, and escape priorities
pwander = 1
pavoid = 2
pescape = 3
'current priority level
plevel = 0
'next direction to go
ndir=0
ldir=99
'Timers for behaviors
tout = 0
'Directions & IRPD input
fwd=0
rev=3
left=1
right=2
nogo=99
'Set object addresses
lmotor$ = "47J"
rmotor$ = "53J"
sonar$ = "41J"
irpd$ = "127J"
'timeouts, in seconds
towander = 1
toavoid = 1
toescape = 3
'Motor speed to use
speed$ = "32N"
'====Main program====
open "com1:",9600 as #5 else bolo
print #5,"\0V"
tout=0
plevel=0
while (1=1)
    call dowander()
    call doavoid()
    call doescape()
    call setdir(ndir)
wend
bolo:
print "Error opening com port"
end
'====behaviors====
```

```
sub dowander()
    if plevel <= pwander then
        if plevel < pwander then tout=0
        plevel = pwander
        if tout > 0 then
            if timer-tout >= towander then
                twander=0
                plevel = 0
            endif
        else
            ndir = rnd(3)
            tout = timer
        endif
    endif
return
sub doavoid()
  tmpa=getirpd()
  if (tmpa > 0) or (plevel = pavoid) then
    if (tmpa > 0) and (plevel <= pavoid) then
      plevel=pavoid
      ndir=tmpa
      tout = timer
    elseif plevel = pavoid then
      if timer-tout >= toavoid then
        tavoid = 0
        plevel = 0
    endif
  endif
return
sub doescape()
    ta=getirpd()
    if (ta=3) then
        longest=0
        call setdir(nogo)
        for n = 1 to 10
            call setdir(left)
            fn wait(1)
            call setdir(nogo)
            d=getsonar()
            if d>longest then
                longest=d
                turn=n
            endif
        next n
        tback=10-turn
        for n=1 to tback
            call setdir(right)
            fn wait(1)
            call setdir(nogo)
        next n
        ndir=fwd
        call setdir(ndir)
        plevel=pwander
        tout=timer
    endif
return
'====Sub programs====
sub setdir(dir)
```

(Continued)

11

OOPIC R SERIAL CONTROL PROTOCOL

```
  if (dir = ldir) then
    goto sddone
  else
    if ndir = fwd then
        cmd1$=lmotor$+speed$
        cmd2$=rmotor$+speed$
    elseif dir = rev then
        cmd1$=lmotor$+"-"+speed$
        cmd2$=rmotor$+"-"+speed$
    elseif dir = left then
        cmd1$=lmotor$+"0N"
        cmd2$=rmotor$+speed$
    elseif dir = right then
        cmd1$=lmotor$+speed$
        cmd2$=rmotor$+"0N"
    else
        cmd1$=lmotor$+"0N"
        cmd2$=rmotor$+"0N"
    endif
    print #5,"\0"+cmd1$
    call getstat()
    print #5,"\0"+cmd2$
    call getstat()
    ldir=dir
  endif
sddone:
return
sub getirpd()
    print #5,"\0"+irpd$+"M"
    for x=1 to 50
    next x
    s=fn serial(5)
    getirpd=val(left$(get$(#5,s),(s-1)))
end sub
sub getsonar()
    print #5,"\0"+sonar$+"M"
    for x=1 to 50
    next x
    s=fn serial(5)
    getsonar=val(left$(get$(#5,s),(s-1)))/64
end sub
sub getstat()
    for x=1 to 50
    next x
    s = fn serial(5)
    if s>0 then dummy$=get$(#5,s)
return
run
```

Listing 11-9

This program looks less pretty because I've left fewer spaces and comments than usual. Because a memo is limited to 4000 characters, you don't want to use up any more space than you have to when writing it. Still I include a few comments and descriptive variable names to help.

The three behaviors are called dowander(), doavoid(), and doescape(). They have been assigned the priorities 1, 2, and 3, respectively, and higher-level behaviors interrupt lower-level ones when they need to. This is basic subsumption behavioral programming. None of the behaviors know anything about any of the other ones; they only know about other priorities that may be active if those priorities are higher than the behavior can preempt (or subsume, to use the lingo). Each behavior has a timer associated with it such that when its time is up, it releases control of the motors. This allows each behavior to be exited and reentered repeatedly and gives other behaviors time to see if they need to act.

At the end of the behavior loop is the *setdir()* subroutine, which takes the winning *ndir* value (the direction to go next) and sends the instructions to the motors. setdir()only sends new instructions to the OOPic R board if the direction has changed since the last time it was called. This helps to avoid too much traffic over the serial port that is not necessary. This type of programming involves simple state machines and can give the feeling of running a multitasking system because it appears that all the behaviors are running at once.

The subroutines below the behaviors handle the SCP mechanisms needed to set motor speed and direction, as well as read the IRPD and SONAR inputs. The following is a listing of all the routines in the *irobot.bas* program and what they do:

- **Dowander()**   This implements the Wander behavior. It chooses a random direction and goes that way for the amount of time set in the *towander* variable. It has the lowest priority.
- **Doavoid()**   This implements the Avoid behavior. It looks at the IRPD object results and chooses a direction that avoids trouble. It continues in that direction for the time set in the *toavoid* variable. It's the next highest priority.
- **Doescape()**   This implements the Escape behavior and has the highest priority. It circles in place looking with SONAR for the avenue with the least obstruction. It then turns to that location and goes forward. It then sets the priority to the Wander behavior and exits.
- **Setdir()**   This sets the motor directions and handles the SCP commands to the OOPic R servos. It does not reissue instructions to go in the same direction it is already headed in, which reduces activity on the SCP bus.
- **Getirpd()**   This routine handles the SCP required to read the IRPD hardware object on the OOPic R board.
- **Getsonar()**   This routine handles the SCP dialog to read the SONAR hardware object on the OOPic R board.
- **Getstat()**   This is the routine that makes sure the last instruction has been completed and gets the status back. Nothing is done with this status, but that can be changed.
- **Main**   This is just the first part of the program. It sets up the initial conditions and loops forever, calling all the behaviors.

I don't like "magic" numbers embedded in the code, so at the top of this program is a listing of all the global variables and constants used in the program.

It's pretty fun to watch this little guy run around, get into trouble, and use the SONAR to accurately find a way out. Some obvious enhancements would be a set of bumpers and even more proximity detectors, or even mounting the SONAR on a servo so it can scan without turning the whole robot.

11

OOPIC R SERIAL CONTROL PROTOCOL

Preventing Your PDA from Shutting off While Your Robot Is Running

One more thing: Did you notice your PDA shut off after a short while? This is because the PalmOS only enables your PDA to remain on for a maximum of three minutes without human activity. You can get around this problem, though, by using the following command periodically in your program:

```
fn set_auto_off(n)
```

where (n) is the number of seconds between 0 and 300. Simply include this function call in your main loop and use it periodically to keep your PDA from shutting down.

Now What? Where to Go from Here

This chapter was a lot of fun to write, mostly because it was a lot of fun to experiment with SCP on the OOPic R. The combination of the wide variety of connectors on the OOPic R board meant I didn't have to do a bunch of soldering to get the pins or the circuit I wanted. The ease with which SCP could be programmed made for a great experience.

Figure 11-6 What a mess.

I'm hoping that these examples stimulate your imagination and allow you to pick up where I left off with your remote-control applications and robotics in particular. The robot in this chapter, for instance, would greatly benefit from some more hardware like bumpers, more sensors, some wheel encoders, and, of course, a compass. Using the combination of the OOPic R and the PalmOS PDA, a lot of power can be brought to bear on a project. The PDA has full floating-point and trig functionality, whereas the OOPic has all those simple-to-use hardware and processing objects that simplify the interface to your main programs. The combination has proved to be powerful.

The information in this chapter is not only useful for robotics, but also for supporting products based on the OOPic B.2.X+ firmware in the field. You can build applications on your PDA that will enable you to download new code on site without lugging around your laptop. It is utterly reliable, as reliable as the OOPic IDE is, and is much more convenient. I can think of lots of robotics competitions where I would have *loved* to have kept several versions of the code that were easy to download right there on the playing field!

HotPaw isn't the only PDA language out there. You can find more Basic and even C interpreters, but few are as inexpensive. Still, feel free to look around and see what you can find now that you know you can do this kind of interfacing easily.

This is the end of this first book about OOPic programming and interfacing, and although it was fun to write, I am happy to complete it. As you can see from Figure 11-6, my lab needs a good housecleaning now.

Have fun with your OOPic microcontrollers; they are easy to use and powerful if you're imaginative. I hope that I've helped you realize both of those happy situations!

11

OOPIC R SERIAL CONTROL PROTOCOL

OOPIC A.2.X OBJECTS

CONTENTS AT A GLANCE

Keys
  IOGroup Selections

Hardware Objects
  oA2D
  oDio1
  oDio4
  oDio8
  oDio16
  oDio16X
  oI2C
  oKeypad
  oPWM
  oSerial
  oSerialPort
  oServo
  oTimer(X)

Processing Objects
  oConverter(C)
  oCounter
  oDataStrobe
  oDDELink
  ODebounce
  oEvent(C)(X)
  oFanout(C)
  oGate(C)
  oIndex(C)
  oMath(I)(O)(C)
  oOneShot
  oSrvSync
  oRandomizer(O)(C)
  oRTC
  oWire

Variable Objects
  oBit
  oBuffer
  oByte
  oEEPROM
  oNibble
  oRam
  oWord

User-Definable Objects
  oUserClass

System Objects
  OOPic
  PIC

This appendix covers the objects that appear in the A.2.X firmware. This list is just a reference for the properties and methods of the objects; it is *not* a full description of the objects and how to use them. For detailed object information, see the referenced chapters or the *O*OOPic web site (www.oopic.com) and refer to the manual pages. Some objects were updated in later firmware releases (B.X.X) and those changes are reflected here to avoid duplication of objects and any possible confusion.

A note about the *Operate* property : When this property is cvFalse (0), it doesn't mean the object's output just disappears. It means it stops changing state based on its current inputs; the object continues to hold the last value and set of flags it had when Operate property was cvTrue (1).

Keys

Abbreviations are used throughout this appendix and Appendix B, "OOPic B.2.X Objects," as shown in Table A-1. The common aliases defined in the compiler are outlined in Table A-2.

TABLE A-1   KEY TO PROPERTY TYPES

*	Default property. The property read when no property is given.
F	Flag. This is a bit that is linkable; a flag is also a bit.
W	Word. Sixteen bits, two bytes.
B	Byte. Eight bits.
N	Nibble. Four bits.
Bi	Bit. Not always a flag.
FP	Flag pointer. A pointer to a flag in another object (your choice).
NP	Number pointer. A pointer to a number in another object.
AP	Address pointer. Useful only to the oRam and oDDELink objects.
BP	Buffer pointer. A special case pointer for oIndex and oRTC.
S	String. Most objects have this property.

TABLE A-2 COMMON ALIASES DEFINED IN THE COMPILER

ALIASES FOR 0	ALIASES FOR 1
cvOff	cvOn
cvLow	cvHigh
cvFalse	cvTrue
cvOutput	cvInput
cvPressed	cvUnpressed

IOGroup Selections

Many objects require 8 bits of I/O to be specified, some 4 bits. Table A.3 shows the I/O lines chosen by each IOGroup. Table A.4 shows the nibble (4 bits) chosen by specific IOGroup and Nibble settings.

TABLE A-3 IOGroup DESCRIPTION

0	*Input/output* (I/O) lines 1 through 7. Not normally assignable in groups.
1	I/O lines 8 through 15 (programmable pull-ups on these).
2	I/O lines 16 through 23 (exclusive use by some objects).
3	I/O lines 24 through 31.

TABLE A-4 IOGroup PLUS THE NIBBLE SELECTION MATRIX

IOGROUP	NIBBLE	DESCRIPTION
0	0	Object is disabled.
0	1	In B.2.0 onward, the oDIO4 object uses the three I/O Lines 5 through 7.
1	0	I/O lines 8 through 11.
1	1	I/O lines 12 through 15.
2	0	I/O lines 16 through 19.
2	1	I/O lines 20 through 23.
3	0	I/O lines 24 through 27.
3	1	I/O lines 28 through 31.

A

OOPIC A.2.X OBJECTS

Hardware Objects

The following objects are based on OOPic hardware.

oA2D

The oA2D object is a basic 8-bit *analog-to-digital* (A2D) channel (refer to Chapter 5, "Analog-to-Digital and Hobby Servos [Project: Push My Finger]"). Its properties are listed in Table A-5.

Memory Size: 3 Bytes

TABLE A-5 oA2D PROPERTIES	
Address (AP)	The address of the object in RAM (0–127).
IOLine (N)	Assigns the I/O line to use. Essentially, 1 through 7 are I/O lines 1 through 7; however, on the A.X.X firmware, 1 through 3 are I/O lines 1through 3, but 0 is I/O line 4. On B.2.X firmware, I/O lines 1 through 7 are A2D lines.
MSB (F)	The *most significant bit* (MSB) of the *Value* property. If this bit is 1, Value is equal to or over half of the reference voltage. If it is 0, then it is under half.
Operate (F)	Must be 1 to enable the object.
String (S)	The string representation of Value.
Value (B)*	The result of the A2D conversion.

ODio1

oDio1 is a 1-bit I/O digital I/O port (refer to Chapter 4, "Your First OOPic Program, OOPic I/O [Project #1: Das Blinken Light]"). Its properties are shown in Table A-6. Its methods are shown in Table A-7.

Memory Size: 1 Byte

TABLE A-6 ODIO1 PROPERTIES	
Address (AP)	The address of the object in RAM (0–127).
Direction (Bi)	0 means (cvOutput) is an output.
	1 means (cvInput) is an input.
I/O Line (B)	OOPic I/O lines 1 through 31; 0 means disabled.
NonZero (F)	Returns 0 if Value is equal to 0; it returns 1 otherwise.
String (S)	The string representation of Value.
Value(Bi)*	The number written to or read from the I/O lines specified.

TABLE A-7

oDio1 METHODS

Clear	Clears the Value property to 0.
Invert	Inverts the bits in the Value property.
Set	Sets the Value property to 1.

oDio4

This is a 4-bit digital I/O port (refer to Chapter 4, "Your First OOPic Program, OOPic I/O [Project #1: Das Blinken Light]"). Its properties are shown in Table A-8. Its methods are outlined in Table A-9.

Memory Size: 1 Byte

TABLE A-8   oDio4 PROPERTIES

Address (AP)	The address of the object in RAM (0–127).
Direction (Bi)	0 means (cvOutput) is an output. 1 means (cvInput) is an input.
IOGroup (N)	The I/O group to use. See IOGroup Selection Key.
Nibble(Bi)	0 means use the lower 4 bits of IOGroup. 1 means use the upper 4 bits of IOGroup.
String (S)	The string representation of the Value property.
Value(N)*	The number written to or read from the I/O lines specified.

TABLE A-9   oDio4 METHODS

Clear	Clears the Value property to 0.
Dec	Decrements the Value property by 1.
Inc	Increments the Value property by 1.
Invert	Inverts the bits in the Value property.
LShift	Shifts the bits in the Value property to the left.
RShift	Shifts the bits in the Value propertyto the right.
Set	Sets the Value property to 15.

ODio8

oDio8 is an 8-bit digital I/O port (refer to Chapter 4, "Your First OOPic Program, OOPic I/O [Project #1: Das Blinken Light]"). Its properties and methods are illustrated in Tables A-10 and A-11.

Memory Size: 1 Byte

TABLE A-10 oDio8 PROPERTIES	
Address (AP)	The address of the object in RAM (0–127).
Direction (Bi)	0 means (cvOutput) is an output. 1 means (cvInput) is an input.
IOGroup (N)	The I/O group to use (refer to beginning of appendix).
String (S)	The string representation of the Value property.
Value(B)*	The number written to or read from the I/O lines specified.

TABLE A-11 oDio8 METHODS	
Clear	Clears the Value property to 0.
Dec	Decrements the Value property by 1.
Inc	Increments the Value property by 1.
Invert	Inverts the bits in the Value property.
LShift	Shifts the bits in the Value property left.
RShift	Shifts the bits in the Value property right.
Set	Sets the Value property to 255.

oDio16

This is a 16-bit digital I/O port (refer to Chapter 4, "Your First OOPic Program, OOPic I/O [Project #1: Das Blinken Light]"). Its properties and methods are shown in Tables A-12 and A-13.

Memory Size: 3 Bytes

TABLE A-12 oDio16 PROPERTIES

Address (AP)	The address of the object in RAM (0–127).
Direction (Bi)	0 means (cvOutput) is an output. 1 means (cvInput) is an input.
IOGroup (Bi)	0 means the object is disabled. 1 indicates you should use I/O lines 8 through 15 and 24 through 31.
String (S)	The string representation of the Value property.
Value(W)*	The number written to or read from the I/O lines specified.

TABLE A-13 oDio16 METHODS

Clear	Clears the Value property to 0.
Dec	Decrements the Value property by 1.
Inc	Increments the Value property by 1.
Invert	Inverts the bits in the Value property.
LShift	Shifts the bits in the Value property left.
RShift	Shifts the bits in the Value property right.
Set	Sets the Value property to 65535.

oDio16X

oDio16X multiplexes 2 sets of 16-bit ports, 1 input and 1 output using I/O lines 5 through 7 and 8 through 15. This command is obsolete in A.2 and B.2 firmware.

oI2C

This is an *inter-IC* (I2C) bus object for use on the local I2C bus/programming port (refer to Chapter 9, "OOPic I2C and Distributed DDE Programming [Project: Remote Control]"). Its properties are outlined in Table A-14.

A

OOPIC A.2.X OBJECTS

Memory Size: 5 Bytes

TABLE A-14 oI2C PROPERTIES	
Address (AP)	The address of the object in RAM (0–127).
Location (B/W)	In Mode 0, this is a 16-bit address usually used in *electrically erasable programmable read-only memory* (EEPROM) that follows the I2C address byte.
	In Mode 1, this is the 8-bit register address usually used in devices that are register based.
	Not used in Mode 2.
Mode (N)	0 (cv23Bit) means the I2C uses Location as the two address bytes that follow the device address byte.
	1 (cv10Bit) means I2C uses the Location property as the single address byte that follows the device address byte, usually to select a register.
	2 (cv7Bit) means the I2C uses the 7-bit I2C address mode for simple 1-byte writes or reads, usually for device commands.
Node (B)	The I2C address of the device (refer to Figure 9-2).
NoInc (F)	0 (cvFalse) increments the Location property each time the Value property is read or written.
	1 (cvTrue) does not increment the Location property each time the Value property is read or written.
	This is only used with Mode 0 or 1.
String (S)	Although listed, this property is not supported in oI2C.
Value (B/W)*	When Width is 0, bytes are read from or written to the device.
	When Width is 1, words are read from or written to the device.
Width (Bi)	0 (cv8Bit) means 8 bits are read and written at a time (1 byte).
	1 (cv16Bit) means 16 bits are read and written at a time (2 bytes).

oKeypad

This is a 16-key keypad hardware object (refer to Chapter 7, "OOPic Events, Keypads, and Serial I/O [Project: A Mini-terminal]"). Its properties are shown in Table A-15.

Memory Size: 4 Bytes

TABLE A-15 oKeypad PROPERTIES	
Address (AP)	The address of the object in RAM (0–127).
Mode (Bi)	0 returns the encoded row and column. 1 returns phone-pad values.
Operate (F)	Must be 1 for this object to respond to input.
Received (F)	0 means no key is being pressed. 1 means a key is pressed and a new value is stored in the Value property.
String (S)	The Value property represented as a string.
Value (N)*	Value encoded as of the last key press.

oPWM

A hardware object that generates *pulse width modulation* (PWM), two of these oPWMs are possible (refer to Chapter 10, "OOPic Robotics and URCP [Project: A Robot That Toes the Line]"). Its properties and methods are outlined in Tables A-16 and A-17.

Memory Size: 3 Bytes

TABLE A-16 oPWM PROPERTIES	
Address (AF)	The address of the object in RAM (0–127).
IOLine (Bi)	0 uses I/O line 18. 1 uses I/O line 17.
Operate (F)	Must be 1 for this object to respond to input.
Period (B)	Selects the active time period for the PWM cycle (out of 256 possible).
Prescale (N)	Sets the clock divider for PWM hardware. 0 means the base frequency is 5 MHz. 1 means the base frequency is 1.25 MHz (system clock divided by 4). 2 means the base frequency is 312.5 KHz (system clock divided by 16). 3 is the same as 2.
Value (B)*	This sets the on time for the PWM signal defined by Period.

A

TABLE A-17   oPWM METHODS

Clear	Clears the Value property to 0.
Dec	Decrements the Value property by 1.
Inc	Increments the Value property by 1.
LShift	Shifts the bits in the Value property left.
RShift	Shifts the bits in the Value property right.
Set	Sets the Value property to 255.

oSerial

This ses the *Universal Asynchronous Receiver Transmitter* (UART) as a bidirectional serial port (refer to Chapter 7, "OOPic Events, Keypads, and Serial I/O [Project: A Mini-terminal]"). Its properties are listed in Table A-18.

Memory Size: 4 Bytes

TABLE A-18   oSerial PROPERTIES

Address (AF)	The address of the object in RAM (0–127).
Baud (N)	0 (cvMidi) sets UART at 31,250 baud (MIDI).
	1 (cv1200) sets UART at 1200 baud.
	2 (cv2400) sets UART at 2400 baud.
	3 (cv9600) sets UART at 9600 baud.
Mode (Bi)	0 means the serial port operates asynchronously.
	1 means the serial port operates synchronously.
Operate (F)	Must be 1 for this object to respond to input.
Received (F)	Is 1 when a byte of data has been received; it is cleared to 0 when that data byte has been read (the value has been read).
String (S)	The string representation of the data in the Value property.
Transmitting (F)	Is 1 when the UART is transmitting data; it is 0 otherwise.
Value (B)*	The data to be transmitted or the data that has been received.

oSerialPort

oSerialPort uses the hardware UART as a serial port with input buffer and flow control (refer to Chapter 7, "OOPic Events, Keypads, and Serial I/O [Project: A Mini-terminal]"). Its properties are shown in Table A-19.

Memory Size: 10 Bytes

TABLE A-19	OSERIALPORT PROPERTIES
Address (AF)	The address of the object in RAM (0–127).
Baud (N)	0 (cvMidi) sets UART at 31,250 baud (MIDI). 1 (cv1200) sets UART at 1200 baud. 2 (cv2400) sets UART at 2400 baud. 3 (cv9600) sets UART at 9600 baud. 4 (cv19200) sets UART at 19,200 baud. 5 (cv4800) sets UART at 4800 baud. 6 (cv50000) sets UART at 50,000 baud.
Mode (Bi)	0 means the serial port operates asynchronously (default). 1 means the serial port operates synchronously.
Operate (F)	Must be 1 for this object to respond to input.
Received (F)	Is 1 when a byte of data has been received. It is cleared to 0 when that data byte has been read (Value has been read).
ReceivedOut (FP)	Links to an I/O line for flow control.
String (S)	The string representation of the data in the Value property.
Transmitting F)	Is 1 when the UART is transmitting data; it is 0 otherwise.
Value (B)*	The data to be transmitted or the data that has been received.

oServo

This is a hardware object used to control a hobby servo (refer to Chapter 5, "Analog-to-Digital and Hobby Servos [Project: Push My Finger]"). Its properties are outlined in Table A-20.

Memory Size: 4 Bytes

TABLE A-20	oServo PROPERTIES
Address (AF)	The address of the object in RAM (0–127).
Center (B)	Used to move the servo's mechanical center position. 22 is the 1.5-millisecond servo pulse center when Value equals 32.
InvertOut (F)	1 reverses the servo's direction. 0 is the normal servo direction.
IOLine (B)	Chooses the I/O line (1–31) for the object.
Operate (F)	Must be a 1 to enable the servo.
Refresh (F)	When set to 1, it doubles the refresh rate (to about every 17 milliseconds).
Value (B)*	The position to move the servo to, from 0 to 63.

A

OOPIC A.2.X OBJECTS

oTimer(X)

This is the hardware 16-bit timer (refer to Chapter 6, "OOPic Timers, Clocks, LCDs, and SONAR [Project: SONAR Ping]"). (The X variant is B.X.X only.) The oTimer(X) properties are shown in Table A-21.

Memory Size: 1 Byte (oTimerX: 3 bytes)

TABLE A-21 oTimer(X) PROPERTIES	
Address (AF)	The address of the object in RAM (0–127).
ExtClock (Bi)	0 uses the internal 5 MHz clock. 1 uses I/O lines 16 or 16 and 17 for the clock.
ExtXtal (Bi)	0 uses external clock input on I/O line 16 only. 1 uses external crystal circuitry on I/O lines 16 and 17.
MSB (F)	MSB value of oTimer. This bit cycles once for every count to 65,535. Only in OOPic B.X.X or later.
Operate (F)	Must be a 1 to enable the oTimer object to start counting.
PreScale (N)	Specifies the divisor to the clock signal. 0 divides by 1 (5 MHz clock). 1 divides by 2 (2.5 MHz clock). 2 divides by 4 (1.25 MHz clock). 3 divides by 8 (625 KHz clock).
String(S)	The Value property represented as a string.
Value (W)*	The current count of the timer.

Processing Objects

These objects modify data or control other objects.

oConverter(C)

This converts one type of data into another for output on I/O lines. (Refer to Chapter 8, "OOPic Interfacing and Electronics [and Steppers and Seven-Segment LEDs]"). (The C variant is B.X.X only.) Its properties are shown in Table A-22.

Memory Size: 3 Bytes (oConverterC: 5 bytes)

TABLE A-22	oConverter(C) PROPERTIES
Address (AF)	The address of the object in RAM (0–127).
Blank (F)	If this bit is set to 1, the output is cleared to 0s instead of being updated. 0 will have a normal operation.
ClockIn (FP)	Links to the flag that clocks this object (oConverterC only).
Input (NP)	Links to the Value property that will be used as the input to the conversion chosen.
InvertC (F)	If this bit is set to 1, the clock sense will be inverted; that means a 0 will clock the object, not a 1 (oConverterC only).
InvertOut (F)	If this bit is set to 1, the bits of the converted value are inverted before written to the *Output* object.
Mode (N)	0 (cv7Seg) is the binary to/from seven-segment display. 1 (cvPhase) is the binary to eight-phase stepper motor. 2 (cvSin) is binary to sin. 3 (cvDecimal) is binary to or from decimal. The same as 74150 IC. Binary:011 equals Decimal:00000100; Binary:111 equals Decimal:10000000.
Operate (F)*	Must be set to 1 for this object to respond to input.
Output (NP)	Links to the Value property of the target object that receives the result of the conversion.

oCounter

oCounter is an object that counts event transitions. Its properties are shown in Table A-23.

Memory Size: 5 Bytes

TABLE A-23	oCounter PROPERTIES
Address (AP)	The address of the object in RAM (0–127).
ClockIn1 (FP)	Links to an object's Flag property used as a clocking signal to increment or decrement the object's Value property linked to the Output property.
ClockIn2 (FP)	Links to an object's Flag property used as the ClockIn1's complement when the Mode is set to cvPhase.

(continued)

A

OOPIC A.2.X OBJECTS

TABLE A-23 oCounter PROPERTIES (Continued)

Direction (F)	Direction	Mode	Description
	0	0	Counts positively (increment).
	0	1	Output is incremented.
	1	0	Counts negatively (decrement).
	1	1	Output is decremented.
Input (NP)	Links to an object whose Value property is used as a count limit value.		
	Each time a clock tick increments the Output object's Value property, the new Value is tested against the Input object's Value. If it is greater, the Value of the Output object is cleared to 0.		
Input (NP)	Each time a clock tick decrements the Output object's Value, the new Value is tested against 0. If it is lower, the Value property of the Output object is set to the Input object's Value.		
Mode (N)	0 (cvCount) means that each cycle of the Flag property pointed to by ClockIn1 is considered to be a clock tick and the Value of the Output object is incremented or decremented in the direction specified by the Direction property.		
	1 (cvPhase) means that, if Mode is set to cvPhase or 1, the flags pointed to by ClockIn1 and Clockin2 are quadrant-encoded and the Value property of the Output object is incremented or decremented in the direction specified by quadrant change. The Direction property is updated to reflect the direction of the change.		
Operate (F)*	Must be set to 1 for this object to respond to input.		
Output (NP)	Links to an object whose Value property is incremented or decremented for each clock tick.		
Tick (F)	0 means the Value property of the Output object is incremented or decremented by 1.		
	1 means the Value property of the Output object is cleared.		

oDataStrobe

This clocks data out, either as bytes or nibbles to an output port (refer to Chapter 6, "OOPic Timers, Clocks, LCDs, and SONAR [Project: SONAR Ping]"). Its properties are shown in Table A-24.

Memory Size: 5 Bytes

TABLE A-24	ODATASTROBE PROPERTIES
Address (AP)	The address of the object in RAM (0–127).
Mode (Bi)	0 (cv8Bit) means a *single 8-bit transfer and a single data strobe.* 1 (cv4Bit) means two 4-bit transfers with the high-order nibble being transferred first and using two data strobes.
Nibble (F)	Indicates which half of a two-nibble transfer is currently taking place. 0 is the lower nibble. 1 is the upper nibble.
OnChange (Bi)	Specifies how the data strobe is generated. 0 is a strobe every time the Value property is written. 1 is a strobe every time the Value property changes.
Operate (F)	Must be 1 for this object to respond to input.
Output (NP)	Links to the I/O lines (either oDio8 or oDio4) being used.
Result (F)	1 means the data strobe is active; it is 0 otherwise.
String (S)	The Value property represented as a string. Multiple bytes are allowed.
Strobe (FP)	Links to the strobe I/O line being used (E on LCD devices).
Value (B)*	The data that is to be transferred

oDDELink

This links multiple OOPics together over the network I2C bus (refer to Chapter 9, "OOPic 12C and Distributed DDE Programming [Project: Remote Control]"). Its properties are shown in Table A-25.

Memory Size: 8 Bytes

TABLE A-25	ODDELINK PROPERTIES
Address (AP)	All objects have an Address property. This is what will be referred to by a master's Location property.
Direction (F)	0 (cvSend) means data is copied from the local object specified by the Input property to the remote object specified by the Node and Location properties. 1 (cvReceive) means data is copied from the remote object specified by the Node and Location properties to the local object specified by the Output property. This defines only the master's actions.

(continued)

TABLE A-25 oDDELink PROPERTIES *(Continued)*

Input (NP)	A link to a local object whose default property value is sent to the remote object's default property.
Location (B)	The Address property of the remote slave object.
Node (B)	The Node property of the remote OOPic. (0 means inactive.)
Operate (F)	Must be 1 (cvTrue) for this object to be active.
Output (NP)	A link to a local object whose default property value is updated with the data received from the remote object.
Sync (F)*	Sends or retrieves data to or from the remote object, just once. This bit is cleared when a sync is completed.
Transmitting (F)	Is 1 when this OOPic is transferring data; it is 0 otherwise.

oDebounce

oDebounce is used to remove switch bounce from buttons (refer to Chapter 4, "Your First OOPic Program, OOPic I/O [Project #1: Das Blinken Light]"). Its properties are listed in Table A-26.

Memory Size: 5 Bytes

TABLE A-26 ODebounce PROPERTIES

Address (AP)	The address of the object in RAM (0–127).
Input (FP)	A link to a flag denoting the switch connection input.
InvertIn (F)	0 means when the Input property is 1, the Result property is 1 after the debounce. 1 means when Input is 0, the Result is 1 after the debounce.
InvertOut (F)	0 means Result is copied to the flag linked to the Output property. 1 means Result is inverted and then copied to the flag linked to the Output property.
Operate (F)*	When set to 1, the object is enabled; otherwise, it is disabled.
Output (FP)	A link to a flag that is updated with the debounced value.
Period (B)	The number of 1/60 time increments that take place before accepting the switch change.
Result (F)	The result of the debounce operation.

oEvent(C)(X)

This command triggers event code with a flag property (refer to Chapter 7, "OOPic Events, Keypads, and Serial I/O [Project: A Mini-terminal]"). (C and X variants are B.X.X only.) Its properties are shown in Table A-27.

Memory Size: 3 Bytes (oEventC and oEventX: 5 bytes), (oEventXC: 7 bytes)

TABLE A-27	oEvent(C)(X) PROPERTIES
Address (AP)	The address of the object in RAM (0–127).
ClockIn (FP)	Points to a property that determines when this object performs its operation.
InvertC (F)	0 means a transition from 0 to 1 triggers this object. 1 means a transition from 1 to 0 triggers this object (B.X.X only).
Operate (F)	When set (cvTrue), the event is triggered; if this bit is still set upon completion of the subprogram, the event is not triggered again until this bit transitions back to 0 and then to 1 again (default property). (oEventXC only: The event can only be clocked when this bit is set to 1; otherwise, the clockIn trigger has no effect.)
Priority (N)	The range is 0 to 15. Lower numbers are higher priority (B.X.X only).

oFanout(C)

This command connects one flag output to several flag inputs. (The C variant is B.X.X only.) Its properties are shown in Table A-28.

Memory Size: 2 + the number of outputs (oFanoutC: 4 + the number of outputs)

TABLE A-28	oFanout(C) PROPERTIES
Address (AP)	The address of the object in RAM (0–127).
ClockIn (FP)	Links to the flag bit used to clock (signal) this object to perform an operation (B.X.X only).
Input (FP)	Links to an object's Flag property that will be copied to the Output flags.
InvertC (F)	When set to 1, this inverts the logical sense of ClockIn; that is, a 0 clocks the object instead of a 1 (B.X.X only).
InvertOut1..4 (Bi)	Inverts that Output flag's logic.
Operate (F)*	Must be 1 (cvTrue) for this object to be active.
Output11..4 (FP)	Links to another object's input flag.
Width (N)	(Read-only) The configured number of output links.

oGate(C)

oGate(C) logically combines multiple flag outputs and link to a flag input (see Chapter 3, "OOPic Object Standard Properties"). (C variant is B.X.X only.) Table A-29 outlines the oGate(C) properties.

Memory Size: 2 + the number of inputs (oGateC: 4 + the number of inputs)

TABLE A-29	oGate(C) PROPERTIES
Address (AP)	The address of the object in RAM (0–127).
ClockIn (FP)	Links to the flag bit that is used to clock (signal) this object to perform an operation (B.X.X only).
Exclusive (Bi)	When set to 1, this makes the inputs of the OR gate exclusive; the logic table is shown in Chapter 3.
Input1..8 (FP)	Links to flag properties of an object that will be used as an input to the object.
InvertC (F)	When set to 1, this inverts the logical sense of ClockIn; that is, a 0 clocks the object instead of a 1 (B.X.X only).
InvertIn1..8 (F)	When a selected input's InvertIn flag is set to 1, the logical sense of that input is inverted; a 1 becomes a 0 and vice versa.
InvertOut (F)	When this property is set to 1, the output of the object is logically inverted.
Operate (F)*	This must be set to 1 (cvTrue) in order for the object to respond to input.
Output (FP)	Links to a flag property in an object that will be affected by the output of the object's logical operation.
Result (F)	This is the result of the logical operations defined by the input of the oGate object. This value reflects the result *before* the output is inverted if that property is set.
Width (N)	Returns the number of inputs defined when this object is dimensioned.

oIndex(C)

oIndex(C) provides indexing functions for oBuffer objects. (C variant is B.X.X only.) The properties are shown in Table A-30.

Memory Size: 4 Bytes (oIndexC: 6 bytes)

TABLE A-30	oIndex(C) PROPERTIES
Address (AP)	The address of the object in RAM (0–127).
Array (BP)	Links to a oBuffer object from which values are returned or stored.
ClockIn (FP)	Links to the flag bit used to clock (signal) this object to perform an operation (B.X.X only).
Direction (F)	0 means (cvRetrieve) retrieves the Index value from the Array object and places it into the Value property of the Unit object.
	1 means (cvStore) stores the Value property of the Unit object into the Index position of the Array object.
Index (NP)	Links to an object whose Value property holds the position within the Array object.
InvertC (Bi)	When set to 1, this inverts the logical sense of ClockIn; that is, a 0 clocks the object instead of a 1 (B.X.X only).
Operate (F)*	This must be set to 1 (cvTrue) in order for the object to respond to input.
Unit (NP)	Links to an object whose Value property is stored in or returned from the Array object.

oMath(I)(O)(C)

This command provides various math functions to a *virtual circuit* (VC) (refer to Chapter 5, "Analog-to-Digital and Hobby Servos [Project: Push My Finger]"). (I, O, and C variants are B.X.X only.) The oMath(I)(O)(C) properties are shown in Table A-31.

Memory Size: 4 Bytes (oMathI and oMathO: 5 bytes, and oMath*C: 6 bytes)

TABLE A-31	OMATH(I)(O)(C) PROPERTIES
Address (AP)	The address of the object in RAM (0–127).
Input1 (NP)	Links to an output Value property to use from another object.
ClockIn (FP)	Links to the flag bit used to clock (signal) this object to perform an operation (B.X.X only).
Input2 (NP)	Links to an output Value property to use from another object.
InvertC (Bi)	Used by a clocked oMath object to invert the logic of the clock property (B.X.X only).

(continued)

TABLE A-31   oMath(I)(O)(C) PROPERTIES *(Continued)*

Mode (N)	Selects one of eight math operations: 0: cvAdd adds Input1 to Input2. 1: cvSubtract subtracts Input2 from Input1. 2: cvLShift bit shifts Input1 left a number of times given by Input2. 3: cvRShift bit shifts Input1 right a number of times given by Input2. 4: cvAND logically ANDs Input1 and Input2. 5: cvOR logically ORs Input1 and Input2. 6: cvXOr logically exclusive ORs Input1 and Input2. 7: cvLatch Input1 is copied to Output.
Negative (F)	Is 1 if the result of Mode 0 through 3 is a negative number.
NonZero (F)	Is 1 if the result of any operation is nonzero.
Operate (F)*	Must be 1 for this object to respond to input.
Output (NP)	Links to an input Value property of another object.
Value (B)	An optional value when used with oMathI or oMathO variants in B.X.X or later firmware (B.X.X only).

oOneShot

This provides a logical one-shot function (refer to Chapter 9, "OOPic 12C and Distributed DDE Programming [Project: Remote Control]").Its properties are shown in Table A-32.

Memory Size: 4 Bytes

TABLE A-32   oOneShot PROPERTIES

Address (AP)	The address of the object in RAM (0–127).
Input (FP)	Links to the flag that will be the trigger level for the one shot.
InvertIn (F)	1 inverts the logic of the input, so that a 0 level will be seen as the trigger. 0 means normal operation, and a 1 level is required to trigger the one shot.
InvertOut (Bi)	Inverts the logic of the trigger before storing the result of the operation and linking to the Output property.
Operate (F)*	Must be 1 (cvTrue) for this object to be active.
Output (FP)	Links to the flag that get the one-shot pulse.
Result (F)	The result of the one-shot operation as a value, not a link.

oSrvSync

oSrvSync provides a method of synchronizing multiple hobby servos for such things as walking robots. The properties are shown in Table A-33.

Memory Size: 6 Bytes

TABLE A-33 oSrvSync PROPERTIES	
Address (AP)	The address of the object in RAM (0–127).
Input (NP)	Links to the object whose Value property specifies the position of the first servo.
Operate (F)*	Must be 1 for this object to respond to input.
Output (NP)	Links to an array of oServo objects.
RCJoint (F)	Specifies if the even-numbered servos are to be used as knee joints on a legged robot. This means that the RCSpan value won't be added to these servos. 0 means the servos are continuous. 1 means the servos are paired with the second in each pair used as the knee joint.
RCMax (F)	The maximum value that the servos are allowed to reach before the value is rolled over. 0 means the maximum servo value is 15. 1 means the maximum servo value is 31.
RCRise (F)	The value that the knee joint servos will be set to when the leg is in the upright position. 0 means the rise value is 15. 1 means the rise value is 31.
RCSpan (N)	Specifies the distance between consecutive servos; each servo in order adds RCSpan to the one preceding it (0–63).
Sync (F)	When set to 1, it tells the object to perform the synchronization.

oRandomizer(O)(C)

This command generates random numbers. Its Output property can link to either a word or byte property. (O and C variants are B.X.X only.) The oRandomizer(O)(C) properties are shown in Table A-34.

A

OOPIC A.2.X OBJECTS

Memory Size: 4 Bytes (oRandomizer*C: 6 bytes)

TABLE A-34	oRandomizer(O)(C) PROPERTIES
Address (AP)	The address of the object in RAM (0–127).
ClockIn (FP)	Links to the flag bit used to clock (signal) this object to perform an operation (B.X.X only).
InvertC (Bi)	Used by a clocked oRandomizer object to invert the logic of the clock property (B.X.X only).
Operate (F)*	Must be 1 for this object to respond to input.
Output (NP)	Links to an object whose Value property is randomized.
Result (B)	Bit 15 of the randomized Seed property.
Result2	Bit 7 of the randomized Seed property.
Seed (B)	The value this object will use as a starting point for randomization.
Value (B)	An optional value when used with oRandomizerO variants in B.X.X or later firmware. It is used instead of the Output property to hold the random number in this object instead of linking it (B.X.X only).

oRTC

This is the OOPic RTC object. To function, it requires linking to the oBuffer object, described later (refer to Chapter 6, "OOPic Timers, Clocks, LCDs, and SONAR [Project: SONAR Ping]"). Its properties are shown in Table A-35.

Memory Size: 4 Bytes

TABLE A-35	oRTC PROPERTIES
Address (AP)	The address of the object in RAM (0–127).
ClockIn1 (FP)	Links to a flag in another VC supplying clock ticks.
Direction (F)	0 = Increment clock. 1 = Decrement clock.
Operate (F)*	Must be 1 for this object to respond to input.
Output (BP)	Links to an oBuffer object that holds the RTC values.
PM (F)	0 = AM. 1 = PM.
Tick (Bi)	0 means each tick increments the one's position of the Seconds field. 1 means each tick increments the one's position of the 1/60-second field.

oWire(C)

oWire is a special case oGate object with a single input and a single output (refer to Chapter 4, "Your First OOPic Program, OOPic I/O [Project #1: Das Blinken Light]"). Its properties are shown in Table A-36.

Memory Size: 5 Bytes

TABLE A-36	oWire PROPERTIES (UNCLOCKED VERSION)
Address (AP)	The address of the object in RAM (0–127).
ClockIn (FP)	Links to the flag bit that will be used to clock (signal) this object to perform an operation. (B.X.X only)
Input (FP)	A link to the signal input.
InvertC (Bi)	Used by clocked oRandomizer object to invert the logic of the clock property. (B.X.X only)
InvertIn (F)	0 means when the Input is 1, the Result is 1. 1 means when the Input is 0, the Result is 1.
Operate (F)*	When set to 1, the object is enabled; otherwise, it is disabled.
Output (FP)	A link to a flag that will be updated with the input value.
Result (F)	The result of the input bit evaluation.

Variable Objects

These objects are used to store data.

oBit

A 1-bit variable object, oBit just holds one bit. Its properties and methods are shown in Tables A-37 and A-38.

Memory Size: 1 Byte

TABLE A-37	oBit PROPERTIES
Address (AP)	The address of the object in RAM (0–127).
NonZero (F)	Returns 0 if Value is equal to 0; it returns 1 otherwise.
String (S)	The string representation of the Value property.
Value(Bi)*	The number written to or read from the I/O lines specified.

TABLE A-38   oBit METHODS

Clear	Clears the Value property to 0.
Invert	Inverts the bits in the Value property.
Set	Sets the Value property to 1.

oBuffer

oBuffer creates an array of bytes that can be used to store a string. The Location property auto-increments when storing a string, and each byte must be manually addressed via Location when storing values. Location 1 is the first byte. Tables A-39 and A-40 outline the oBuffer properties and methods.

Memory Size: 4 + the number of bytes configured (new oBuffer(10): 10 bytes)

TABLE A-39   oBuffer PROPERTIES

Address (AP)	The address of the object in RAM (0–127).
Location (B)	The index within the buffer referencing a specific byte.
NonZero (F)	Returns 0 if the Value property is equal to 0; it returns 1 otherwise.
RTCString (S)	The string representation of the BCD values stored in oBuffer by the oRTC object.
String (S)	The string representation of the Value property, starting at the Location specified.
Value (B)*	The number stored in Location of the buffer.
Width (B)	The number of bytes configured in this object (read only).

TABLE A-40   oBuffer METHODS

Clear	Clears the Value property to 0.
Dec	Decrements the Value property by 1.
Inc	Increments the Value property by 1.
Invert	Inverts the bits in the Value property.
LShift	Shifts the bits in the Value property left one position.
RShift	Shifts the bits in the Value property right one position.
Set	Sets the Value property to 255.

oByte

oByte is a variable object holding 1 byte of data. Its properties are shown in Tables A-41 and A-42.

Memory Size: 1 Byte

TABLE A-41   oByte PROPERTIES

Address (AP)	The address of the object in RAM (0–127).
MSB (F)	Returns a 1 if the MSB is 1; it returns 0 otherwise.
NonZero (F)	Returns 0 if the Value property is equal to 0; it returns 1 otherwise.
Signed (Bi)	Setting this property to 1 causes this object to be treated as an 8-bit signed number whose value is from -128 to 127. If set to 0, this object's value is from 0 to 255.
String (S)	The string representation of the Value property.
Value(B)*	The number that is stored.

TABLE A-42   oByte METHODS

Clear	Clears the Value property to 0.
Dec	Decrements the Value property by 1.
Inc	Increments the Value property by 1.
Invert	Inverts the bits in the Value property.
LShift	Shifts the bits in the Value property left one position.
RShift	Shifts the bits in the Value property right one position.
Set	Sets the Value property to 255.

oEEPROM

This is the EEPROM object for use on the local I2C bus and programming port. This object is identical to oI2C but adds a method (refer to Chapter 9, "OOPic I2C and Distributed DDE Programming [Project: Remote Control]"). Its properties and methods are outlined in Tables A-43 and A-44.

Memory Size: 5 Bytes

TABLE A-43   oEEPROM PROPERTIES

Address (AP)	The address of the object in RAM (0–127).
Location (B/W)	In Mode 0, this is a 16-bit address usually used in EEPROMs that will follow the I2C address byte.
	In Mode 1, this is the 8-bit register address usually used in devices that are register based.
	Not used in Mode 2.
Mode (N)	0 (cv23Bit) means the I2C uses Location as the two address bytes that follow the device address byte.
	1 (cv10Bit) means I2C uses Location as the single address byte that follows the device address byte, usually to select a register.
	2 (cv7Bit) means the I2C uses the 7-bit I2C address mode for simple 1-byte writes or reads, usually for device commands.
Node (B)	The I2C address of the device.
NoInc (F)	0 (cvFalse) means the Location property is incremented each time the Value property is read or written.

(continued)

	1 (cvTrue) means the Location property is not incremented each time the Value property is read or written. This is only used with Mode 0 or 1.
String (S)	Although listed, this property is not supported in oI2C.
Value (B/W)*	When Width equals 0, bytes are read from or written to the device. When Width equals 1, words are read from or written to the device.
Width (Bi)	0 (cv8Bit) means 8 bits are read and written at a time (1 byte). 1 (cv16Bit) means 16 bits are read and written at a time (2 bytes).

TABLE A-44   oEEPROM METHODS

Data (Value1, Value2 . . .)	Stores data into EEPROM above a program's maximum used address location. Bytes, words, and quoted strings are all supported as data and are only stored in the program's EEPROM.

oNibble

This is a variable object holding 4 bits (a half-byte) of data. Its properties and methods are shown in Tables A-45 and A-46.

Memory Size: 1 Byte

TABLE A-45   oNibble PROPERTIES

Address (AP)	The address of the object in RAM (0–127).
NonZero (F)	Returns 0 if Value is equal to 0; it returns 1 otherwise.
String (S)	The string representation of the Value property.
Value(N)*	The number written to or read from the I/O lines specified.

A

OOPIC A.2.X OBJECTS

TABLE A-46   oNibble METHODS

Clear	Clears the Value property to 0.
Dec	Decrements the Value property by 1.
Inc	Increments the Value property by 1.
Invert	Inverts the bits in the Value property.
LShift	Shifts the bits in the Value property left one position.
RShift	Shifts the bits in the Value property right one position.
Set	Sets the Value property to 15.

oRam

oRam enables access to any RAM location in the OOPic memory space. It is one of two objects that use the Address property of other objects; oDDELink is the other object (refer to Chapter 3, "OOPic Object Standard Properties"). Its properties and methods are shown in Tables A-47 and A-48.

Memory Size: 3 Bytes

TABLE A-47   oRam PROPERTIES

Address (AP)	The address of the object in RAM (0–127).
Location (W)	The address within RAM that is to be accessed.
String (S)	The string representation of the Value property.
Value(B)*	The number written to or read from the I/O lines specified.

TABLE A-48   oRam METHODS

Clear	Clears the Value property to 0.
Dec	Decrements the Value property by 1.
Inc	Increments the Value property by 1.
Invert	Inverts the bits in the Value property.
LShift	Shifts the bits in the Value property left one position.
RShift	Shifts the bits in the Value property right one position.
Set	Sets the Value property to 255.

oWord

oWord is a variable object holding one word or 16 bits of data. Its properties and methods are shown in Tables A-49 and A-50.

Memory Size: 1 Byte

TABLE A-49 oWord PROPERTIES	
Address (AP)	The address of the object in RAM (0–127).
MSB (F)	Returns a 1 if the MSB is 1; it returns 0 otherwise.
NonZero (F)	Returns 0 if Value is equal to 0; it returns 1 otherwise.
Signed (Bi)	Setting this property to 1 causes this object to be treated as an 8-bit signed number whose value is -128 to 127. If set to 0, this object's value will be 0 to 255.
String (S)	The string representation of the Value property.
Value(W)*	The number written to or read from the I/O lines specified.

TABLE A-50 oWord METHODS	
Clear	Clears the Value property to 0.
Dec	Decrements the Value property by 1.
Inc	Increments the Value property by 1.
Invert	Inverts the bits in the Value property.
LShift	Shifts the bits in the Value property left one position.
RShift	Shifts the bits in the Value property right one position.
Set	Sets the Value property to 65,535.

User-Definable Objects

Only one object is included in this section.

oUserClass

oUserClass enables the definition of OO-style class files to be used in a program (refer to Chapter 3, "OOPic Object Standard Properties"). To create a user class, simply write a program with a main() function and any other subroutines. The main() function becomes the constructor method and all other subroutines become methods. oUserClass objects have a couple of restrictions:

■ You can't reference the OOPic object from within the user class.
■ You can't reference any method of the user class from within the class.

This is how you create a class:

```
Dim LCD as new oUserClass("dlclcd.osc")
```

You can reference methods (subroutines) or attributes (variables) in the standard OO way, such as using LCD.Init, to execute an Init() method within an oUserClass called LCD.

System Objects

System objects are not declared and are intrinsic to the OOPic operating system.

OOPic

This is the object that runs the user program and coordinates all VC activities (refer to Chapter 3, "OOPic Object Standard Properties"). Its properties are shown in Table A-51.

Memory Size: 10 Bytes (after which 86 bytes are left for other objects)

TABLE A-51 OOPic PROPERTIES	
Delay (W)	A write to this property delays a number (0–65,535) multiplied by 1/100's of a second before executing the next instruction.
ExtVRef (Bi)	This is discussed in Chapter 5, "Analog-to-Digital and Hobby Servos (Project: Push My Finger)." When set to 1, this sets the internal voltage reference to the voltage appearing on I/O line 4 not to exceed 5V.
Hz1 (F)	This read-only bit has a duty cycle of once per second; actually, it's closer to 0.99957 Hz if you want to be picky.
Hz60 (F)	This read-only bit has a duty cycle of $1/60$ of a second. For a 60 Hz clock, this is useful for clocking your RTC objects.
Node (B)	0 means no I2C node is assigned; 1 through 127 is the DDELink I2C node number of this OOPic. This is used for IDE debugging and for DDELink communications between OOPic controllers.
	(continued)

Operate (F)	If you set this bit to 1, the OOPic goes into low-power halt mode, effectively turning it off (default property).
Pause (F)	Similar to Operate, setting this bit to 1 pauses the OOPic, and it stops executing instructions.
PullUp (Bi)	Setting this bit to 1 enables the internal pull-up resistors on I/O lines 8 through 15. This is useful if you want buttons on those inputs and don't want to use your own pull-up resistors. This mode is used by oKeypad and oKeypadX.
Reset (F)	When set to 1, this causes an OOPic software reset.
StartStat (N)	This read-only property returns the reason for the last OOPic reset: 0 = Standard power on; the OOPic just turned on. 1 = Reset by the hardware reset line. 2 = Reset caused by a power brown-out (low voltage on Vcc). 3 = Reset caused by the watch dog timer or the Reset property being set.

PIC

This object defines all the internal PIC registers for the PIC used with that OOPic variant. In general, they are all the same names and fields (refer to Chapter 1, "OOPic Family Values," and Chapter 3, "OOPic Object Standard Properties". This is for advanced users that know their PICs. The following PIC settings should not be changed:

FSR	PIE1
INDF	PIE2
INTCON	PIR1
OPTION	PIR2
PCON	STATUS
PCL	TMR0
PCLATH	TRISA
PORTA	

Refer to the PICMicro datasheet for details about how to use the properties of the PIC processor used by your OOPic variant (as shown in Chapter 1, "OOPic Family Values," and here in Table A-52). The properties are encoded somewhat from what you will find in the PICMicro datasheets; for instance, the BRG register in the PIC16F877 is the SPBRG_REG property. The PIC object costs no object memory to use.

TABLE A-52 PIC PROPERTIES

CON0_ADCS (N)	PIE1_TMR2IE (Bi)
ADCON0_ADON (Bi)	PIE1_TXIE (Bi)
ADCON0_CHS (N)	PIE2_CCP2IE (Bi)
ADCON0_GODONE (Bi)	PIR1_ADIF (Bi)
ADCON1_PCFG (N)	PIR1_CCP1IF (Bi)
ADRES_REG (N)	PIR1_PSPIF (Bi)
CCP1CON_CCP1M (N)	PIR1_RCIF (Bi)
CCP1CON_CCP1X (Bi)	PIR1_SSPIF (Bi)
CCP1CON_CCP1Y (Bi)	PIR1_TMR1IF (Bi)
CCP2CON_CCP2M (N)	PIR1_TMR2IF (Bi)
CCP2CON_CCP2X (Bi)	PIR1_TXIF (Bi)
CCP2CON_CCP2Y (Bi)	PIR2_CCP2IF (Bi)
CCPR1H_REG (B)	PORTA_REG (B)
CCPR1L_REG (B)	PORTB_REG (B)
CCPR2H_REG (B)	PORTC_REG (B)
CCPR2L_REG (B)	PORTD_REG (B)
FSR_REG (B)	PORTE_REG(N)
INDF_REG (B)	PR2_REG (B)
INTCON_GIE (Bi)	RCREG_REG (B)
INTCON_INTE (Bi)	RCSTA_CREN (Bi)
INTCON_INTF (Bi)	RCSTA_FERR (Bi)
INTCON_PEIE (Bi)	RCSTA_OERR (Bi)
INTCON_RBIE (Bi)	RCSTA_REX9D (Bi)
INTCON_RBIF (Bi)	RCSTA_RX9 (Bi)
INTCON_TOIE (Bi)	RCSTA_SPEN (Bi)
INTCON_TOIF (Bi)	RCSTA_SREN (Bi)
OPTION_INTEDG (Bi)	SPBRG_REG (B)
OPTION_PS (N)	SSPADD_REG (B)
OPTION_PSA (Bi)	SSPBUF_REG (B)
OPTION_RBPU (Bi)	SSPCON_CKP (Bi)
OPTION_T0SC (Bi)	SSPCON_SSPEN (Bi)
OPTION_T0SE (Bi)	SSPCON_SSPM (N)
PCL_REG (B)	SSPCON_SSPOV (Bi)
PCLATH_REG (B)	SSPCON_WCOL (Bi)
PCON_BOR (Bi)	SSPSTAT_BF (Bi)
PCON_POR (Bi)	SSPSTAT_DA (Bi)
PIE1_ADIE (Bi)	SSPSTAT_P (Bi)
PIE1_CCP1IE (Bi)	SSPSTAT_RW (Bi)
PIE1_PSPIE (Bi)	SSPSTAT_S (Bi)
PIE1_RCIE (Bi)	SSPSTAT_UA (Bi)
PIE1_SSPIE (Bi)	

(continued)

STATUS_C (Bi)
STATUS_DC (Bi)
STATUS_IRP (Bi)
STATUS_PD (Bi)
STATUS_RP (N)
STATUS_TO (Bi)
STATUS_Z (Bi)
T1CON_T1CKPS (N)
T1CON_T1OSCEN (Bi)
T1CON_T1SYNC (Bi)
T1CON_TMR1CS (Bi)
T1CON_TMR1ON (Bi)
T2CON_T2CKPS (N)
T2CON_TMR2ON (Bi)
T2CON_TOUTPS (N)
TMR0_REG (B)
TMR1H_REG (B)
TMR1L_REG (B)

TMR2_REG (B)
TRISA_REG (B)
TRISB_REG (B)
TRISC_REG (B)
TRISD_REG (B)
TRISE_IBF (Bi)
TRISE_IBOV (Bi)
TRISE_OBF (Bi)
TRISE_PSPMODE (Bi)
TRISE_REG (N)
TXREG_REG (B)
TXSTA_BRGH (Bi)
TXSTA_CSRC (Bi)
TXSTA_SYNC (Bi)
TXSTA_TRMT (Bi)
TXSTA_TX9 (Bi)
TXSTA_TX9D (Bi)
TXSTA_TXEN (Bi)

A

OOPIC A.2.X OBJECTS

OOPIC B.2.X OBJECTS

These are the additional objects that appear in the B.2.X (OOPic II) firmware. The OOPic II *greatly* expanded the number of high-level objects available to the user. The objects in this appendix are *only* the additional objects; this is not a full list of all the objects in the OOPic II. This list is just a reference for the properties and methods of the objects; it is not a full description of the object and how to use it. For detailed object information, see the referenced chapters or the OOPic web site (www.oopic.com) and refer to the manual pages.

The same abbreviations, aliases, and selection matrices are used in this appendix as in Appendix A, "OOPic A.2.X Objects." Refer to Appendix A for them when needed.

A note about the *Operate* property: When the Operate property is cvFalse (0), this doesn't mean the object's output just disappears. It means it stops changing state based on its current inputs. The object continues to hold the last value and set of flags that it had when Operate was cvTrue (1).

A note about the *UnSigned* property: When UnSigned is set to 1, this does not mean the Value property changes somehow into a positive number. This property is a flag to the OOPic that tells the operating system whether or not to extend the sign bit when the value is moved to a larger variable (as in a byte being assigned to a word).

New Hardware Objects

These are additional objects that access hardware devices.

oA2D10

oA2D10 is a 10-bit *analog-to-digital* (A2D) object. The two *least significant bits* (LSBs) are only valid in B.2.X+ Firmware; in B.2.X firmware, they read 00 (refer to Chapter 5, "Analog-to-Digital and Hobby Servos [Project: Push My Finger]"). Its properties are shown in Table B-1.

Memory Size: 3 Bytes

TABLE B-1 OA2D10 PROPERTIES	
Address (AP)	The address of the object in RAM (0–127).
IOLine (N)	Assigns the *input/output* (I/O) line to use. Essentially, 1 through 7 are I/O lines 1 through 7.
MSB (F)	The *most significant bit* (MSB) of the Value property. If this bit equals 1, then Value is equal to or over half of the reference voltage. If it is 0, then it is under half.
Operate (F)	Must be 1 to respond to input.
String (S)	The string representation of the Value property.
Value (W)*	The result of the A2D conversion.

oA2DX

oA2DX is an 8-bit A2D object that allows a center point with signed values above and below a 0 setting (refer to Chapter 5, "Analog-to-Digital and Hobby Servos [Project: Push My Finger]"). Its properties are shown in Table B-2.

Memory Size: 5 Bytes

TABLE B-2 oA2DX PROPERTIES	
Address (AP)	The address of the object in RAM (0–127).
Center (B)	Adjusts the center's compare point by -128 to 127.
IOLine (N)	Selects the A2D line to use, which is within 1 through 4 or 1 through 7 depending upon which OOPic is being used. 0 selects no A2D line.
Limit (Bi)	When set to 1, this limits the Value property from -128 to +127, no matter what the center is set to.
Negative (F)	Is set whenever the Value property is a negative number.
NonZero (F)	Is 0 whenever the voltage is 0.
Operate (F)	Must be 1 for the object to respond to input.
Unsigned (Bi)	When set to 1, the percentage is from 0 to 255 and is not signed.
Value (W)*	The adjusted A2D result.

oBumper4

This object reads and returns a 4-bumper digital input as a *Uniform Robotic Control Protocol* (URCP) value (see Chapter 10, "OOPic Robotics and URCP [Project: A Robot That Toes the Line]"). Its properties are shown in Table B-3.

Memory Size: 4 Bytes

TABLE B-3 oBumper4 PROPERTIES	
Address (AP)	The address of the object in RAM (0–127).
IOGroup (N)	The 8-bit I/O group to use (refer to Appendix A).
Nibble (Bi)	0 means use the lower 4 bits of the selected I/O group. 1 means use the upper 4 bits of the selected I/O group.
Operate (F)	Must be set to 1 for this object to function.
Received (F)	Is 1 if *any* valid bumper press is sensed; it is 0 otherwise.
Unsigned (Bi)	0 means the Value property is interpreted as a signed number. 1 means the Value property is interpreted as an unsigned number.
Value (B)*	The URCP directional value returned between -128 and +127.

Assuming you use the same bumper positions, this matrix shows the URCP value returned. Assign the bumpers in bit order such that bit 0 is the front and bit 2 is the rear in clockwise order (as shown in Table B-4) or counterclockwise.

TABLE B-4   oBUMPER4 BUMPER ENCODING

I/O INPUTS	BUMPER	POSITION	SIGNED VALUE	UNSIGNED VALUE	RECEIVED
0001	---x	Front	0	0	1
0011	--xx		32	32	1
0010	--x-	Right	64	64	1
0110	-xx-		96	96	1
0100	-x--	Back	−128	128	1
1100	xx--		−96	160	1
1000	x---	Left	−64	192	1
1001	x--x		−32	224	1
Other			Unchanged	Unchanged	0

oBumper8

This object reads and returns an 8-bumper digital input as a URCP value (refer to Chapter 10, "OOPic Robotics and URCP [Project: A Robot That Toes the Line]"). Its properties are shown in Table B-5.

Memory Size: 4 Bytes

TABLE B-5   oBumper8 PROPERTIES

Address (AP)	The address of the object in RAM (0–127).
IOGroup (N)	The 8-bit I/O group to use (refer to Appendix A).
Operate (F)	Must be set to 1 for this object to function.
Received (F)	Is 1 if *any* valid bumper press is sensed; it is 0 otherwise.
Unsigned (Bi)	0 means the Value property is interpreted as a signed number. 1 means the Value property is interpreted as an unsigned number.
Value (B)*	The URCP directional value returned between −128 and +127.

Assuming you use the same bumper positions, this matrix shows the URCP value returned. Three and four bumper switch presses will be read too, as averages of their respective values (those that make sense!). In other words, add the values of the adjacent bumpers; that will be the value that is reported. Left and right are arbitrarily assigned. Assign bumpers in bit order either clockwise (as shown in Table B-6) or counterclockwise such that bit 0 is the front bumper and bit 4 is the rear bumper.

TABLE B-6   oBUMPER8 BUMPER ENCODING

I/O INPUTS	BUMPERS	POSITION	SIGNED VALUE	UNSIGNED VALUE	RECEIVED
00000001	-------x	Front	0	0	1
00010001	------xx		16	16	1
00010000	------x-	Front right	32	32	1
00010010	-----xx-		48	48	1
00000010	-----x--	Right	64	64	1
00100010	----xx--		80	80	1
00100000	----x---	Back right	96	96	1
00100100	---xx---		112	112	1
00000100	---x----	Back	−128	128	1
01000100	---xx---		−112	114	1
01000000	-x------	Back left	-96	160	1
01001000	-xx-----		−80	176	1
00001000	-x------	Left	−64	192	1
10001000	xx------		−48	208	1
10000000	x-------	Front left	−32	224	1
10000001	x------x		−16	240	1
Other			Unchanged	Unchanged	0

oButton

This object interacts with a button–*light-emitting diode* (LED) combination, such as those found on the OOPic R. The LED resistor should be equal to the pull-up resistor (refer to Chapter 1, "OOPic Family Values"). Its properties are shown in Table B-7.

Memory Size: 4 Bytes

TABLE B-7	oButton PROPERTIES
Address (AP)	The address of the object in RAM (0–127).
InvertIn (Bi)	When InvertIn is 0 and
	Mode is 0, a button pressed Value = cvOn (1), and when released Value = cvOff (0).
	Mode is 1: the state toggles when the button is pressed.
	When InvertIn is 1 and Mode is 0, the button pressed Value = cvOff (0), and when released Value = cvOn (1).
	Mode is 1, the state toggles when the button is released.
IOLine (B)	The I/O line used for the button or button-LED combination.
Mode (Bi)	0: The button is treated as a momentary contact pushbutton.
	1: The button is treated as a toggle switch.
Option (Bi)	0: The LED is controlled by the Value property.
	1: The LED is controlled by the ValueL property.
Style (N)	LED appearance when turned on:
	0: Solid on.
	1: Blinks at 1 Hz.
	2: Blinks at 4 Hz.
	3: Blinks twice at 4 Hz, off for half-second, and repeats.
Value (F)*	Reflects the current state of the button.
ValueL (F)	Optionally (Mode = 1) sets the state of the LED.

oCompassDN

oCompassDN reads the values of the Dinsmore 1490 Compass module and reports in URCP (refer to Chapter 10, "OOPic Robotics and URCP [Project: A Robot That Toes the Line]"). Its properties are shown in Table B-8.

Memory Size: 4 Bytes

TABLE B-8	oCompassDN PROPERTIES
Address (AP)	The address of the object in RAM (0–127).
IOGroup (N)	The 8-bit I/O group to use (refer to Appendix A).
Nibble (Bi)	0 means use the lower 4 bits of the selected I/O group.
	1 means use the upper 4 bits of the selected I/O group.
Operate (F)	Must be set to 1 for this object to respond to input.
Unsigned (Bi)	0 means the Value property is interpreted as a signed number.
	1 means Value is interpreted as an unsigned number.
Value (B)*	URCP directional value returned between −128 and +127.

oCompassVX

This object reads the values of the Vector 2X Compass module and reports in URCP. Its properties are shown in Table B-9.

Memory Size: 9 Bytes

TABLE B-9 oCompassVX PROPERTIES	
Address (AP)	The address of the object in RAM (0–127).
IOGroup (N)	I/O group to use (refer to Appendix A). Only uses the lower 5 bits of the IOGroup; the upper bits can be used in other objects.
NonZero (F)	Is 0 if Value property is 0; it is 1 otherwise.
Operate (F)	Must be set to 1 for this object to respond to input.
Scale (Bi)	0 means the Value property read is scaled to URCP ranges. 1 means the Value property is not scaled and ranges from 0 to 359 (per the V2X normal output).
Unsigned (Bi)	0 means the Value property is interpreted as a signed number. 1 means Value is interpreted as an unsigned number.
Value (W)*	Either URCP byte or V2X word value, depending on the Scale setting.

oDCMotor

This is the LMD18200-style DC motor controller object (refer to Chapter 10, "OOPic Robotics and URCP [Project: A Robot That Toes the Line]"). Its properties are shown in Table B-10.

Memory Size: 5 Bytes

TABLE B-10 oDCMotor PROPERTIES	
Address (AP)	The address of the object in RAM (0–127).
Brake (F)	Sets (1) or clears (0) the brake line to the H-bridge chip.
Direction (F)	If the Unsigned property is 0, this bit reflects the motor direction: 0 is used for positive Value settings. 1 is used for negative Value settings. If Unsigned is 1, you set or clear this bit to determine motor direction.

(continued)

TABLE B-10   oDCMotor PROPERTIES (Continued)

InvertOutB (Bi)	Enables you to change the output level of the Brake functionality. 0 is standard logic; 0 = brake pin goes low. 1 is inverted logic; 0 = 1 = brake pin goes high.
InvertOutD (Bi)	When set to 1, the motor control line logic is reversed. 0 means IOLineD reflects the setting of the Direction property. 1 means IOLineD logic is inverted with respect to the Direction property.
IOLineB (B)	Brake control line to the H-bridge.
IOLineD(B)	Direction control line to the H-bridge.
IOLineP (Bi)	The I/O line to use for the *pulse width modulation* (PWM) signal. 0 is IOLine 18. 1 is IOLine 17.
Mode (Bi)	Configures how the PWM line is set when the Brake line is asserted. 0 (cvOff): When Brake equals 1, PWM output line is low. 1 (cvOn): When Brake equals 1, PWM output line is high.
Operate (F)	The bit must be 1 in order for the object to respond to input.
Period (B)	The number of clock tics that define the PWM time period.
PreScale (N)	Sets the clock divider for PWM hardware: 0 means the base frequency is 5 MHz. 1 means the base frequency is 1.25 MHz (system clock divided by 4). 2 means the base frequency is 312.5 KHz (system clock divided by 16). 3 is the same as 2 above.
Unsigned (Bi)	0 means the Value range is -128 to +127. 1 means the Value range is 0 to 255.
Value (B)*	This sets the on time for the PWM signal in clock tics.

oDCMotor2

oDCMotor2 controls an L298, L293, TI754410, or any similar DC Motor H-bridge. Its mproperties are shown in Table B-11.

Memory Size: 5 Bytes

TABLE B-11 oDCMotor2 PROPERTIES	
Address (AP)	The address of the object in RAM (0–127).
Brake (F)	When this bit is set, the object sets all control lines to the brake condition. This means that both motor outputs are at the same state and the PWM line is not cycling. *Mode* and *InvertOutB* define the operation of this state. Check your motor driver documentation for how to get an active brake, and then configure this object to match it.
Direction (F)	If Unsigned is 0, this bit reflects the motor direction: 0 for positive Value settings. 1 for negative Value settings. If Unsigned is 1, set or clear this bit to determine the motor direction.
InvertOutB (Bi)	Enables you to change the output level of the Brake functionality: 0 means both motor lines are logic 1. 1 means both motor lines are logic 0.
InvertOutD (Bi)	When set to 1, the motor control line logic is reversed. This is useful if you can't unsolder and swap your IOLine1 and IOLinet2 control lines to change this logic. 0: When Direction equals 0, IOLine1 equals 1, IOLine2 equals 0, and vice versa. 1: When Direction equals 0, IOLine1 equals 0, IOLine2 equals 1, and vice versa.
IOLine1 (B)	One of the two direction control lines to the H-bridge.
IOLine2 (B)	The other direction line to the H-bridge.
IOLineP (Bi)	The I/O line to use for the PWM signal (connected to enable): 0: IOLine 18. 1: IOLine 17.
Mode (Bi)	Sets how the PWM line is set when the Brake line is handled. 0 (cvOff) means when Brake is 1, PWM output is 0. 1 (cvOn) means when Brake is 1, PWM output is 1. If you choose incorrectly, don't worry; you may never notice. One way causes the motor to stop immediately; the other stop the motor gradually.
Operate (F)	The bit must be 1 in order for the object to respond to input.
Period (B)	The number of clock tics that define the PWM time period.

(continued)

TABLE B-11   oDCMotor2 PROPERTIES *(Continued)*

PreScale (N)	Sets the clock divider for PWM hardware: 0 means the base frequency is 5 MHz. 1 means the base frequency is 1.25 MHz (system clock divided by 4). 2 means the base frequency is 312.5 KHz (system clock divided by 16). 3 is the same as 2.
Unsigned (Bi)	0 means the Value range is from -128 to +127. 1 means the Value range is 0 to 255.
Value (B)*	This sets the on time for the PWM signal in clock tics.

oDCMotorMT

This object controls a MondoTronics-style H-bridge where the PWM signal is inverted (refer to Chapter 10, "OOPic Robotics and URCP [Project: A Robot That Toes the Line]"). Its properties are shown in Table B-12.

Memory Size: 5 Bytes

TABLE B-12   oDCMotorMT PROPERTIES

Address (AP)	The address of the object in RAM (0–127).
Brake (F)	Sets (1) or clears (0) the brake line to the H-bridge.
Direction (F)	If Unsigned is 0, this bit reflects the motor direction: 0 is used for positive Value settings. 1 is used for negative Value settings. If Unsigned is 1, you set or clear this bit to determine motor direction.
InvertOutB (Bi)	Enables you to change the output level of the Brake functionality. 0 is standard logic; 0 = 0 on brake pin. 1 is inverted logic; 0 = 1 on brake pin.
InvertOutD (Bi)	When set to 1, the motor control line logic is reversed: 0 means IOLineD reflects the setting of the Direction property. 1 means IOLineD logic is inverted with respect to the Direction property.
IOLineB (B)	The brake control line to the H-bridge.

(continued)

IOLineD(B)	The direction control line to the H-bridge.
IOLineP (Bi)	The I/O line to use for the PWM signal: 0 is IOLine 18. 1 is IOLine 17.
Mode (Bi)	Configures how the PWM line is set when the Brake line is asserted: 0 (cvOff) means when Brake is 1, PWM output is 0. 1 (cvOn) means when Brake is 1, PWM output is 1.
Operate (F)	The bit must be 1 in order for the object to respond to input.
Period (B)	The number of clock tics that define the PWM time period.
PreScale (N)	Sets the clock divider for PWM hardware: 0 means the base frequency is 5 MHz. 1 means the base frequency is 1.25 MHz (system clock divided by 4). 2 means the base frequency is 312.5 KHz (system clock divided by 16). 3 is the same as 2 above.
Unsigned (Bi)	0 means the Value range is −128 to +127. 1 means the Value range is 0 to 255.
Value (B)*	This sets the on time for the PWM signal in clock tics.

oDCMotorWZ

oDCMotorWZ controls a Wirz 203 motor driver board (refer to Chapter 10, "OOPic Robotics and URCP [Project: A Robot That Toes the Line]"). Its memory size is 5 bytes. This object is functionally identical to the oDCMotor object, so just ignore the Brake properties because the Wirz 203 board doesn't have any.

oFreq

oFreq outputs a frequency on IOLine 21 whose minimum frequency is about 38 Hz and whose maximum frequency is about 9765 Hz. Use the following formula to find the divisor to use (the result *must* be less than 65,279):

$$65535 - \left(\frac{2500000}{freq} \right)$$

The oFreq properties are shown in Table B-13.

Memory Size: 2 Bytes

TABLE B-13 oFreq PROPERTIES	
Address (AP)	The address of the object in RAM (0–127).
Operate (F)	The bit must be 1 in order for the object to respond to input.
Value (W)*	The divisor calculated as previously for desired frequency. 0 means no output.

oIRPD1

oIRPD1 reports the status of a single reflection-type *IR proximity detector* (IRPD). Its properties are shown in Table B-14.

Memory Size: 1 Byte

TABLE B-14 oIRPD1 PROPERTIES	
Address (AP)	The address of the object in RAM (0–127).
I/O line (B)	OOPic I/O lines 1 to 31; 0 means disabled.
Value (F)*	0 means nothing is detected. 1 means the object reflection is detected.

oIRPD2

oIRPD2 reports the status of a two-output (left and right) IRPD, such as the Lynxmotion IRPD board. Its properties are shown in Table B-15.

Memory Size: 2 Bytes

TABLE B-15 oIRPD2 PROPERTIES	
Address (AP)	The address of the object in RAM (0–127).
Center (F)	Reflection seen from both LEDs' transmissions.
InvertIn (Bi)	0 means a 1 on the IOLineS input is a reflection-detected signal. 1 means a 0 on the IOLineS input is a reflection-detected signal.
InvertOut (Bi)	0 means a 1 is required on IOLineL and IOLineR to select that side. 1 means a 0 is required on IOLineL and IOLineR to select that side.
IOLineL (B)	Left-side LED transmission selection IOLine.
IOLineS (B)	IR sensor input IOLine.

(continued)

IOLineR (B)	Right-side LED transmission selection IOLine.
Left (F)	Reflection seen when the left side (IOLineL) is selected.
NonZero (F)	0 means nothing is detected. 1 means something is detected.
Operate (F)	The bit must be 1 in order for the object to respond to input.
Right (F)	Reflection seen when the right side (IOLineR) is selected.
Value (N)*	0 means nothing is detected on either side. 1 is a left-side detection. 2 is a right-side detection. 3 means both sides detected something.

oIRRange

oIRRange reads and interprets URCP ranging information from a Sharp GP2D12-type range finder. It reports in 64 steps per foot and uses an A2D I/O Line (refer to Chapter 10, "OOPic Robotics and URCP [Project: A Robot That Toes the Line]"). Its properties are shown in Table B-16.

Memory Size: 5 Bytes

TABLE B-16	OIRRANGE PROPERTIES
Address (AP)	The address of the object in RAM (0–127).
Center (B)	Enables you to adjust (-128 to +127) the center point of your ranging unit if your unit does not read 128 at maximum range (nothing seen).
IOLine (N)	I/O lines 1 through 7 can be used. Remember, the OOPic II and II+ offer seven potential A2D lines, whereas OOPic I has only four.
Operate (F)	Must be set to 1 for this object to respond to input.
Value (B)*	Distance to the nearest object in URCP format; 0 to 128 is the range.

oJoystick

This object reports the button press information from an Atari-style button-press joystick (not one with potentiometers in it). Reports are in URCP format, just like the oBumper4 object. Its properties are shown in Table B-17. In fact, this object *is* the oBumper4 object, just aliased to another name to avoid confusion.

Memory Size: 4 Bytes

TABLE B-17   oJoystick PROPERTIES

Address (AP)	The address of the object in RAM (0–127).
IOGroup (N)	The 8-bit I/O group to use (refer to Appendix A).
Nibble (Bi)	0 uses the lower 4 bits of the selected I/O group.
	1 uses the upper 4 bits of the selected I/O group.
Operate (F)	Must be 1 for this object to respond to inputs.
Received (F)	Is 1 if *any* valid button press is sensed; it is 0 otherwise.
UnSigned (Bi)	0 means that the Value property is interpreted as a signed number.
	1 means that Value is interpreted as an unsigned number.
Value (B)*	The URCP directional value between −128 and +127.

oKeypad

oKeypad uses IOLines 8 to 15 to scan a 4×4 keypad matrix and returns the key pressed by the formula:

$$(((Row - 1) \times 4) - (Column - 1))$$ (Refer to Chapter 7, "OOPic Events, Keypads, and Serial I/O [Project : A Mini-terminal]")

Its properties are shown in Table B-18.

Memory Size: 4 Bytes

TABLE B-18   oKeypad PROPERTIES

Address (AP)	The address of the object in RAM (0–127).
Mode (Bi)	0 return the encoded row and column.
	1 returns phone-pad values.
Operate (F)	Must be 1 for this object to respond to inputs.
Received (F)	0 means no key is pressed.
	1 means the key is pressed, and a new value is stored in the Value property.
String (S)	The Value property represented as a string.
Value (N)*	The Value property encoded as of the last key press.

oKeypadX

oKeypadX decodes up to an 8×8 keyboard to be scanned for key presses. I/O lines 8 though 15 are used for the columns, and any I/O group of 8 lines can be used for the rows. Any I/O lines that are masked "off" in the matrix can be used for other OOPic I/O (refer to Chapter 7, "OOPic Events, Keypads, and Serial I/O [Project: A Mini-terminal]"). Its properties are shown in Table B-19.

ColMask or RowMask: Bit 0 is the LSB I/O line of the I/O group selected; a 0 in any bit-mask place means it will be ignored.

Memory Size: 6 Bytes

TABLE B-19	oKeypadX PROPERTIES
Address (AP)	The address of the object in RAM (0–127).
ColMask (B)	Bit mask determining which I/O lines are to be read.
IOGroup (N)	1 means the oKeypadX object uses IOGroup 1 for the row outputs.
	3 means the oKeypadX object uses IOGroup 3 for the row outputs.
	0 and 2 are not supported and cause odd things to happen.
Operate (F)	Must be 1 for this object to respond to inputs.
Received (F)	0 means no key is being pressed.
	1 means a key is pressed, and a new value is stored in the Value property.
RowMask (B)	Bit mask determining which I/O lines are to be scanned.
String (S)	The Value property represented as a string.
Value (B)*	The Value property encoded as of the last key press.

oLCD

oLCD controls an LCD that uses the standard 44780 chipset in 4-bit parallel mode (refer to Chapter 6, "OOPic Timers, Clocks, LCDs, and SONAR [Project: SONAR Ping]"). Its properties and methods are shown in Tables B-20 and B-21.

Memory Size: 6 Bytes

TABLE B-20	oLCD PROPERTIES
Address (AP)	The address of the object in RAM (0–127).
IOGroup (N)	The 8-bit I/O group to use (refer to Appendix A).
IOLineE (B)	The I/O line to use for the E line to the LCD.
IOLineRS (B)	The I/O line to use for the RS line to the LCD.
	(continued)

TABLE B-20 **oLCD PROPERTIES** *(Continued)*

Nibble (Bi)	0 uses the lower 4 bits of the IOGroup selection.
	1 uses the upper 4 bits of the IOGroup selection.
Operate (F)	Must be 1 for this object to respond to inputs.
RS (Bi)	Enables you to manually set the RS line to 0 or 1 so you can issue either instructions or data to the LCD.
String (S)	The Value property as a string. Whole strings may be sent here.
Value (B)*	The individual byte value to be sent to the LCD.

TABLE B-21 **oLCD METHODS**

Clear	Clears the display and homes the cursor to line 0, location 0.
Locate(R,C)	Sets the LCD cursor to the row and column specified. (0,0) is the first column and first line.

oLCDSE(T)

oLCDSE(T) controls a Scott Edwards serial LCD board (refer to Chapter 6, "OOPic Timers, Clocks, LCDs, and SONAR [Project: SONAR Ping]"). Its properties and methods are shown in Tables B-22 and B-23.

Memory Size: 5 Bytes

TABLE B-22 **oLCDSE(T) PROPERTIES**

Address (AP)	The address of the object in RAM (0–127).
Baud (N)	2 is the 2400-baud serial rate (default).
	3 is the 9600-baud serial rate.
	1, 4, 5, 6, and 7 are reserved for future expansion.
IOLine (B)	Any I/O line you choose to connect your LCD controller to.
Operate (F)	Must be 1 for this object to respond to inputs.
String (S)	The Value property as a string. Whole strings are sent here.
Value (B)*	The individual byte value to be sent to the LCD.

TABLE B-23	oLCDSE(T) METHODS
Clear	Clears the display and homes the cursor to line 0, location 0.
Locate(R,C)	Sets the LCD cursor to the row and column specified. (0,0) is first column and first line.

oLCDWZ

oLCDWZ controls a Wirz SLI-OEM LCD board (refer to Chapter 6, "OOPic Timers, Clocks, LCDs, and SONAR [Project: SONAR Ping]"). Its properties and methods are shown in Tables B-24 and B-25.

Memory Size: 5 Bytes

TABLE B-24	oLCDWZ PROPERTIES
Address (AP)	The address of the object in RAM (0–127).
Baud (N)	4 is the 19,200-baud serial rate (default). 1, 2, 3, 5, 6, and 7 are reserved for future expansion.
IOLine (B)	Any I/O line you choose to connect your LCD controller to.
Operate (F)	Must be 1 for this object to respond to inputs.
String (S)	The Value property as a string. Whole strings are sent here.
Value (B)*	The individual byte value sent to the LCD.

TABLE B-25	oLCDWZ METHODS
Clear	Clears the display and homes the cursor to line 0, location 0.
Locate(R,C)	Sets the LCD cursor to the row and column specified. (0,0) is the first column and first line.

oMotorMind

oMotorMind controls a Solutions Cubed™ Motormind-B DC motor controller module using the serial communications mode at 2400 baud. Its properties are shown in Table B-26.

Memory Size: 5 Bytes

TABLE B-26	OMOTORMIND PROPERTIES
Address (AP)	The address of the object in RAM (0–127).
IOLine (B)	The I/O line you use to connect to the device (1–31).
Operate (F)	Must be 1 for this object to respond to input.
Value (B)*	The value to send to the Motormind module (0–255).

oPWMX

This is a PWM object that can use any I/O line, not just the hardware PWM I/O lines. The frequency is 555 Hz with 16 settings (refer to Chapter 10, "OOPic Robotics and URCP [Project: A Robot That Toes the Line]"). The oPWMX properties are shown in Table B-27.

Memory Size: 4 Bytes

TABLE B-27	oPWMX PROPERTIES
Address (AP)	The address of the object in RAM (0–127).
IOLine (B)	The I/O line you use to connect to the device (1–31).
Operate (F)	Must be 1 for this object to respond to input.
Value (B)*	The value to send to the Motormind module (0–255).

oQEncode

oQEncode reads and interprets a quadrature encoder (rotational feedback). Its properties are shown in Table B-28.

Memory Size: 6 Bytes

TABLE B-28	oQEncode PROPERTIES
Address (AP)	The address of the object in RAM (0–127).
Direction (F)	0 means the last change to the Value property was an increment. 1 means the last change to the Value property was a decrement.
InvertD (F)	0 means the Value property is incremented when the first input changes to the same state as the second input. 1 means the Value property is decremented when the first input changes to the same state as the second input.

(continued)

IOLine1 (B)	Chooses the I/O line to use for the B input.
IOLine2 (B)	Chooses the I/O line to use for the A input.
Moved (F)	Is set to 1 when the Value property changes. This indication is for one list loop cycle, so it's only for use in a VC.
Operate (F)	Must be 1 for this object to respond to input.
Signed (F)	0 means the Value property can be from 0 to 65,535. 1 means the Value property can be from -32,768 to +32,767.
Value (W)*	The running count of the quadrature encoder signal that increments forward and decrements backwards.

oSerialX

oSerialX enables a serial input or output on any I/O line, with optional flow control (refer to Chapter 7, "OOPic Events, Keypads, and Serial I/O [Project: A Mini-terminal]"). Its properties are shown in Table B-29.

Memory Size: 5 Bytes

TABLE B-29	oSerialX PROPERTIES
Address (AP)	The address of the object in RAM (0–127).
Baud (N)	1 (cv1200) sets UART at 1200 baud. 2 (cv2400) sets UART at 2400 baud. 3 (cv9600) sets UART at 9600 baud. 5 (cv4800) sets UART at 4800 baud.
InvertF (F)	1 means invert flow control logic (set this to be like oSerialPort flow control).
InvertS (F)	1 means invert serial line logic.
IOLineF (B)	Chooses the I/O line to be the flow control line.
IOLineS (B)	Chooses the I/O line to be the serial data line.
Operate (F)	Must be 1 for this object to respond to input.
Value (B)*	The data to be transmitted or the data that has been received.

oServoSE

oServoSE controls a Scott Edwards SSCII serial servo controller (refer to Chapter 5, "Analog-to-Digital and Hobby Servos [Project: Push My Finger]"). Its properties and methods are listed in Tables B-30 and B-31.

Memory Size: 5 Bytes

TABLE B-30 oServoSE PROPERTIES	
Address (AP)	The address of the object in RAM (0–127).
IOLine (B)	Chooses the I/O line (1–31) for the object.
Value (B)*	A write transmits the value, unmodified directly to the SSC II board.

TABLE B-31 oServoSE METHODS	
Position (S, P)	S is the servo = the servo to move (byte value)
	P is the position = the position to move the servo to (byte value).

oServoSP1

oServoSP1 is the URCP control of a hacked hobby servo (refer to Chapter 5, "Analog-to-Digital and Hobby Servos [Project: Push My Finger]"). Its properties are shown in Table B-32.

Memory Size: 6 Bytes

TABLE B-32 oServoSP1 PROPERTIES	
Address (AP)	The address of the object in RAM (0–127).
Center (B)	Can be used to adjust the mechanical center of your servo. If Center equals 0, 1.5 milliseconds would be the center pulse width to the servo if one was sent (it's not). The range is −32 to +31.
InvertOut (F)	1 reverses the servo's direction.
	0 is the normal servo direction.
IOLine (B)	Chooses the I/O line (1–31) for the object.
Operate (F)	Must be a 1 to enable the servo.
Value (B)*	The position to move the servo to, from −128 to +127.

oServoSP2

This object enables URCP control of a hacked hobby servo using a constant pulse width and a varied repetition time period (refer to Chapter 5, "Analog-to-Digital and Hobby Servos [Project: Push My Finger]"). Its properties are shown in Table B-33.

Memory Size: 6 Bytes

TABLE B-33 oServoSP2 PROPERTIES	
Address (AP)	The address of the object in RAM (0–127).
InvertOut (F)	1 reverses the servo's direction. 0 is the normal servo direction.
IOLine (B)	Chooses the I/O line (1–31) for the object.
Operate (F)	Must be a 1 to enable the servo.
Value (B)*	The speed to move the servo.

oServoX

oServoX is the URCP control of an unhacked hobby servo (refer to Chapter 5, "Analog-to-Digital and Hobby Servos [Project: Push My Finger]"). Its properties are shown in Table B-34.

Memory Size: 6 Bytes

TABLE B-34 oServoX PROPERTIES	
Address (AP)	The address of the object in RAM (0–127).
Center (B)	Can be used to adjust the mechanical center of your servo. If Center equals 0, 1.5 milliseconds is the center pulse width to the servo. The range is from -32 to +31.
InvertOut (F)	1 reverses the servo's direction. 0 is the normal servo direction.
IOLine (B)	Chooses the I/O line (1–31) for the object.
Mode (N)	Specifies how the servo responds to values outside its mechanical limit. 0 means the servo splits the inaccessible area into two parts and positions itself to the limit closest to the Value property. 1 means the servo positions itself on the low value side. 2 means the servo positions itself on the high value side. 3 means the servo shuts off.

(continued)

TABLE B-34   oServoX PROPERTIES *(Continued)*

Offset (B)	This specifies the Value property, which the object considers to be mechanical 0 of the servo. For example, if Offset equals 64, then 64 is the center and the Value range is 0 to 127.
Operate (F)	Must be a 1 to enable the servo.
OutOfRange (F)	Will be a 1 to denote when the servo Value property is set outside the mechanical range of the servo.
Refresh (F)	When set to 1, it doubles the refresh rate (to about every 17 milliseconds).
Value (B)*	The position to move the servo to, from -128 to $+127$.

oSonarDV

This object controls and interprets a Devantech SRF04 SONAR board in URCP distance units (refer to Chapter 6, "OOPic Timers, Clocks, LCDs, and SONAR [Project: SONAR Ping]"). Its properties are shown in Table B-35.

Memory Size: 6 Bytes

TABLE B-35   OSONARDV PROPERTIES

Address (AP)	The address of the object in RAM (0–127).
IOLineE (B)	The I/O line to use for the Echo signal line from the SRF04.
IOLineP (B)	The I/O line to use for the Ping signal line to the SRF04.
Operate (F)	A SONAR ping is sent whenever this property is toggled from 0 to 1.
Received (F)	Set to 1 when an echo has been received.
TimeOut (F)	Set to 1 if the board times out (no echo received within time limits).
Transmitting (F)	Set to 1 while waiting for an echo after a ping has been sent.
Value (W)*	URCP distance reading (0–32,768).

oSonarPL

oSonarPL controls and interprets a Polaroid 6500 SONAR board in URCP distance units. Its properties are shown in Table B-36.

Memory Size: 6 Bytes

TABLE B-36 oSonarPL PROPERTIES

Address (AP)	The address of the object in RAM (0–127).
IOLineE (B)	The I/O line to use for the Echo signal line from the 6500.
IOLineP (B)	The I/O line to use for the Ping signal line to the 6500.
Operate (F)	A SONAR ping is sent whenever this property is toggled from 0 to 1.
Received (F)	Set to 1 when an echo has been received.
TimeOut (F)	Set to 1 if the board times out (no echo received within time limits).
Transmitting (F)	Set to 1 while waiting for an echo after a ping has been sent.
Value (W)*	URCP distance reading (0–32768).

oSP0256

oSP0256 controls the venerable SP0256 (that's SP zero 256) speech chip using an 8-bit digital I/O port. Connect the LSB of the I/O group selected to A1 of the chip, and increment up from there in order. You may want to connect the /Reset and SBY pins to other digital I/O lines to reset or check on the voice completion status, but they are not needed. Its properties are shown in Table B-37.

Memory Size: 6 Bytes

TABLE B-37 oSP0256 PROPERTIES

Address (AP)	The address of the object in RAM (0–127).
IOGroup (N)	The 8-bit I/O group to use (refer to Appendix A).
IOLineS (B)	Specifies the I/O line to use for the /ALD (strobe) pin.
IOLineB (B)	Specifies the I/O line to use for the /LRQ (busy) pin.
Operate (F)	Must be 1 for this object to respond to input.
String (S)	Use this property to send a string of commands using the Chr$() function to the SP0256-AL2 chip; it checks the /LRQ (busy) line between bytes.
Value (B)	Use this property to send a single command to the chip; it does not check the /LRQ (busy) line before sending.

oSPI

oSPI controls the synchronous *serial port interface* (SPI) or three-wire data transfers. SPI is essentially a high-speed shift-in or shift-out interface. Its memory size is 6 bytes and its properties are shown in Table B-38.

Memory Size: 6 Bytes

TABLE B-38 oSPI PROPERTIES	
Address (AP)	The address of the object in RAM (0–127).
Direction (F)	0 means data is shifted out LSB first.
	1 means data is shifted out MSB first.
InvertE (Bi)	Inverts the logical output of the enable line.
IOLineC (B)	Selects the I/O line to use for the SPI clock signal.
IOLineE (B)	Selects the I/O line to use for the device enable signal.
IOLineI (B)	Selects the I/O line to use for the SPI input. If set to 0, oSPI uses the IOLineO I/O line for input.
IOLineO (B)	Selects the I/O line to use for the SPI output.
Mode (Bi)	0 means data is sampled before the clock pulse (when IOLineC is low).
	1 means data is sampled after the clock pulse (when IOLineC is high).
Operate (F)	Must be 1 for this object to respond to input.
Rate (N)	The clock rate to use for SPI transfers:
	0: 12 KHz. 2: 360 Hz.
	1: 720 Hz. 3: 240 Hz.
Value (B/W)*	Data read in or sent out. Whether it is bytes or words depends on the Width property.
Width (Bi)	0 means 8-bit data are used.
	1 means 16-bit data are used.

oStepperSP

This is a URCP controllable stepper where either the rate or the number of steps can be linked in a VC (refer to Chapter 8, "OOPic Interfacing and Electronics [and Steppers and Seven-Segment LEDs]"). Its properties are shown in Table B-39.

Memory Size: 10 Bytes

TABLE B-39 oStepperSP PROPERTIES	
Address (AP)	The address of the object in RAM (0–127).
Direction (F)	0 means forward, and 1 means backward. Of course, these are kind of arbitrary.
Free (F)	0 means engaged. When the motor isn't turning, the windings are left energized so the motor holds its position.
	1 means off. The windings are not energized and the motor will turn freely.
Invertout (Bi)	0 means a logic 1 turns a coil on.
	1 means a logic 0 turns a coil on.
IOGroup (N)	The 8-bit I/O group to use (refer to Appendix A).
Mode (Bi)	0 turns the number of steps specified.
	1 turns continuously.
Nibble (Bi)	0 uses the lower 4 bits of the selected I/O group.
	1 uses the upper 4 bits of the selected I/O group.
NonZero (F)	Is 0 when the Value property is 0 (finished stepping).
Operate (F)	Must be set to 1 for this object to respond to inputs.
Phasing (N)	0 means 1 of the 4 coils is active, whereas the other 3 are inactive (Wave).
	1 means 2 of the 4 coils are active, whereas the other 2 are inactive (two-phase).
	2 alternates between 1 and 2 active coils (half-step).
	3 indicates three-phase stepper-motor phasing.
Rate (B)	The step frequency is 1132.246 Hz/(128 − rate).
Steps (W)	Mode 0 sets the number of steps to take in the direction specified at the rate specified.
	Mode 1 keeps track of the number of steps taken.
Value (B)*	From −128 to 127, the URCP speed setting.

oTone

oTone generates low-frequency tones. The frequency of the tone is derived from a 1,132 KHz clock divided by 128 minus the Value property. The minimum frequency is 8.8 Hz and the maximum frequency is 1132 Hz. Its properties are shown in Table B-40.

Memory Size: 5 Bytes

TABLE B-40	OTONE PROPERTIES
Address (AP)	The address of the object in RAM (0–127).
IOLine (B)	The I/O line to use for frequency output.
Operate (F)	Must be set to 1 for this object to respond to inputs.
Value (B)*	The frequency to produce: 1132/(128 − Value).

oTracker

oTracker interprets a three- or four-line sensor input to track the location of a black line on a white background (refer to Chapter 10, "OOPic Robotics and URCP [Project: A Robot That Toes the Line]"). Its properties are shown in Table B-41.

Memory Size: 5 Bytes

TABLE B-41	oTracker PROPERTIES
Address (AP)	The address of the object in RAM (0–127).
Center (F)	1 signifies that the robot is centered on the line.
Direction (F)	0 means the last movement of the line was toward the LSB sensor.
	1 indicates the last movement of the line was toward the MSB sensor.
InvertIn (Bi)	0 indicates a high-logic input, which means the line is under the sensor.
	1 indicates a low-logic input, which means the line is under the sensor.
IOGroup (N)	The 8-bit I/O group to use (refer to Appendix A Keys).
Mode (Bi)	0 means if no line is detected, assume the line is far outside either the left or right sides, depending on the last direction seen. The Value is set to either -32 or 32 depending on the Direction property.
	1 means if no line is detected, assume the line is between the center two sensors. The Value property is set to 0 and the Center property is set to 1.
Moved (F)	If set to 1, the line has moved since the last time Moved was set to 0. It simply means something has changed.
Nibble (Bi)	0 uses the lower 4 bits of the selected IOGroup.
	1 uses the upper 4 bits of the selected IOGroup.
Operate (F)	Must be set to 1 for this object to respond to input.

(continued)

OutOfRange(F)	When this bit equals 1, the line is not under any sensor.
Unsigned (Bi)	When this bit equals 1, don't use URCP; the Value property is unsigned.
Value (B)*	The URCP (or unsigned) directional data.
Width (Bi)	0 uses four sensors for line detection. 1 uses three sensors for line detection.

oUVTronHM

oUVTronHM times and interprets readings from a Hamamatsu UVTron driver board and detector. Its properties are shown in Table B-42.

Memory Size: 6 Bytes

TABLE B-42 oUVTronHM PROPERTIES

Address (AP)	The address of the object in RAM (0–127)
IOLine (B)	The I/O line connected to the UVTron driver board.
NonZero (F)	Set to 1 whenever any pulses are detected from the driver board.
Operate (F)	Must be set to 1 for this object to respond to input.
Value (B)	The number of pulses sensed in the last sensor window.

OVideoIC

OVideoIC controls an Intuitive Circuits OSD232 On-Screen Display Character Overlay Board (Preliminary). Its properties and methods are shown in Tables B-43 and B-44. The most commonly used commands are represented as methods which send data to this object at 4800 baud.

Memory Size: 5 Bytes

TABLE B-43 OVideoIC PROPERTIES

Address (AP)	The address of the object in RAM (0–127).
IOLine (B)	The I/O line used to communicate with the device.
String (S)	Input sent to the display board in string format.
Value (B)*	ASCII data sent to the display board.

TABLE B-44   OVideoIC METHODS

Clear	Clears the screen.
Color (R, B)	Sets the text color. R is red and B is blue.
Fullscreen	Use full-screen mode.
Locate (R, C)	Sets the cursor to the specified location. It uses two arguments, row and column, both of which are 0 based.

New Processing Objects

These are the new processing objects that have been added to B.X.X firmware.

oBus(I)(O)(C)

This object links data from one object default value to another (refer to Chapter 3, "OOPic Object Standard Properties"). Their properties are shown in Table B-45.

Memory Size: 3 Bytes (oBusI and oBusO: 4 bytes, and oBus*C: 5 bytes).

TABLE B-45   oBus(I)(O)(C) PROPERTIES

Address (AP)	The address of the object in RAM (0–127).
ClockIn (FP)	Used by clocked variants of this object for the signal that clocks data through.
Input (NP)	Links to an output Value property to use from another object.
InvertC (F)	Used by clocked variants of this object to determine the clock direction: 0: Clock changes from 0 to 1. 1: Clock changes from 1 to 0.
InvertIn (F)	Input is negated bitwise if set to 1; it is unmodified otherwise. The input is negated bitwise before being used if set to 1; it is unmodified otherwise.
Mode (Bi)	Determines when the copy is performed: 0: Data is copied continuously as long as Operate equals 1. 1: Data is only copied when Operate changes from 0 to 1.
Operate (F)	Must be 1 for this object to respond to inputs.
Output (P)	Links to the input Value property of another object.
Value (B)	A signed 8-bit number optionally used by oBusI and oBusO variants.

oChanged(O)(C)

This object detects when an input has changed. It then updates the new value and keeps looking (refer to Chapter 9, "OOPic I2C and Distributed DDE Programming [Project: Remote Control]"). Their properties are shown in Table B-46.

Memory Size: 3, 4 Bytes (for oChangedO), or 5 (for oChangedC)

TABLE B-46	OCHANGED(O)(C) PROPERTIES
Address (AP)	The address of the object in RAM (0–127).
Changed (F)	Set to 1 if the Input property changed during the last cycle of the object list loop. It is 0 otherwise.
ClockIn (FP)	Links to a flag that clocks the object into action for one object list loop cycle.
Direction (F)	Is 0 if the Input property is greater than the last value saved. Is 1 if Input is less than the last value saved.
Input (NP)	Links to the value being watched for change. It can be any number, bit, nibble, byte, or whatever.
InvertC (F)	0 (cvFalse) means the compare and copy function is performed when the Flag property that the ClockIn property is linked to transitions from 0 to 1. 1 (cvTrue) means the compare and copy function is performed when the Flag property that the ClockIn property is linked transitions from 1 to 0.
Operate (F)*	Must be 1 (cvTrue) for this object to respond to inputs.
Output (NP)	Links to the object holding the value last read from the Input property (such as oChanged or oChangedC).
Value (B)	The 8-bit value of the last input read when not using the Output property (such as oChangedO or oChangedOC).

oClock

oClock provides a low-speed programmable logic clock (refer to Chapter 6, "OOPic Timers, Clocks, LCDs, and SONAR [Project: SONAR Ping]"). Its properties are displayed in Table B-47.

Memory Size: 5 Bytes

TABLE B-47	oClock PROPERTIES
Address (AP)	The address of the object in RAM (0–127).
InvertOut (F)	0 means that the Output and Result properties are the same.
	1 means the Result property is inverted before it appears as part of the Output property.
Mode (Bi)	0 means clock output is a 50 percent duty cycle square wave.
	1 means the clock only pulses once every time period for the duration of one object list loop.
Operate (F)	Must be a 1 to respond to inputs.
Output (FP)	Points to a property that is updated with the value of the Result property.
Rate (B)	Defines the divisor for the 283 Hz clock. The frequency is 283/(256 − Rate).
Result (F)*	The result of the clocking operation.

oCompare(0)(2)(C)

This object compares two numbers or a number versus 0 and sets flags based on the results. Table B-48 describes what compares to what.

If the Input property is greater than the reference + fuzzyness value, the Above property is set. If the Input is less than the reference-fuzzyness value, the Below property is set. If it is neither, then Between is set.

TABLE B-48 oCompare OBJECT COMPARISONS	CONTINUOUS OPERATION	CLOCKED OPERATION
Compare made with one reference < greater than (ReferenceIn1 + Fuzzyness) or less than (ReferenceIn1 − Fuzzyness)	oCompare	oCompareC
Compared to 0	oCompare0	oCompare0C
Compare made with two references, greater than (ReferenceIn1 + Fuzzyness) or less than (ReferenceIn2 − Fuzzyness)	oCompare2	oCompare2C

Their properties are shown in Table B-49.

Memory Size: 4 Bytes (oCompare2: 5 bytes, oCompareC[OC]: 6 bytes, and oCompare2C: 7 bytes

TABLE B-49 oCompare(0)(2)(C) PROPERTIES

Above (F)	The Input property is greater than the reference used.
Address (AP)	The address of the object in RAM (0–127).
Below (F)	The Input property is less than the reference used.
Between (F)	The Input property is between the references used.
ClockIn (FP)	Links to the flag that clocks this object (oCompare*C only).
Fuzzyness (B)	A dead zone that is added to the first reference and is subtracted from the second reference when a comparison is made.
Input (NP)	Links to the object whose default value is compared to limits.
InvertC (Bi)	Inverts the ClockIn logic (the object operates on a 1 to 0 transition).
Operate (F)	Must be set to 1 for this object to respond to inputs.
ReferenceIn1 (NP)	An object whose Value property is used for the greater than and less than comparison in oCompare. It is also used for the greater than comparison only in oCompare2, not in oCompare0.
ReferenceIn2 (NP)	An object whose Value property is used for the less than comparison in oCompare2 and is not used in oCompare or oCompare0.

oCountDown(0)

This object counts down to 0 either using a linked value or an 8-bit value when clocked by an external event. It is useful as a timeout counter; when it reaches 0, it stops (refer to Chapter 7, "OOPic Events, Keypads, and Serial I/O [Project: A Mini-terminal]"). Its properties are shown in Table B-50.

Memory Size: 5 Bytes

TABLE B-50 OCountdown(O) PROPERTIES

Address (AP)	The address of the object in RAM (0–127).
ClockIn (FP)	Links to a flag in another object supplying clock ticks.
NonZero (F)	Reads 0 when the Value property is 0; it reads 1 otherwise.
Operate (F)*	Must be 1 for this object to respond to input.
Output (NP)	Links to the object whose Value property is decremented.
Value (B)	Internal 8-bit Value property to be decremented (oCountDownO).
PreScale (B)	Specifies what to divide the ClockIn frequency by: ClockIn/(PreScale + 1) = Final frequency.

oDivider

oDivider divides an incoming clock by the rate specified. Think of this as a digital divider. Its properties are shown in Table B-51.

Memory Size: 5 Bytes

TABLE B-51	oDivider PROPERTIES
Address (AP)	The address of the object in RAM (0–127).
ClockIn (FP)	Links a flag in another object supplying clock ticks.
InvertC (Bi)	Inverts the normal logic of the ClockIn flag.
InvertOut (Bi)	Inverts the normal logic of the Output flag.
Operate (F)*	Must be 1 for this object to respond to input.
Output (FP)	Links to the object whose flag property receives the final frequency.
Rate (B)	Divisor; 0 means divide by 256.
Result (F)	Result of the divided ClockIn rate.

oFlipFlop(C)

This is a VC equivalent to a set/reset FlipFlop. When oFlipFlopC is used, this object is like single-bit memory. Its properties are shown in Table B-52.

Memory Size: 5 Bytes (oFlipFlopC is 6 bytes)

TABLE B-52	oFlipFlop(C) PROPERTIES
Address (AP)	The address of the object in RAM (0–127).
ClockIn (FP)	Links to a flag that clocks this object (oFlipFlopC only).
Input1 (FP)	Links to an object's Flag property whose value is used to set the Output linked flag.
Input2 (FP)	Links to an object's Flag property whose value is used to set the Output linked flag.
InvertC (Bi)	Inverts the normal logic of the ClockIn flag.
InvertIn1 (Bi)	Inverts the logic of Input1 (0 causes a set).
InvertIn2 (Bi)	Inverts the logic of Input2 (0 causes a reset).
Operate (F)*	Must be 1 for this object to respond to input.
Output (FP)	Optional link to an object's input flag property that is set or reset.
Result (F)	The result of the FlipFlop operation.

oNavCon(I)

This object is designed to take URCP turn information and convert it into URCP motor control speed information for two motors to drive a robot in differential drive mode (refer to Chapter 10, "OOPic Robotics and URCP [Project: A Robot That Toes the Line]"). Its properties are shown in Table B-53.

Memory Size: 7 Bytes (oNavCon(I)C: 8 bytes)

TABLE B-53   ONAVCON(I) PROPERTIES

Address (AP)	The address of the object in RAM (0–127).
Center (B)	Added to Input2, which is used as the turn value to offset a motor inequality bias.
ClockIn (FP)	Links to a flag used to clock this object (oNavConC and oNavConIC only).
Input1 (NP)	Links to a Value property used as the robot speed.
Input2 (NP)	Links to a Value property used as the turn value.
InvertC (Bi)	Inverts the active logic sense of the ClockIn flag when equal to 1 (oNavConC and oNavConIC only).
Limit (Bi)	When equal to 1, the motor values are limited between -128 and $+127$.
Operate (F)	Must be 1 for this object to respond to input (default property for oNavCon and oNavConC).
Output1 (NP)	Links to the motor Value input for one side of the robot.
Output2 (NP)	Links to the motor Value input for the other side of the robot.
Value (B)*	Used by oNavConI as the turn value (default property only then).

oRamp(I)(C)

oRamp(I)(C) implements a form of servo feedback for motor control in URCP notation. The ramp setting is really the setting where feedback changes allow motor speed changes; it is *not* a ramp function, more like a servo motor feedback object that differentiates differences between desired and current settings. Its properties are shown in Table B-54.

Memory Size: 7 Bytes (oRamp*C: 8 bytes)

TABLE B-54	oRamp(I)(C) PROPERTIES
Address (AP)	The address of the object in RAM (0–127).
ClockIn (FP)	Links to a flag used to clock this object (oRamp*C only).
Input1 (NP)	Links to an object measuring the current absolute location.
Input2 (NP)	Links to an object that holds the absolute location desired at the Output property.
InvertC (Bi)	Inverts the logic of the ClockIn object (0 clocks instead of 1).
Limit (F)	0 means Output = Input1 − Input2 (no interpretation). 1 means Output = Input1 − Input2 *unless* the sign of Input1 is opposite of the sign of Input2. Then a motor full reverse is given: If Input1's sign is (+), then Output equals −128. If Input1's sign is (-), then Output equals +127.
Operate (F)	Must be 1 for this object to respond to input.
Output (NP)	Links to an object whose Value setting is to be affected.
Style (N)	After the difference between the two inputs is taken and the result is amplified, this value is filtered to affect the curve of the result.
	0: Ramp is linear throughout the curve, the difference goes to the Output.
	1: Ramp values have less effect closer to the target value and more effect further from the target value by using the Sin() function on the result. This leaves a "dead band" when Input1 is near Input2.
	2: Ramp values have more effect closer to the target value and less effect further from the target value by using the Cos() function on the result.
Value (B)*	Internally held location instead of a link (oRampI and oRampIC only).
Width(N)	The difference between the two inputs is multiplied by shifting the bits the number of positions equal to this setting (0-7). This is basically an amplifier for the difference reading.

oRepeat

oRepeat enables a static event, such as a button held down, to cause multiple events to occur, such as a key auto-repeat (refer to Chapter 7, "OOPic Events, Keypads, and Serial I/O [Project: A Mini-terminal]"). Its properties are shown in Table B-55.

Memory Size: 7 Bytes

TABLE B-55 OREPEAT PROPERTIES	
Address (AP)	The address of the object in RAM (0–127).
Input (FP)	Links to the flag that causes the event.
InvertIn (F)	1 inverts the flag property so a 0 causes the repeat, not a 1.
InvertOut (F)	1 inverts the flag property before sending it to the Output property.
Operate (F)*	Must be 1 for this object to function.
Output (FP)	Links to the object triggered by the event Input flag.
Period (B)	The number of 1/60-second tics to wait before repeating the event. The limit is from 0 to 127.
Rate (B)	The number of 1/60th-second tics to wait between repeats of the event. The limit is from 0 to 63.
Result (F)	The result of this object's work as it occurs.

OOPIC B.2.X+ OBJECTS

CONTENTS AT A GLANCE

The OOPic R, OOPic C, and OOPic II+ add only a few new features to the basic OOPic II. In fact, the firmware is identical to OOPic II except for the hardware differences between the PICMicro 16F877 and PICMicro 16F77 chips.

New Variable Type: sByte (Preliminary)

The *sByte* variable is a persistent variable that is stored into the 16F877's internal *electrically erasable programmable read-only memory* (EEPROM). You use it just like you would a Byte or Word variable; it is *only* a byte. This variable type is new and not fully specified.

New Hardware Objects

This object in OA2D10 is in the B.2.2 Firmware, but it is only fully enabled for 10-bit A2D results in the B.2.2+ Firmware.

oA2D10

oA2D10 is a 10-bit A2D object. Two *least significant bits* (LSBs) are only valid in B.2.X+ Firmware. (See Chapter 5, "Analog-to-Digital and Hobby Servos.") The oA2D10 Table C-1 outlines its properties.

Memory Size: 3 Bytes

TABLE C-1 oA2D10 PROPERTIES	
Address (AP)	The address of the object in *random access memory* (RAM) (0 to 127).
IOLine (N)	Assigns the *input/output* (I/O) line to use. Essentially, 1 through 7 are I/O lines 1 through 7.
MSB (F)	The *most significant bit* (MSB) of the Value property. If this bit equals 1, then Value is equal to or over half of the reference voltage. If it is 0, then it is under half.
Operate (F)	Must be 1 to respond to input.
String (S)	The string representation of the Value property.
Value (W)*	The result of the A2D conversion.

New Processing Objects (Preliminary)

Only one new processing object, *oSequencer*, has been added. This object uses the 16F877's internal EEPROM storage to operate. This object is new and not fully specified.

oSequencer(C)

oSequencer(C) stores a sequence of commands in the internal EEPROM that sequences data to objects using its own form of language. The Table C-2 outlines its properties.

The oSequencer object reads a series of programmed functions from the OOPic's internal EEPROM and executes them. The three different type of functions are:

- **Command** These are used to control the sequencer's program flow.
- **Set Data** This function is used to set value into objects.
- **Set RAM** This function is used to set values into RAM.

Memory Size: 5 Bytes (oSequencer(C) memory size: 6 bytes)

TABLE C-2   oSequencer PROPERTIES

Address (AP)	The address of the object in RAM (0 to 127).
ClockIn (FP)	Link to a flag that will clock the object into action for one object list loop cycle.
InvertC (F)	Used by clocked variants of this object to determine the clock direction. 0: The clock changes from 0 to 1. 1: The clock changes from 1 to 0.
Mode (Bi)	0: The programmed sequence runs when the Operate property is 1 and stops when Operate is 0. 1: The programmed sequence starts running when the Operate property is set to 1 and continues until the end of the sequence.
Operate (F)	The bit must be 1 in order for the object to respond to input.
Value (W)*	Specifies which programmed sequence will be executed next.

The Command function compares the value of the item pointed with the pointer with the 16-bit value. Based on the results, it either resets the program counter, sets the program counter to reevaluate the current command, or skips the next command. Table C-3 describes the command byte 00AASTEE.

TABLE C-3   Command byte: 00AASTEE

00	Header	Used to specify the command function.
AA	Action	00: The program counter will be set to 0, which will restart the sequence. 01: N/A 10: The program counter will be set to reevaluate the current command. 11: The program counter will be set to skip the next command.
S	Signed	0: The 16-bit value is evaluated as an unsigned number. 1: The 16-bit value is evaluated as a signed number
T	Type	0: The pointer points to a flag. 1: The pointer points to an object.
EE	Evaluation	00: Values are equal. 01: Value pointed to is greater than the 16-bit value. 10: Value pointed to is less than the 16-bit value. 11: N/A

This sets a byte of data and specifies a delay before the next instruction is executed.

TABLE C-4 Set Data byte format 01XXXXXX		
01	Header	Used to specify the Set Data function.
XXXXXX	Wait	The duration to wait until executing the next command.

TABLE C-5 Set RAM byte format 1SSSNNNW		
1	Header	Used to specify the Set Property function.
SSS	Shift	The number of bits to shift the value.
NNN	Bits	The number of bits to write.
W	Width	0: Write data in one byte. 1: Write data in two bytes

OOPIC PRODUCTS, ACCESSORIES, AND RESOURCES

The following companies sell Savage Innovations OOPic products and/or accessories and interfaces. I have certainly not found and listed all of the shops that sell the OOPic and OOPic accessory boards and upgrades, but this selection will be a good starting point. Unless otherwise noted, these companies are based in the United States, but they ship internationally. This list also contains several companies outside of the United States.

Acroname, Inc.

This company supplies OOPic kits and other manufacturers' OOPic accessories as well as a wide variety of sensors, motor controllers, and other accessories. Acroname, Inc., sells many hard-to-find robotics supplies.

Acroname, Inc.
4894 Sterling Dr.
Boulder, CO 80301-2350
Phone: (720) 564-0373
Fax: (720) 564-0376
E-mail: info@acroname.com
Web site: www.acroname.com

BG Micro

This company sells the OOPic S boards and accessories as well as motor drivers and electronics.

B.G. Micro
555 N. 5th St., Suite 125
Garland, TX 75040
Phone: (800) 276-2206
Fax: (972) 205-9417
Web site: www.bgmicro.com

Budget Robotics.com

This is a company that specializes in inexpensive robotics and is run by Gordon McComb. They sell a variety of robotics platforms, sensors, and accessories as well as kits featuring the OOPic R board.

Budget Robotics.com
P.O. Box 5821
Oceanside, California 92056
Phone: (760) 941-6632
E-mail: info@budgetrobotics.com
Web site: www.budgetrobotics.com

Scooterbot platform.

DIY Electronics (Malayasia)

This company sells the OOPic as well as many other sensor and robotics kits.

E-mail: inquiry@diyelectronics.com
Web site: www.diyelectronics.com

Electroprint ApS (Denmark)

This company sells the OOPic product line.

Electroprint ApS
Bøgekildevej 21
DK-8361 Hasselager
Denmark
Phone: +45 8628 4566
E-mail: oopic@electroprint.dk
Web sites: www.electroprint.dk and www.oopic.dk

D

PRODUCTS, ACCESSORIES, & RESOURCES

Energise Technology Ltd (UK)

(competition-robotics.com)
PO Box 1178
Swindon
SN25 4ZL
England
Phone: +44 (0)1793 636119
Fax: +44 (0)1793 705772
Web site: www. competition-robotics.com

Esutech Oy (Finland)

This company sells the OOPic product line.

Esutech Oy
Rasikuja 4
78870 Varkus
Finland
Phone: +358 400 920 139
Fax: +358 17 5512 138
E-mail: info@esutech.com
Web site: www.esutech.com/

Hanitech Co. (South Korea)

This company sells the OOPic and many other robots, kits, and supplies.

Hanitech Co.
Phone: +82-2-337-4910
Fax: +82-2-337-492
E-mail: sales@hanitech.co.kr
Web site: www.hanitech.co.kr/

HVW Technologies, Inc. (Canada)

This Canadian company sells the OOPic R and S boards and accessories as well as a complete line of sensors, microcontroller parts and kits, motor controllers, and much more. The company's prices are also very reasonable.

HVW Technologies Inc.
3907-3A St. N.E., Unit 218
Calgary, Alberta T2E 6S7
Canada
Phone: (403) 730-8603
E-mail: info@HVWTech.com
Web site: www.hvwtech.com

Mark III Robot Store

This online store sells the Mark III robot base used in Chapter 10, "OOPic Robotics and URCP (Project: A Robot That Toes the Line)," as well as OOPic II+-compatible boards and OOPic II+ upgrade processors and serial *electrically erasable programmable read-only memory* (EEPROMs).

E-mail: MarkIII@junun.org
Web site: www.junun.org/MarkIII/Store.jsp

Mondotronics (The Robot Store)

Mondotronics sells a variety of robot kits and electronics, including the OOPic line.

Mondotronics
4286 Redwood Highway PMB-N
San Rafael, CA 94903
Phone: (800) 374-5764
Fax: (415) 491-4696
E-mail: info@RobotStore.com
Web site: www.robotstore.com

Oricom Technologies

This company supplies specialized robotics and controller boards, the OOBot40 and OOBot40 II, that are compatible with the OOPic S chips as well as OOPic II+ upgrades.

Oricom Technologies
P.O. Box 68
Boulder, CO 80306
Phone: (303) 449-6428
E-mail: support@oricomtech.com
Web site: www.oricomtech.com

OOBot40 controller.

Primastream SRL (Italy)

This company carries the OOPic II.

Primastream SRL
Phone: +39 06 5005696
Fax: +39 06 5004275
Web site: www.primastream.com/Start.htm

Reynolds Electronics

This company sells the full line of OOPic products as well as other electronics.

Reynolds Electronics
3101 Eastridge Lane
Canon City, CO 81212
Phone: (719) 269-3469
Fax: (719) 276-2853
Web site: www.rentron.com

Robotic Education Products (Australia)

This company carries the full line of OOPic products as well as a host of other robotics supplies.

World of Robotics Online
110 Mt. Pleasant Road Belmont
Geelong, Victoria 3216
Australia
Phone: 1800 000 745 (toll free within Australia) or 61 3 52 419 581
Fax: 61 3 52 419 089
Web site: www.robotics.com.au/

Robot Store HK (Hong Kong)

This shop sells the OOPic product line as well as a variety of other embedded processors and hobby robotics supplies

.Robot Store HK
7th Floor, Fok Wa Mansion
No. 19 Kin Wah Street
North Point,
Hong Kong
Phone: 1852 9752-0677
Fax: 1852 2887-2519
E-mail: info@robotstorehk.com
Web site: www.robotstorehk.com/

D
PRODUCTS, ACCESSORIES, & RESOURCES

SuperDroid Robots

This shop sells all the OOPic products and chip upgrades as well as a selection of accessory cards for both the OOPic S and OOPic R boards. SuperDroid Robots is a subdivision of

Team Half-Life Inc.
2120 Old Frederick Road
Catonsville, MD 21228
Phone: (410) 788-1683
Fax: (410) 788-1683
Web site: www.superdroidrobots.com

Trekker Robot using an OOPic controller

Total Robots Ltd. (UK)

This company sells the full OOPic line as well as many of their own sensors and communications products, which are designed explicitly to work with the OOPic. These products include interface boards, expansion kits, and the *Wireless Control Module* (WCM) and *Fast Wireless Control Module* (FWCM) boards, which are wireless networking boards designed to interface to the OOPic via I2C. They also sell a wide variety of robotics platforms.

Total Robots Ltd.
Global House
Ashley Avenue
Epsom, Surrey KT18 5AD
United Kingdom
Phone: +44 (0)208 823 9220
Fax: +44 (0)208 823 9240
E-mail: enquiry@totalrobots.com
Web site: www.totalrobots.com

The following section lists some companies (other than those listed previously) that sell accessories that interface with the OOPic and that offer other useful electronics.

All Electronics

All Electronics sells surplus electronics parts. You will also find a selection of small DC motors.

All Electronics
P.O. Box 567
Van Nuys, CA 91408-0567
Phone: (888) 826-5432
Web site: www.allelectronics.com

Alltronics

Alltronics sells surplus electronics components. Their inventory usually includes a few DC motors suitable for robotics uses.

Alltronics
P.O. Box 730
Morgan Hill, CA 95038-0730
Phone: (408) 847-0033
Web site: www.alltronics.com

AWC Electronics

Al Williams Electronics makes the popular PAK series coprocessors. The OOPic has objects designed to work with these (oI2C and oSPI in particular).

Al Williams Electronics
310 Ivy Glen
League City, TX 77573
Phone: (281) 334-4341
Fax: (281) 754-4462
Web site: www.al-williams.com/index.htm

D

PRODUCTS, ACCESSORIES, & RESOURCES

Devantech Ltd. (UK)

This company is the home of the SRF04 and SRF08 SONAR units as well as electronic compasses and more. Many other shops distribute their wares.

Robot Electronics
Unit 2B Gilray Road
Diss, Norfolk IP22 4EU
England
Phone: +44 (0)1379 640450 or 644285
Fax: +44 (0)1379 650482
E-mail: sales@robot-electronics.co.uk
Web site: www.robot-electronics.co.uk

Electronic Goldmine

Electronic Goldmine sells all kinds of surplus electronics parts and assemblies.

Electronic Goldmine
P.O. Box 5408
Scottsdale, AZ 85261
Phone: (800) 445-0697
Web site: www.goldmine-elec.com

GotRobots.com

This is the home of the Ziggy robot, which is a very cool hexapod. It is designed for and run with the OOPic micro. It is not cheap, but it is amazing.

E-mail: nick@gotrobots.com
Web site: www.gotrobots.com/ziggy/purchase.shtml

The Ziggy Robot

Jameco Electronics

It was hard to determine where to place this one. You can buy electronics, motors, steppers, drivers, and robotic specialty hardware from this company.

Jameco Electronics
1355 Shoreway Road
Belmont, CA 94002-4100
Phone: (800) 831-4242
Web site: www.jameco.com

Lynxmotion

Although they are primarily known for their robot kits, including various wheeled platforms and a variety of walker configurations, Lynxmotion also sells a wide variety of sensors that are designed to work with OOPic objects such as the *Tracker Sensor* for use with the *oTracker* object.

Lynxmotion
P.O. Box 818
Pekin, IL 61555-0818
 hone: (866) 512-1024
Web site: www.lynxmotion.com

Magnevation

This company manufactures OOPic interface components, sensors, displays, and other accessories. Their products include the Logic Status Board, H-Bridge drivers, many cables, and OOPic add-ons.

Magnevation
P.O. Box 217
Capshaw, AL 35742
Fax: (801) 469-0838
E-mail: sales@magnevation.com
Web site: www.magnevation.com/

ServoCity

This is a great site for low-cost servos, gears, chain links, and so on for small robots.

ServoCity
620 Industrial Blvd.
Winfield, KS 6715
Phone: (620) 221-0421
Fax: (620) 221-0858
E-mail: sales@ServoCity.com
Web site: www.servocity.com

Today's Toys and Technology

This is my online company, which currently sells sensors. In the near future, it will also sell OOPic-related expansion boards, the (now famous) buffered parallel port programmer cable, and more.

E-mail: dlc@frii.com
Web site: www.techtoystoday.com

ASCII CODES

Because ASCII codes come up so often in this book, a reference table has been provided. This table includes the decimal and hexadecimal values for all 128 ASCII codes. Most of these you probably do not care about, but you should be interested in some codes such as *Carriage Return* (CR), *Line Feed* (LF), and the letters and numbers.

DEC	HEX	CODE	DEC	HEX	CODE	DEC	HEX	CODE
0	00	NUL	43	2B	+	86	56	V
1	01	SOH	44	2C	'	87	57	W
2	02	STX	45	2D	-	88	58	X
3	03	ETX	46	2E	.	89	59	Y
4	04	EOT	47	2F	/	90	5A	Z
5	05	ENQ	48	30	0	91	5B	[
6	06	ACK	49	31	1	92	5C	\
7	07	BEL	50	32	2	93	5D	]
8	08	BS	51	33	3	94	5E	^
9	09	HT	52	34	4	95	5F	_
10	0A	LF	53	35	5	96	60	`
11	0B	VT	54	36	6	97	61	a
12	0C	FF	55	37	7	98	62	b
13	0D	CR	56	38	8	99	63	c
14	0E	SO	57	39	9	100	64	d
15	0F	SI	58	3A	:	101	65	e
16	10	DLE	59	3B	;	102	66	f
17	11	XON	60	3C	<	103	67	g
18	12	DC2	61	3D	=	104	68	h
19	13	XOFF	62	3E	>	105	69	i
20	14	DC4	63	3F	?	106	6A	j
21	15	NAK	64	40	@	107	6B	k
22	16	SYN	65	41	A	108	6C	l
23	17	ETB	66	42	B	109	6D	m
24	18	CAN	67	43	C	110	6E	n
25	19	EM	68	44	D	111	6F	o
26	1A	SUB	69	45	E	112	70	p
27	1B	ESC	70	46	F	113	71	q
28	1C	FS	71	47	G	114	72	r
29	1D	GS	72	48	H	115	73	s
30	1E	RS	73	49	I	116	74	t
31	1F	US	74	4A	J	117	75	u
32	20	SPACE	75	4B	K	118	76	v
33	21	!	76	4C	L	119	77	w
34	22	"	77	4D	M	120	78	x
35	23	#	78	4E	N	121	79	y
36	24	$	79	4F	O	122	7A	z
37	25	%	80	50	P	123	7B	{
38	26	&	81	51	Q	124	7C	|
39	27	'	82	52	R	125	7D	}
40	28	(	83	53	S	126	7E	~
41	29	)	84	54	T	127	7F	DEL
42	2A	*	85	55	U			

INDEX

SOFTWARE AND INFORMATION LICENSE

The software and information on this diskette (collectively referred to as the "Product") are the property of The McGraw-Hill Companies, Inc. ("McGraw-Hill") and are protected by both United States copyright law and international copyright treaty provision. You must treat this Product just like a book, except that you may copy it into a computer to be used and you may make archival copies of the Products for the sole purpose of backing up our software and protecting your investment from loss.

By saying "just like a book," McGraw-Hill means, for example, that the Product may be used by any number of people and may be freely moved from one computer location to another, so long as there is no possibility of the Product (or any part of the Product) being used at one location or on one computer while it is being used at another. Just as a book cannot be read by two different people in two different places at the same time, neither can the Product be used by two different people in two different places at the same time (unless, of course, McGraw-Hill's rights are being violated).

McGraw-Hill reserves the right to alter or modify the contents of the Product at any time.

This agreement is effective until terminated. The Agreement will terminate automatically without notice if you fail to comply with any provisions of this Agreement. In the event of termination by reason of your breach, you will destroy or erase all copies of the Product installed on any computer system or made for backup purposes and shall expunge the Product from your data storage facilities.

LIMITED WARRANTY

McGraw-Hill warrants the physical diskette(s) enclosed herein to be free of defects in materials and workmanship for a period of sixty days from the purchase date. If McGraw-Hill receives written notification within the warranty period of defects in materials or workmanship, and such notification is determined by McGraw-Hill to be correct, McGraw-Hill will replace the defective diskette(s). Send request to:

Customer Service
McGraw-Hill
Gahanna Industrial Park
860 Taylor Station Road
Blacklick, OH 43004-9615

The entire and exclusive liability and remedy for breach of this Limited Warranty shall be limited to replacement of defective diskette(s) and shall not include or extend any claim for or right to cover any other damages, including but not limited to, loss of profit, data, or use of the software, or special, incidental, or consequential damages or other similar claims, even if McGraw-Hill has been specifically advised as to the possibility of such damages. In no event will McGraw-Hill's liability for any damages to you or any other person ever exceed the lower of suggested list price or actual price paid for the license to use the Product, regardless of any form of the claim.

THE McGRAW-HILL COMPANIES, INC. SPECIFICALLY DISCLAIMS ALL OTHER WARRANTIES, EXPRESS OR IMPLIED, INCLUDING BUT NOT LIMITED TO, ANY IMPLIED WARRANTY OF MERCHANTABILITY OR FITNESS FOR A PARTICULAR PURPOSE. Specifically, McGraw-Hill makes no representation or warranty that the Product is fit for any particular purpose and any implied warranty of merchantability is limited to the sixty day duration of the Limited Warranty covering the physical diskette(s) only (and not the software or information) and is otherwise expressly and specifically disclaimed.

This Limited Warranty gives you specific legal rights; you may have others which may vary from state to state. Some states do not allow the exclusion of incidental or consequential damages, or the limitation on how long an implied warranty lasts, so some of the above may not apply to you.

This Agreement constitutes the entire agreement between the parties relating to use of the Product. The terms of any purchase order shall have no effect on the terms of this Agreement. Failure of McGraw-Hill to insist at any time on strict compliance with this Agreement shall not constitute a waiver of any rights under this Agreement. This Agreement shall be construed and governed in accordance with the laws of New York. If any provision of this Agreement is held to be contrary to law, that provision will be enforced to the maximum extent permissible and the remaining provisions will remain in force and effect.